MORAL GROUND

Carly Lettero, *Research Associate*
Carol Mason, *Production Associate*
Frank L. Moore, *Information Manager*

MORAL GROUND

Ethical Action for a Planet in Peril

Edited by
Kathleen Dean Moore
and
Michael P. Nelson

Foreword by
Desmond Tutu

TRINITY UNIVERSITY PRESS
SAN ANTONIO

Published by Trinity University Press
San Antonio, Texas 78212

Copyright © 2010 by Kathleen Dean Moore and Michael P. Nelson

All rights reserved. No part of this book may be reproduced in any form or by any electronic or mechanical means, including information storage and retrieval systems, without permission in writing from the publisher.

Front cover design by the Office of Paul Sahre
Dust jacket composition by Anne Richmond Boston
Book design by BookMatters, Berkeley

Trinity University Press strives to produce its books using methods and materials in an environmentally sensitive manner. We favor working with manufacturers that practice sustainable management of all natural resources, produce paper using recycled stock, and manage forests with the best possible practices for people, biodiversity, and sustainability. The press is a member of the Green Press Initiative, a nonprofit program dedicated to supporting publishers in their efforts to reduce their impacts on endangered forests, climate change, and forest dependent communities.

The paper used in this publication meets the minimum requirements of the American National Standard for Information Sciences—Permanence of Paper for Printed Library Materials, ANSI Z39.48-1992.

Library of Congress Cataloging-in-Publication Data

Moral ground : ethical action for a planet in peril / edited by Kathleen Dean Moore and Michael P. Nelson ; foreword by Desmond Tutu.
   p. cm.
SUMMARY: "An anthology bringing together the testimony of over eighty theologians, religious leaders, scientists, elected officials, business leaders, naturists, activists, and writers to present a diverse and compelling call to honor humans' moral responsibility to the planet in the face of environmental degradation and global climate change"—*Provided by publisher.*
978-1-59534-085-6 (paperback: alk. paper)
978-1-59534-105-1 (e-book)
  1. Environmental ethics. 2. Environmental responsibility. I. Moore, Kathleen Dean. II. Nelson, Michael P., 1966–
GE42.M65 2010
179'.1—dc22   2010021362

For Lem, Theo, and Zoey

*For Heather,
provider of unflinching and unwavering support of
this and every wide- and wild-eyed endeavor*

*The publisher gratefully acknowledges
the support of the Germeshausen Foundation
and the Kendeda Fund in the publication of this book.*

Contents

Foreword Desmond Tutu — xiii

INTRODUCTION — xv

Do we have a moral obligation to take action to protect the future of a planet in peril?

1. **Yes, for the survival of humankind.** — 1

    James Gustave Speth The Limits of Growth — 3
    Daniel Quinn The Danger of Human Exceptionalism — 9
    The Dalai Lama A Question of Our Own Survival — 15
    E. O. Wilson The Fate of Creation Is the Fate of Humanity — 21
    Sheila Watt-Cloutier The Inuit Right to Culture Based on Ice and Snow — 25
    Barack Obama The Future I Want for My Daughters — 30
    Alan Weisman Obligation to Posterity? — 32
    ETHICAL ACTION — 38

2. **Yes, for the sake of the children.** — 40

    Oren Lyons Keepers of Life — 42
    Scott Russell Sanders We Bear You in Mind — 45
    Gary Snyder For the Children — 50
    John Paul II and the Ecumenical Patriarch Bartholomew I Steering the Earth Toward Our Children's Future — 51
    Hylton Murray-Philipson A Letter to My Boys — 55
    Derrick Jensen You Choose — 60
    ETHICAL ACTION — 65

CONTENTS

3. **Yes, for the sake of the Earth itself.** ... 67

   Brian Turner   Sky ... 69
   Holmes Rolston III   A Hinge Point of History ... 70
   F. Stuart Chapin III   The Planet Is Shouting but Nobody Listens ... 75
   Thich Nhat Hanh   The Bells of Mindfulness ... 79
   Robin Morris Collin   Restoration and Redemption ... 82
   Kate Rawles   A Copernican Revolution in Ethics ... 88
   ETHICAL ACTION ... 96

4. **Yes, for the sake of all forms of life on the planet.** ... 98

   Dave Foreman   Wild Things for Their Own Sakes ... 100
   Carly Lettero   Spray Glue Goes. Maggots Stay. ... 103
   Shepard Krech III   Ornithophilia ... 108
   Gary Paul Nabhan   Heirloom Chile Peppers and Climate Change ... 115
   David Quammen   Imagining Darwin's Ethics ... 119
   Robert Michael Pyle   Evening Falls on the Maladaptive Ape ... 123
   ETHICAL ACTION ... 129

5. **Yes, to honor our duties of gratitude and reciprocity.** ... 131

   Ecumenical Patriarch Bartholomew I   To Commit a Crime Against the Natural World Is a Sin ... 133
   Nirmal Selvamony   Sacred Ancestors, Sacred Homes ... 137
   Robin W. Kimmerer   The Giveaway ... 141
   Courtney S. Campbell   From the Mountain, a Covenant ... 146
   Deborah Bird Rose   So the Future Can Come Forth from the Ground ... 154
   Ursula K. Le Guin   A Conference in Time ... 158
   ETHICAL ACTION ... 163

CONTENTS

6. **Yes, for the full expression of human virtue.**   165

    Brian Doyle   A Newt Note   167

    John Perry   Worship the Earth   169

    Bill McKibben   Something Braver Than Trying to Save the World   174

    Massoumeh Ebtekar   Peace and Sustainability Depend on the Spiritual and the Feminine   178

    Dale Jamieson   A Life Worth Living   183

    Thomas L. Friedman   Who We Really Are   189

    ETHICAL ACTION   192

7. **Yes, because all flourishing is mutual.**   194

    George Tinker   An American Indian Cultural Universe   196

    Fred W. Allendorf   No Separation Between Present and Future   202

    Jonathan F. P. Rose   A Transformational Ecology   207

    Mary Catherine Bateson   Why Should I Inconvenience Myself?   211

    Angayuqaq Oscar Kawagley   Extra! Extra! New Consciousness Needed   217

    Edwin P. Pister   Just a Few More Yards   220

    Kimberly A. Wade-Benzoni   Why Sacrifice for Future Generations?   225

    Jesse M. Fink   Hope and the New Energy Economy   230

    ETHICAL ACTION   239

8. **Yes, for the stewardship of God's creation.**   241

    Sallie McFague   A Manifesto to North American Middle-Class Christians   243

    Marcus J. Borg   God's Passion in the Bible: The World   250

    Seyyed Hossein Nasr   Our Obligation to Tomorrow   254

    Tri Robinson   The Biblical Mandate for Creation Care   260

CONTENTS

    Martin S. Kaplan   Will Religions Guide Us on Our Dangerous Journey?   263
    ETHICAL ACTION   267

9. **Yes, because compassion requires it.**   268

    Libby Roderick   Winter Wheat   269
    Wangari Maathai   We Are Called to Help the Earth to Heal   271
    Ming Xu and Xin Wei   An Invisible Killer   275
    James Garvey   Climate Change Is a Moral Problem for You, Right Now   279
    Sulak Sivaraksa   From Engagement to Emancipation   283
    Quincy Troupe   The Architecture of Language, Parts 9 and 10   286
    ETHICAL ACTION   290

10. **Yes, because justice demands it.**   292

    Carl Pope   Ethics as if Tomorrow Mattered   294
    Michael M. Crow   Sustainability as a Founding Principle of the United States   301
    Steve Vanderheiden   Climate Change and Intergenerational Responsibility   306
    Lauret Savoy   Still an American Dilemma   312
    Ismail Serageldin   There Is a Tide   317
    Peter Singer   A Fair Deal   321
    Carl Safina   The Moral Climate   324
    ETHICAL ACTION   327

11. **Yes, because the world is beautiful.**   329

    Bernd Heinrich   Our Edens: Ecological Homes   331
    John A. Vucetich   Wolves, Ravens, and a New Purpose for Science   337
    Hank Lentfer   Get Dirty, Get Dizzy   343

Alison Hawthorne Deming   The Feasting   347
ETHICAL ACTION   352

12. Yes, because we love the world.   354

J. Baird Callicott   Changing Ethics for a Changing World   356
bell hooks   Touching the Earth   363
Katie McShane   Love, Grief, and Climate Change   369
Stephen R. Kellert   For the Love and Beauty of Nature   373
Bron Taylor   Earth Religion and Radical Religious Reformation   379
Wendell Berry   A Promise Made in Love, Awe, and Fear   387
Kathleen Dean Moore   The Call to Forgiveness at the End of the Day   390
ETHICAL ACTION   394

13. Yes, to honor and celebrate the Earth and Earth systems.   395

Thomas Berry   The Great Work   396
N. Scott Momaday   An Ethic of the Earth   400
Curt Meine   Spring's Hopes Eternal   403
Linda Hogan   Dawn for All Time   407
Mary Evelyn Tucker and Brian Swimme   The Universe Story and Planetary Civilization   410
ETHICAL ACTION   417

14. Yes, because our moral integrity requires us to do what is right.   419

Ernest Partridge   Moral Responsibility Is the Price of Progress   421
Terry Tempest Williams   Climate Change: What Is Required of Us?   429
David James Duncan   Being Cool in the Face of Global Warming   434
Paul B. Thompson   Everything Must Go   440

CONTENTS

Joerg Chet Tremmel The No-Man's-Land of Ethics — 444
José Galizia Tundisi The Advocacy Responsibility of the Scientist — 448
Barbara Kingsolver How to Be Hopeful — 452
Michael P. Nelson To a Future Without Hope — 458
Paul Hawken The Most Amazing Challenge — 463
ETHICAL ACTION — 469

Acknowledgments — 471
Credits — 473
List of Contributors — 475

Foreword

To the people of all the world:

The essays in this book tell us the moral ground we stand on. They are a clear call to action. We are called to understand that climate change is a moral challenge, not simply an economic or technological problem. We are called to honor our duties of justice, to prevent the enormities of climate change, as the price of the lifestyles of the privileged is paid by millions of poor people, in the loss of their livelihoods and their lives. We are called to honor our duties of compassion, to prevent the suffering of millions of innocent people, especially the hungry children.

Leading environmental scientists predict that as many as 185 million Africans will die this century as the direct result of climate change. Many more will face untold suffering in other parts of the world. As I write, famine is increasing. Flooding is increasing, as are the disease and insecurity caused by water scarcity around the world. Climate change is real. It has begun.

The countries that are the least responsible for causing climate change are paying the heaviest price. The average U.K. citizen produces nearly fifty times as much carbon dioxide as any citizen in the developing world. And in the United States the production of carbon dioxide is significantly higher. This is a serious injustice.

As an African, I urgently call on ordinary people in rich countries to act as global citizens, not as isolated consumers. We must listen to our consciences, not to governments who speak only about economic markets. These markets will cease to exist if climate change is allowed to develop to climate chaos.

We have a big problem to solve. Climate change is a global threat that will affect my generation surely, but will prove to have a devastating effect on my children, not to mention my grandchildren and great-grandchildren. All scientific prognoses show that the continent

FOREWORD

of Africa will be severely harmed if we do not act now. The consequences could be conflicts and instability, which we must avoid at any price.

Our experience in South Africa confirms that if we act on the side of justice, we have the power to turn tides. Industry, government, civil society, and you and I—we can all make a difference. Raise your voice. I urge you, sisters and brothers, to work together with campaigners in the global south and call for strong climate change laws in your own countries in the north as well as internationally. Do not fly in the face of the poor by allowing the emissions produced by endless and unnecessary business flights to keep growing. Insist on an 80 percent cut in your national emissions and hold your governments to account.

In matters of climate change, as in all our lives, our obligation is clear: we must do unto others as we would wish them to do unto us.

Thank you for caring. Thank you for acting.

Archbishop Emeritus Desmond Tutu

INTRODUCTION

Toward a Global Consensus for Ethical Action

Kathleen Dean Moore and Michael P. Nelson

> The key thing is the sense of universal responsibility; that is the real source of strength, the real source of happiness. If our generation exploits everything available—the trees, the water, and the minerals—without any care for the coming generations or the future, then we are at fault, aren't we?
>
> HIS HOLINESS, THE DALAI LAMA

> The problems that we see every day in the markets represent a *massive* opportunity to stop and think where this is all heading. What sort of people are we?
>
> HYLTON MURRAY-PHILIPSON,
> Canopy Capital, UK

It's late autumn as we begin to write. Crows perch on the electric lines that connect the houses. A teenager swings his car into the driveway and opens the door, releasing a wave of music. The little girl who lives next door shuffles past wearing her mother's high-heeled shoes. Her mother sits on their doorstep in the slanting light of late afternoon, watching her daughter. In the next block, a neighbor adjusts his ear protectors and yanks the cord on his leaf blower. Students walk by, detouring into the street to avoid the roaring engine and the blast of leaves. There will be rain this evening and maybe snow by morning.

The scene feels odd, almost fictional, the way life goes on. It seems almost as if we were watching a herd of dinosaurs grazing on giant fern-trees, oblivious to the shadow of the asteroid that will strike Earth and forever change the conditions under which they will live—or die. A dinosaur lifts his head and hisses at another who is stripping a frond off the same tree. A mother bends her long neck to nudge the eggs nestled under her flank.

But there is no reptilian oblivion among the neighbors and the people passing by. Scientists have done a heroic job of alerting citizens to the world-altering dangers of cascading extinctions, climate change, and other environmental calamities. People know that we already approach tipping points where climate change will trigger irreversible alteration in the conditions under which we will live—or die. They know that by the terrible working out of a few wrong decisions, this generation, our own, will see the end of the Holocene era, when Earth achieved its greatest richness and variety of life, and the beginning of a new epoch of unpredictable change and loss. They know that what is about to hit us is a human creation and that immediate human action is all that can alter its direction, on the chance that calamity will hit the Earth only a glancing blow.

The neighbor switches off the leaf blower and begins to stuff dead leaves into a black plastic bag. The mother walks down the sidewalk to put a little coat on her child.

What will move people to act to save their beloved worlds? Clearly, information is not enough. What is missing is the moral imperative, the conviction that assuring our own comfort at terrible cost to the future is not worthy of us as moral beings.

This book is a call to that moral affirmation. It is a call to people to honor their moral responsibility to the future to strive to avert the worst consequences of the environmental emergencies and leave a world as rich in life and possibility as the world we live in. We have three immediate goals: (1) With the testimony in this book, we aim to demonstrate a global *ethical* consensus among the moral and intellectual leaders of the world that climate action is a moral responsibility, just as scientists demonstrated with the 2002 report of the Union of Concerned Scientists, a global *scientific* consensus that the climate crisis is real, dangerous, and upon us. (2) With the essays in this book, we aim to empower readers with a wide variety of arguments demonstrating that we are called as moral beings to environmental action, so that no matter their religion, their worldview, or their position in life, readers will find reasons here—good reasons, powerful reasons—to respond. (3) Finally, we call people to take themselves seriously as moral agents, to reclaim the right to live the lives they believe in, to live as people of integrity—conscientious, compassionate, joyous—and so to take away the ability of the powerful to destroy the Earth.

The Importance of the Ethical

Scientists continue to provide evidence that environmental degradation and consequent global climate change are profoundly dangerous to humans and to other life on Earth. Activists also are doing their jobs well, urging us to act in ways that will avert these harms to the extent that can be done. But to the surprise and frustration of both scientists and activists, we do not act, we do not create policy, and we casually opt out of efforts to avert future harms—all the while thinking of ourselves as good and sensible people.

How can this be?

Here's one answer: a piece is largely missing from the public discourse about climate change, namely an affirmation of our moral responsibilities in the world that the scientists describe. No amount of factual information will tell us what we ought to do. For that, we need moral convictions—ideas about what it is to act rightly in the world, what it is to be good or just, and the determination to do what is right. Facts and moral convictions together can help us understand what we ought to do—something neither alone can do.

Western society is very good at facts. We aren't as good at values. Our press is filled with scientific reports and predictions about the ecological systems our lives and other lives depend on. Moral discussion is notably sparse. We hope this book will invigorate the discussion of the values and moral aspirations that equally shape our lives, answers to the fundamental questions of the human condition: What is a human being? What is a human's relation to the Earth? How then ought we to live?

This fusion of facts and values is exactly our goal: to articulate explicitly the missing moral premise of arguments that can compel us from terrifying facts to powerful obligations and effective actions.

The factual premise   If we do not act soon, anthropogenic environmental changes will bring serious harms to the future.

The moral premise   We have a moral obligation to avert harms to the future, so as to leave a world as rich in life and possibility as the world we inherited.

The conclusion   Therefore, we have a moral obligation to act, and act now.

The facts are in. The normative premise is still under discussion. The stakes of this discussion are high. Once this argument is complete and the premises are agreed upon, one cannot withdraw from obligation and at the same time profess to be a moral person.

How to Call Forth Discussion of the Ethical?

Our first step as editors was to invite people to join the ethical discussion that would take place in the book. Cultures around the world express moral wisdom in a multitude of ways—in stories, scientific reports, poetry, scripture, economic analysis, literary essays, philosophical argument, letters, edicts, music, and prayers. We wanted to honor all forms of moral discourse. Moreover, there is moral wisdom in people of many different ages, professions, locations, and worldviews. We wanted to hear from them, lots of them—dozens of the world's moral and intellectual leaders.

So we began to call upon people we deeply respect. We asked them to write for the book. Does the world have an obligation to the future to avert the worst effects of a lurching climate and impoverished environment? If yes, why? What explains our obligation? What argument can convince us of our responsibilities?

Almost all of these extraordinary, generous people agreed to write, setting aside their own projects, diverting energy from their political campaigns or their own climate actions or their fight against illness. No writer accepted any payment; indeed, none was offered. If there are proceeds from *Moral Ground*, they will go to climate action. From every continent save Antarctica, the contributors sent in beautiful, moving work, expressions of hope for the future and faith in what is best in us. Each of the writers is deeply and painfully aware that saving a fully thriving future will require not just good science and new technology, but also the greatest exercise of the moral imagination that the world has ever seen.

The Arguments

One moral message sounds loud and clear from almost every writer: as a world, we are undergoing a fundamental change in our understanding of who we are and what role we play in the world. We might once have thought that we were separate from the Earth—in power and in control. We might have thought the Earth was created for our use alone and drew all its value from its usefulness to us, or that we had no obligations except to ourselves, as individuals or as a species. But ecological science and almost all of the world's religions renounce that worldview as simply false and deeply dangerous. Rather, humans are part of intricate, delicately balanced systems of living and dying that have created a richness of life greater than the world has ever seen. Because we are part of the Earth's systems, we are utterly dependent on their thriving.

The writers in this book tell us clearly that we cannot be concerned with humans and human futures without at the same time being directly concerned for all that supports those humans and human futures, because these things are not as distinct as we might have thought—not ecologically distinct or ethically. There's a merging of an ecological and spiritual awareness here that might account for the emergence of a new ethical awareness.

This new ethical awareness includes and goes beyond our obligations to humans, toward the other inhabitants of the world, and to the world itself. We hear from these writers—nearly all of them—that while humans and human interests certainly matter, they are not the only things that matter. A moral life will also honor the interests—and the beauty and mystery—of all the Earth.

Given this, do we have a moral obligation to strive to avert the worst effects of the environmental emergencies? Yes, the writers in this book say. Yes, for many different reasons. We have grouped the essays into sections, according to the reasons they offer:

1. *Moral arguments based on the consequences of acting or failing to act*

Some of the reasons refer to the consequences of acting, or failing to act. Ethicists call this sort of argument "consequentialist," because it is by the act's results that we judge it to be right or wrong. An act is right if it increases or enhances what we value, wrong if it reduces or destroys what we think of as good. One primary good is human well-being. So yes, we have an obligation to avert the worst effects of the environmental emergencies, for the survival of humankind, a fundamental value indeed (section 1). And, yes, we have an obligation, for the good of the children to come, for what could we value more than our children's happiness (section 2)? Yes, we have an obligation to all the Earth; because we are all mutually interdependent, human thriving depends on the thriving of all life (section 7).

But it's not only human well-being that we value—our flourishing, and that of our children. We also value the great, good Earth and all its life. So yes, we must avert the worst effects of the environmental emergencies for the sake of the sparrows and the seagrass, for newborn whales and tons of krill, for fish like confetti on coral reefs, for lingonberries and the pawprints of bears, for the sake of all living things and their thriving (section 4).

And it's not just the individual species at risk, but the great systems, the dance of air, water, Earth, and the fire of life. So yes, we have an obligation to act for the survival of the Earth as a self-regulating, living system (section 3).

Isn't this what it means to live a moral life?—to do what you can to create or protect what you believe is good and beautiful and of great worth?

2. *Moral arguments based on doing what is right*

In affirming a moral obligation to the future, other writers examine not the consequences of acting or failing to act, but the nature of the act itself. As we go about our lives, making moral decisions, are we fulfilling our obligations as moral beings? Ethicists call this sort of argument "deontological," from the Greek word *deon*, which means "duty." Some of the writers argue that God calls us to the stewardship of divine creation and that failure to protect the world he so lovingly created is a failure in our moral duty to God (section 8). Others argue that we have duties of justice. We have duties to honor the rights of all species, both present and future (section 13). And we have duties to respect human rights, duties of intergenerational and international justice (section 10). Alternatively, because the gifts of the Earth are freely given, some writers argue, we are called as recipients of great gifts of life and nourishment and beauty to honor duties of gratitude and reciprocity (section 5).

3. Moral arguments based on virtue

Ethics is certainly about how to act. But in an important sense, ethics is also about who to be. What is the best a human being can be (or, what is beastly or appalling in human nature)? What virtues should form our character? What motivations should shape our acts? Ethicists call arguments based on these sorts of considerations "virtue ethics."

So yes, some writers claim, the full expression of human virtue requires that we honor our obligations to the future (section 6). Others claim that we must act from compassion, recognizing the suffering that environmental degradation imposes on others of all species, of all times (section 9). Moreover, we must protect all of life because the Earth is beautiful, and what sort of thug would destroy what is beautiful or fail to grieve for its loss (section 11)? We love the world; we are delighted and comforted by its wonders and rhythms; love calls us to acts of caring (section 12). And if this is the way the world is—beautiful, extraordinary, astonishing in its detail and in the wide expanse of its creation—aren't we called to honor and celebrate it in everything we do (section 13)?

In the final set of essays (section 14), the moral arguments ask us to think about personal integrity, perhaps the keystone virtue. Integrity is a kind of wholeness, the consistency between belief and action. As we make decisions about how we live in an endangered world, do our actions match up with what we most deeply believe is right and good? Or do we act unthinkingly or allow ourselves to be pushed to live in ways we think are wrong? When all is said and done, will each of us be able to say that we lived a life we believed in, conscientiously refusing what is wrong and destructive, exhibiting in our life choices what is compassionate and just? As hope rapidly fails that we might be able to avert the coming ecological collapses, to live a life that embodies our values may be the strongest ethical stance, some writers argue. To do what is right, even if it does no good; to celebrate and to care, even if it breaks your heart.

What we find, when we look at the writings in this book together, is that the discussion of our obligation to the future has transcended argument. This gathering of reasons has become a global celebration of the foresight, compassion, and humility that are part of the heritage of the human being. Perhaps this is reason enough to hope that we might summon the wisdom to save our lives and what makes our lives worth living.

The Call to a Moral Life in a Time of Environmental Emergency

What the authors collectively argue is that no matter what moral theory one might turn to, one will find a strong affirmation of our obligation to future generations to preserve or restore a thriving, healthy world filled with beauty and possibility. Where is there room for escape from this obligation? What is the name of the virtue that would allow us to believe that we could care only about ourselves, act as if present or future beings had no pull on us, and then turn around and assert our goodness?

Certainly there is no room for escape from our obligations in an appeal to the divine. What sort of deity would sanction disregard for the future and still command our praise? And really, how can it be prudent, or beneficial to any being alive or imagined, to discount the future? The world's wisdom makes the same demand of us: we have a moral obligation to avert harms to the future.

INTRODUCTION

A dissenter might argue in response that the concept of obligations to the future makes no sense at all. The future does not exist. How can we have a responsibility to nothing at all? Or even if future beings can be imagined to exist, how can we know what their interests are, or what they might ask of us? How can there be obligations to abstractions?

On the contrary, we would reply, people show strong loyalty to abstractions every time they act in ways that honor conceptual ideals such as freedom, liberty, and prosperity. Loyalty to abstractions is the stuff of this world. Arguments about responsibility to the future do indeed require us to imagine a world that does not yet exist. They further require us to be morally moved by that imagined vision. Loyalty to what we can only imagine is what hope is made of. If any among us does not have the ability to imagine the pain or rejoicing of people other than themselves, then perhaps a first step in moral development is to practice that imagining. This is the work of moral education.

The fulfillment of our responsibilities requires something extraordinary of us. We must develop the moral imagination that will allow us to see ourselves in the places of our progeny, to see the world through their wide eyes, to imagine their sorrow and their hopes—not so different from our own. Indeed, we must develop empathy with wild creatures. We must nourish in ourselves the humility to escape an infantile egoism, even though this humility is often foreign to us, requiring an unfamiliar renunciation of the self, requiring us to understand that the universe was not created for our particular species and generation alone and that we are not the measure of all its worth. We must find a way to enlist all the powers of the human mind and heart. These are not only our rational powers, but the additional gifts that define us as fully human—our abilities to care, to grieve, to yearn, to celebrate, to fear, to analyze, to dream, to hope, to love, to have faith. Above all, we must learn to listen to the Earth, which resounds with a wisdom we can hear if we try, wisdom born of the longest reaches of time and space, red leaves falling from oak trees, soft rain, children coming home at dusk.

Our hope is that the writers in this book will inspire this change, that they will begin a new dialogue about who we are when we are at our best, and what we must do to be worthy of our gifts. Then we can imagine ourselves as not only clever, not only powerful, but also deeply humane. Then we can find the moral ground for living on Earth respectfully, responsibly, and joyously.

KDM, Corvallis, Oregon
MPN, Bell Oak, Michigan

I

> *Do we have a moral obligation to take action to protect the future of a planet in peril?*
>
> **Yes, for the survival of humankind.**

Start here: An act is right if the consequence of that act is to enhance and protect what we value. It is wrong if its effect is the reverse.

Then ask: What is the highest value? Surely the highest value is human well-being, since without humans there is no valuing at all. If we should care about anything, if there should be a foundational value on which we can stand, it is that humankind is worth preserving. It follows that an act is right if the consequence of that act is to enhance and protect human life. It is wrong if its effect is the reverse.

So: What enhances and protects human life? What will destroy it? There is no escaping the fact that our survival (collective and individual) is entwined with the survival of the planet's ecosystems. Our lives depend utterly on the food, air, water, and weather made possible by healthy ecosystems. When they are poisoned, we are poisoned. When they die, we die. Various human cultures demonstrate this connection in a profound fashion: some pueblo Indians of North America believe humans first emerged into this world through an earthly navel, a *sinapu*; in the Ojibwa tradition, Muskrat remade the world after a great flood by bringing soil to the surface of the world; in the Judeo-Christian tradition, God breathed life into the soil to create the first human, Adam, from the Hebrew *Adamah*, meaning "of the Earth." And now climate scientists tell us that the changes we are making to the planet threaten to undermine the very basis of human life on Earth. The biological and cultural connection between the survival of the planet and the survival of humankind is not only broad and deep—it is inescapable.

It follows that: Because we have an obligation not to destroy our own kind, our own selves, our own cultures, we have a corresponding moral obligation not to destroy the ecological and geological foundations of our lives and the future of humankind of Earth.

The Limits of Growth

James Gustave Speth

Human activities are imposing enormous costs on the Earth's climate and other life-support systems, as the remarkable charts on pages 4 and 5 reveal. Whether one looks at the loss of forests, fisheries, species, or climatic stability, the level of environmental destruction is very high and, in almost all cases, rising rapidly.

The exponential expansions of human populations and, even more, economic activity are the main drivers of these momentous transformations. For all the material blessings economic progress has provided, its impact on the natural world must be counted in the balance as tragic loss.

How serious has the threat to the environment become? The human presence is now so large that all we have to do to destroy the planet's climate and ecosystems and leave a ruined world to our children and grandchildren is to keep doing exactly what we are doing today, with no growth in the human population or the world economy. Just continue to release greenhouse gases at current rates, just continue to

---

JAMES GUSTAVE SPETH is the Carl W. Knobloch Jr. Dean of the School of Forestry and Environmental Studies and the Sara Shallenberger Brown Professor in the Practice of Environmental Policy at Yale University. His recent publications include *The Bridge at the Edge of the World: Capitalism, the Environment, and Crossing from Crisis to Sustainability* (2008) and *Red Sky at Morning: America and the Crisis of the Global Environment* (2005). He has been awarded the National Wildlife Federation's Resources Defense Award, the Natural Resources Council of America's Barbara Swain Award of Honor, a Special Recognition Award from the Society for International Development, the Lifetime Achievement Award of the Environmental Law Institute, and the Blue Planet Prize.

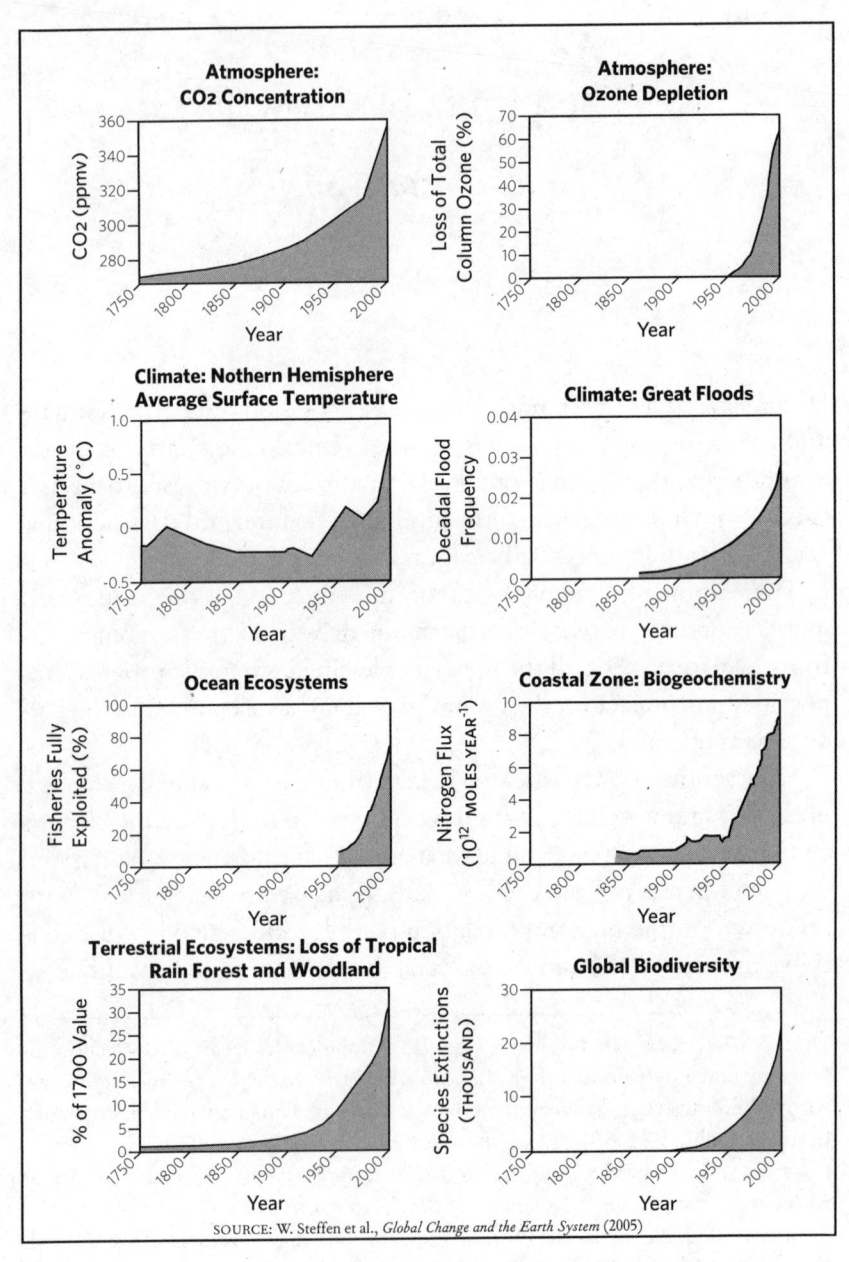

The charts above show degradation from 1750 to 2000 in various Earth systems.

SOURCE: W. Steffen et al., *Global Change and the Earth System* (2005)

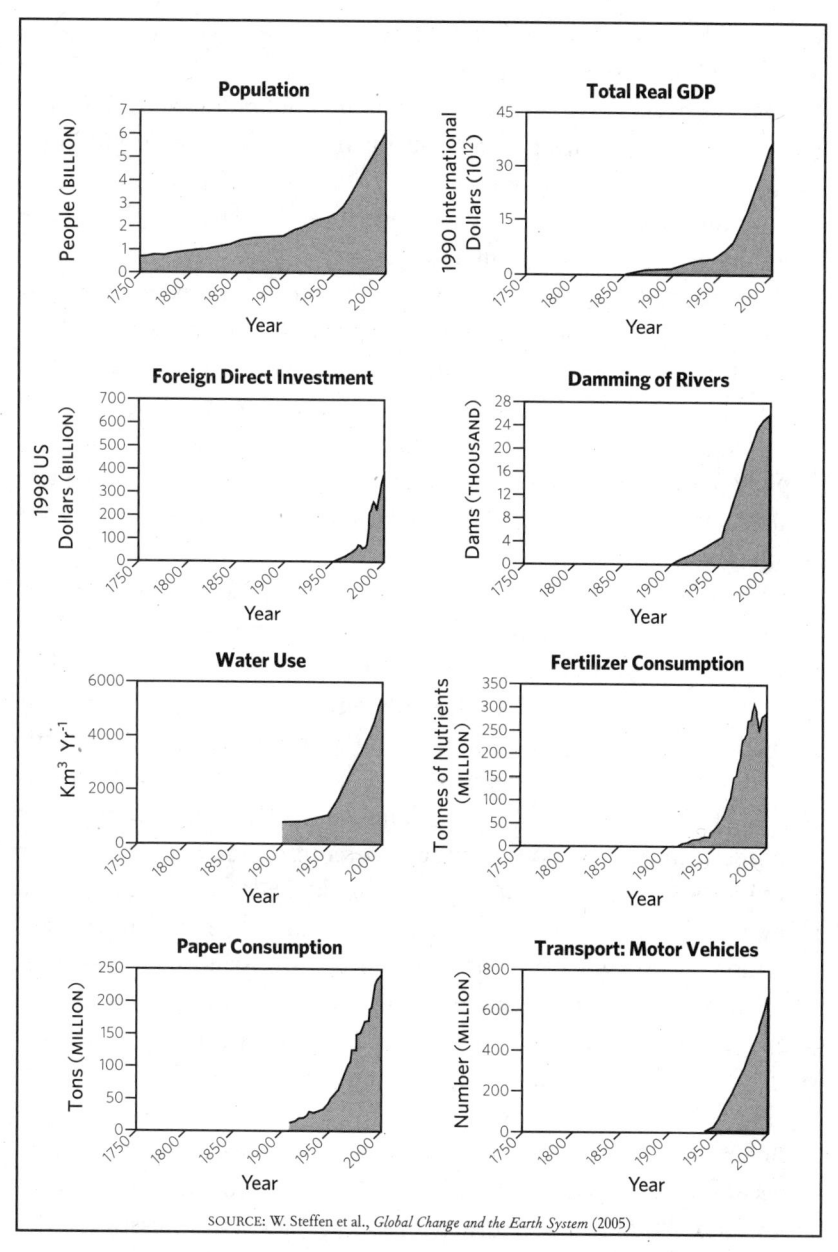

SOURCE: W. Steffen et al., *Global Change and the Earth System* (2005)

The charts above document exponential expansion of human activity (from 1750 to 2000) responsible for the rapid rates of environmental destruction shown on the opposite page.

impoverish ecosystems and release toxic chemicals at current rates, and the world in the latter part of this century won't be fit to live in. But as the charts show, human activities are not holding at current levels—they are accelerating, dramatically. It took all of history to build the $7 trillion world economy of 1950; today economic activity grows by that amount every decade. At current rates of growth, the world economy will double in size in less than two decades. We are thus facing the possibility of an enormous increase in environmental deterioration, just when we need to move strongly in the opposite direction.

Most Americans with environmental concerns have worked within the political mainstream, but it is now clear that approach has not been succeeding. We have been winning battles, including some very important ones, but losing the planet. It is time for the environmental community—indeed, everyone—to develop a deeper critique of what is going on.

Economic growth may be the world's secular religion, but for much of the world it is a god that is failing today—underperforming for most of the world's people and, for those of us in affluent societies, creating more problems than it is solving. The never-ending drive to grow the overall U.S. economy undermines families, jobs, communities, the environment, a sense of place and continuity, even mental health. It fuels a ruthless international search for energy and other resources, and it rests on a manufactured consumerism that is not meeting the deepest human needs. Psychological studies consistently show that materialism is toxic to happiness and that more income and more possessions do not in fact lead to lasting increases in a sense of well-being or satisfaction with life. What does make people happy are warm personal relationships, and giving rather than getting. There may be limits to growth, but I am more concerned with the limits of growth.

Before it is too late, America should begin to move to a post-growth society where working life, the natural environment, our communities, and the public sector are no longer sacrificed merely for the sake of an increase in GDP; where the illusory promises of ever-more growth no longer provide an excuse for neglecting to deal generously with compelling social needs; and where a truly democratic politics is no longer hostage to the primacy of powerful economic interests.

The new environmentalism must embrace a profound challenge to consumerism and commercialism and the lifestyles they offer, a healthy skepticism of growth-mania and a redefinition of what society should be striving to grow, a challenge to corporate dominance and a redefinition of the corporation and its goals, a commitment to deep change in both the functioning and the reach of the market, and a powerful assault on the anthropocentric and contempocentric values that currently dominate.

We live and work in a system of political economy—today's capitalism—that cares profoundly about profits and growth, but it cares about society and the natural world only to the extent that it is required to do so. It is up to us—we, the people—to inject values such as sustainability and justice into this system, and government is the primary vehicle we have for this task. But we mainly fail at it because our politics are too enfeebled and the resistance of vested interests too strong. Our best hope for real change is a fusion of those concerned about environment, social justice, and political democracy into one powerful progressive force.

Environmentalists must join with social progressives in addressing the crisis of inequality now unraveling America's social fabric and undermining democracy. It is a crisis of soaring executive pay, huge incomes and increasingly concentrated wealth for a small minority, occurring simultaneously with poverty rates near a thirty-year high, stagnant wages despite rising productivity, declining social mobility and opportunity, record levels of people without health insurance, failing schools, increased job insecurity, swelling jails, shrinking safety nets, and the longest work hours among the rich countries. In an America with such vast social insecurity, where half the families just get by, economic arguments, even misleading ones, will routinely trump environmental goals.

Environmentalists must also join with those seeking to reform politics and strengthen democracy. America's gaping social and economic inequality poses a grave threat to democracy. What we are seeing is the emergence of a vicious circle: income disparities shift political access and influence to wealthy constituencies and large businesses, which further imperils the potential of the democratic process to act to correct the growing income disparities. Corporations have been the principal economic actors for a long time; now they are the principal political actors as well. Neither environment nor society fares well under corptocracy. Environmentalists need to embrace public financing of elections, regulation of lobbying, nonpartisan congressional redistricting, and other political reform measures as core to their agenda. Today's politics will never deliver environmental sustainability.

The point of departure for many of us has been the momentous environmental challenge we face. But today's environmental reality is linked powerfully with other realities, including growing social inequality and neglect and the erosion of democratic governance and popular control. So we must now mobilize our spiritual and political resources for transformative change on all three fronts. We are all communities of shared fate. We will rise or fall together.

The Danger of Human Exceptionalism

Daniel Quinn

One day some ten years ago I was riding in a taxi to a vacant site under the Brooklyn Bridge. The driver asked what was taking me to such an unusual spot. I explained that I was making an appearance on a show for Granada Television devoted to threats to the human future. He was obviously puzzled, perhaps even a bit incredulous. What in the world was I talking about? Although he seemed intelligent and otherwise normally informed, he had never heard the first word about any such threat.

That, as I say, was about ten years ago. A great deal has changed since then. It would, I think, be hard to find anyone in the United States who was completely unaware of threats we face—even if he or she were unworried by them. These include global warming, the increasingly rapid depletion of fossil fuels (fundamentally imperiling food production as well as our global economy), the increasing pollution and depopulation of our oceans, and the rapid disappearance of arable land and topsoil, to mention just a few.

If we manage to lay each of these threats to rest, another will remain that is even more dire, yet somehow less troubling to the generality of mankind. This is the fact, acknowledged by the vast majority of biolo-

---

DANIEL QUINN is an environmental writer who is best known for his book *Ishmael* (1992), which won the Turner Tomorrow Fellowship Award. His short fiction has appeared in the *Quarterly, Asylum, Magic Realism*, and elsewhere. In 1998 Quinn collaborated with environmental biologist Alan D. Thornhill in producing *Food Production and Population Growth*, a video that elaborates on the ideas presented in his publications. His most recent books include *Work, Work, Work* (2006) and *If They Give You Lined Paper, Write Sideways* (2007).

gists, that we are in the midst of a mass extinction equal in intensity to the one 65 million years ago that swept away the dinosaurs. It must be understood that species extinction is not in itself an abnormal event; as a normal event, it is called "background extinction." But today's extinction rate is estimated to be a thousand times greater than this normal background extinction—a *thousand* times!

"We are indeed experiencing the greatest wave of extinctions since the disappearance of the dinosaurs," said Ahmed Djoghlaf, head of the UN convention on biological diversity. "Extinction rates are rising by a factor of up to 1,000 above natural rates. Every hour, three species disappear. Every day, up to 150 species are lost. Every year, between 18,000 and 55,000 species become extinct," he said. "The cause: human activities."[1]

The cause: human activities. This generalization is all but universally accepted, but it is also all but universally understood to refer to activities that can be moderated or extinguished by vigilance and vigorous regulation, as was done in the case of, say, the American bald eagle. The absurdity of this notion is evident from the fact that fewer than two thousand species are formally considered "endangered," whereas it's estimated that more than this number fall into extinction *every two weeks*—most of them unknown (and therefore not even *on* any endangered list).

To understand the true cause of this mass extinction, we must begin with these facts. Humans, like all other living things, are composed of biomass. To make human beings, you must have some biomass to make them out of; you can't make them out of moonbeams or shadows.

In 1965 the total of human biomass was about 50 million tons. Today, forty-five years later, it is double that amount, about 100 million tons. In other words, just to make it super-clear, the biomass of the human species has gained about 50 million tons in the last forty-five years. The question I'm leading up to is: Where did this biomass come from? The answer: It came from other members of the community of life.

---

1. Quoted in "UN Urges World to Slow Extinctions: 3 Each Hour," by Alister Doyle, Environment Correspondent, 22 May 2007, Reuters Limited.

How do I know that? I know that because there is nowhere else to get it. The biomass of this planet is pretty much a constant (since the amount of land, sea, air, and sunlight are all pretty much constants). We can destroy the Earth's biomass, through the process known as desertification, but it's virtually impossible for us to add to that biomass. About the best we can hope to do is to hold back the deserts; reclaiming them is all but impossible.

So: The biomass we have added to the human race in the past forty-five years has been taken, little by little and day by day by day, from the species around us. How? When we cut down or burn down a forest to turn the land into cropland or pasture land, the biomass that formerly belonged to the tens of thousands of birds, bugs, foxes, squirrels, lizards, and all other wildlife is destroyed—as thoroughly as they can be destroyed. When we grow our crops or pasture our cattle on that land, we make sure that none of that biomass goes back to the creatures that formerly lived there. We protect it carefully, because we want it for ourselves. It is our understanding that we *need* this biomass for ourselves and that we have a *right* to take it for ourselves—in ever-increasing amounts.

And what is the effect of this? I certainly don't mean that clearing a thousand acres of forest destroys whole species—though it's not impossible that it may do so. What I mean is that clearing a thousand acres of forest *diminishes* tens of thousands of species. And when we clear the adjacent thousand acres, we *further* diminish those species. And this is in fact what we do—to thousands of acres every day. And the result is that inevitably species *do* become extinct—as many as 150 of them *per day*.

I want be very clear that this is no accident. It doesn't happen through inadvertence. It doesn't come about as a side effect of driving our cars or running our factories. It comes about through the *deliberate* destruction of wildlife habitats to feed our ever-growing population. Though we may not think of it this way, we are literally turning 150 species a day into human tissue.

I have said that it is *our understanding* that we *need* to take this biomass for ourselves—this year more than last year and next year more than this year. Why do we need to do this? Because humans are unlike all other species. Other species—*all* other species—grow to the limit of their food resources and then cease to grow. If the food available to them decreases, then their population declines. If the food available to them increases, then they grow.

This is a picture of the living community: millions of species constantly growing and declining as the species around them grow and decline. To put it simply: When the rabbit population grows, the fox population that feeds on them grows. As the fox population grows, the rabbit population declines. And as the rabbit population declines, the fox population necessarily declines with it. And as the fox population declines, the rabbit population recovers and grows. And as the rabbit population grows, the fox population recovers and grows. And so on, through millions of generations and species, living together, some growing as others decline, some declining as others grow.

Except for humans.

Humans alone grow no matter what. The human population will continue to grow whether or not you make more food available. To make the human population grow, you don't need to grow more food: it will grow willy-nilly.

This is the theory of human exceptionalism, which was added to science by Thomas Malthus. "Population, when unchecked, increases in a geometrical ratio. Subsistence increases only in an arithmetical ratio."[2] Absolutely no known species population behaves in this way—except ours, if we take the word of a late-eighteenth-century Anglican country curate. (I find it curious that Darwin's theory has been controverted since its first appearance in print, but Malthus's has gone virtually unchallenged for two centuries.)[3]

If casting doubt on a theory is a theory, then it is my theory that there is utterly no reason to suppose that humans, who assuredly evolved from creatures who were not exceptional and who lived unexceptionally as humans for some 3 million years, are now exceptions. If this theory is correct, then our population (like all others) will inevitably grow if more food is continually made available to us—as it has been for the past ten thousand years, during the agri-

2. T. R. Malthus, *An Essay on the Principle of Population* (1798).
3. For an exception, see Russell Hopfenberg's "Human Carrying Capacity Is Determined by Food Availability" and, with David Pimentel, "Human Population Numbers as a Function of Food Supply" (www.panearth.org/publications.htm).

cultural revolution, which continues to spread to the present day into the last remaining wilderness areas, devouring our neighbors in the community of life. And if my theory is correct, then our population would become stable if our food production were to become stable.

There remains, of course, our perennial excuse for increasing food production: we must increase it to feed the starving millions. Except the fact is that, though I have heard this excuse and seen the annual increase in food production for the past sixty years, the starving millions have never gotten fed—ever. It is said in defense of the excuse that the excess food simply doesn't reach them, as if the roads and the railway tracks fall short of them, as if ships carrying food cannot reach their ports. This is nonsense. If food doesn't reach the starving millions, it is because they are too poor to *make* it reach them. End poverty and there will be no starving millions; everyone knows that there is food enough for them—if they could afford to buy it. There are no starving *rich* people anywhere in the world.

A 2004 document issued by the Population Division of the UN's Department of Economic and Social Affairs gives three estimates of global population at the end of this century: a low estimate of 7.4 billion, a medium estimate of 8.9 billion, and a high estimate of 10.6 billion.[4] Since we added .7 billion to our population in just the first nine years of this century, I find it hard to believe that it will take us another 90 years to add just another .7 billion. But it doesn't really matter. The real question is whether we can survive another 90 years in a period of mass extinctions—a period in which between 1.6 and 5 million species are likely to become extinct, even if extinction rates remain stable (which is scarcely credible, given an ever-increasing human population).[5]

We are like people living in the penthouse of a hundred-story building. Every day we go downstairs and at random knock out 150 bricks to take upstairs to increase the size of our penthouse. Since the building below consists of millions of bricks, this seems harmless enough . . . for a single day. But for thirty thousand days?

---

4. http://secint24.un.org/esa/population/publications/longrange2/WorldPop2300final.pdf.

5. It isn't possible to relate these numbers to the total number of species on this planet, since authoritative estimates of this number range from 2 million to 100 million.

Eventually—inevitably—the streams of vacancy we have created in the fabric of the walls below will come together to produce a complete structural collapse.

When this happens—if it is *allowed* to happen—we will join the general collapse, and our lofty position at the top of the structure will not save us.

A Question of Our Own Survival

The Dalai Lama

Just as we should cultivate more gentle and peaceful relations with our fellow human beings, we should also extend that same kind of attitude toward the natural environment. Morally speaking, we should be concerned for our whole environment.

This, however, is not just a question of morality or ethics, but also a question of our own survival. For this generation and for future generations, the environment is very important. If we exploit the environment in extreme ways, we may receive some benefit today, but in the long run, we will suffer, as will our future generations. When the environment changes, the climatic condition also changes. When the climate changes dramatically, the economy and many other things change. Our physical health will be greatly affected. Again, conservation is not merely a question of morality, but a question of our own survival.

SIDNEY PIBURN,
The Dalai Lama[1]

THE DALAI LAMA is the head of state and the spiritual leader of Tibet. In 1989 he was awarded the Nobel Peace Prize for his nonviolent struggle for the liberation of Tibet. He has traveled to more than sixty-two countries with his message of peace, nonviolence, interreligious understanding, universal responsibility, and compassion. He has authored more than seventy-two books, most recently *In My Own Words: An Introduction to My Teachings and Philosophy* (2008).

---

1. Dalai Lama and Sidney Piburn (compiler and editor), "An Ethical Approach to Environmental Protection," in *The Dalai Lama: A Policy of Kindness* (Snow Lion Publications, 1997), 107.

Scientific predictions of environmental change are difficult for ordinary human beings to comprehend fully. We hear about hot temperatures and rising sea levels, increasing cancer rates, vast population growth, depletion of resources, and extinction of species. Human activity everywhere is hastening to destroy key elements of the natural ecosystems all living beings depend on.

These threatening developments are individually drastic and together amazing. The world's population has tripled in this century alone and is expected to double or triple in the next. The global economy may grow by a factor of five or ten, including with it extreme rates of energy consumption, carbon dioxide production, and deforestation. It is hard to imagine all these things actually happening in our lifetime and in the lives of our children. We have to consider the prospects of global suffering and environmental degradation unlike anything before in human history.

I think, however, there is good news in that now we will definitely have to find new ways to survive together on this planet. In this century we have seen enough war, poverty, pollution, and suffering. According to Buddhist teaching, such things happen as the result of ignorance and selfish actions, because we often fail to see the essential common relation of all beings. The Earth is showing us warnings and clear indications of the vast effects and negative potential of misdirected human behavior.

To counteract these harmful practices we can teach ourselves to be more aware of our own mutual dependence. Every sentient being wants happiness instead of pain. So we share a common basic feeling. We can develop right action to help the Earth and each other based on a better motivation. Therefore, I always speak of the importance of developing a genuine sense of universal responsibility. When we are motivated by wisdom and compassion, the results of our actions benefit everyone, not just our individual selves or some immediate convenience. When we are able to recognize and forgive ignorant actions of the past, we gain the strength to constructively solve the problems of the present.

We should extend this attitude to be concerned for our whole environment. As a basic principle, I think it is better to help if you can, and if you cannot help, at least try not to do harm. This is an especially suitable guide when there is so much yet to understand about the complex interrelations of diverse and unique ecosystems. The Earth is our home and our mother. We need to respect and take care of her. This is easy to understand today.

We need knowledge to care for ourselves, every part of the Earth and the life upon it, and all of the future generations as well. This means that education about the environment is of great importance to everyone. Scientific learning and technological progress are essential for improving the quality of life in the modern world. Still more important is the simple practice of getting to know and better appreciate our natural surroundings, and ourselves, whether we are children or adults. If we have a true appreciation for others and resist acting out of ignorance, we will take care of the Earth.

In the biggest sense, environmental education means learning to maintain a balanced way of life. All religions agree that we cannot find lasting inner satisfaction based on selfish desires and acquiring the comforts of the material things. Even if we could, there are now so many people that the Earth would not sustain us for long. I think it is much better to practice enjoying simple peace of mind. We can share the Earth and take care of it together rather than trying to possess it, destroying the beauty of life in the process.

Ancient cultures that have adapted to their natural surroundings can offer special insights about how human societies can be structured to exist in balance with the environment. For example, Tibetans are uniquely familiar with life on the Himalayan Plateau. This has evolved into a long history of a civilization that took care not to overwhelm and destroy its fragile ecosystem. Tibetans have long appreciated the presence of wild animals as symbolic of freedom. A deep reverence for nature is apparent in much of Tibetan art and ceremony. Spiritual development thrived despite limited material progress. Just as species may not adapt to relatively sudden environmental changes, human cultures also need to be treated with special care to ensure survival. Therefore, learning about the useful ways of people and preserving their cultural heritage is also a part of learning to care for the environment.

I try always to express the value of having a good heart. This simple aspect of human nature can be nourished to great power. With a good heart and wisdom you have right motivation and will automatically do what needs to be done. If people begin to act with genuine compassion for everyone, we can still protect each other and the natural environment. This is much easier than having to adapt to the severe and incomprehensible environmental conditions projected for the future.[2]

As a boy studying Buddhism, I was taught the importance of a caring attitude toward the environment. Our practice of nonviolence applies not just to human beings but to all sentient beings—any living thing that has a mind. Where there is a mind, there are feelings such as pain, pleasure, and joy. No sentient being wants pain; all want happiness instead. I believe that all sentient beings share those feelings at some basic level.

In Buddhism practice we get so used to this idea of nonviolence and the ending of all suffering that we become accustomed to not harming or destroying anything indiscriminately. Although we do not believe that trees or flowers have minds, we treat them too with respect. Thus we share a sense of universal responsibility for both mankind and nature.

Our belief in reincarnation is one example of our concern for the future. If you think that you will be reborn, you are likely to say to yourself, I have to preserve such and such because my future reincarnation will be able to continue with these things. Even though there is a chance you may be reborn as a creature, perhaps even on a different planet, the idea of reincarnation gives you reason to have direct concern about this planet and future generations.

Peace and survival of life on Earth as we know it are threatened by human activities that lack a commitment to humanitarian values. Destruction of nature and natural resources results from ignorance, greed, and lack of respect for the Earth's living things. This lack of respect extends even to the Earth's human descendants, the future generations who will inherit a vastly degraded planet if world peace doesn't become a reality and if destruction of the natural environment continues at the present rate.

---

2. The Dalai Lama, "A Universal Task," *EPA Journal: A Magazine on National and Global Environmental Perspectives* 17, no. 4 (September/October 1991).

Our ancestors viewed the Earth as rich and bountiful, which it is. Many people in the past also saw nature as inexhaustibly sustainable, which we now know is the case only if we care for it. It is not difficult to forgive destruction in the past that resulted from ignorance. Today, however, we have access to more information. It is essential that we reexamine ethically what we have inherited, what we are responsible for, and what we will pass on to coming generations.

Clearly this is a pivotal generation. Global communication is possible, yet confrontations take place more often than meaningful dialogues for peace. Our marvels of science and technology are matched, if not outweighed, by many current tragedies, including human starvation in some parts of the world and extinction of other life-forms. Exploration of outer space takes place at the same time that the Earth's own oceans, seas, and freshwater areas grow increasingly polluted and their life-forms are still largely unknown or misunderstood. Many of the Earth's habitats, animals, plants, insects, and even microorganisms that we know as rare may not be known at all by future generations. We have the capability and the responsibility. We must act before it is too late.[3]

We have to accept this. If we unbalance Nature, humankind will suffer. Furthermore, as people alive today, we must consider future generations: a clean environment is a human right like any other. It is therefore part of our responsibility toward others to ensure that the world we pass on is as healthy as when we found it, if not healthier. This is not quite such a difficult proposition as it might sound. For although there is a limit to what we as individuals can do, there is no limit to what a universal response might achieve. It is up to us as individuals to do what we can, however little that may be.

This is where, as a Buddhist monk, I feel that belief in the concept of karma is very useful in the conduct of daily life. Once you believe in the connection between motivation and its effects, you will become more alert to the effects that your own actions have upon yourself and others.

---

3. The Dalai Lama, "Universal Responsibility and the Environment," in *My Tibet* (Berkeley and Los Angeles: University of California Press, 1990), 79–80.

Now, although I have found my own Buddhist religion helpful in generating love and compassion, I am convinced that these qualities can be developed by anyone, with or without religion. I further believe that all religions pursue the same goals: those of cultivating goodness and bringing happiness to all human beings. Though the means might appear different, the ends are the same.

With the ever-growing impact of science on our lives, religion and spirituality have a greater role to play in reminding us of our humanity. There is no contradiction between the two. Each gives us valuable insights into the other. Both science and the teachings of the Buddha tell us of the fundamental unity of all things.

Finally, I would like to share with my readers a short prayer, one that gives me great inspiration and determination:

> For as long as space endures,
> And for as long as living beings remain,
> Until then may I, too, abide
> To dispel the misery of the world.[4]

---

4. The Dalai Lama, "Universal Responsibility and the Good," in *Freedom in Exile: The Autobiography of the Dalai Lama* (New York: HarperOne, 1991), 269–271.

The Fate of Creation Is the Fate of Humanity

E. O. Wilson

Somehow and somewhere back in history humanity lost its way. We evolved in a biologically rich world over tens of thousands of generations. We destroyed most of that biological richness in order to improve our lives and generate more people. Billions more people, to the peril of the Creation. I would like to offer the following explanation of the human dilemma:

According to archaeological evidence, we strayed from Nature with the beginning of civilization roughly ten thousand years ago. That quantum leap beguiled us with an illusion of freedom from the world that had given us birth. It nourished the belief that the human spirit can be molded into something new to fit changes in the environment and culture, and as a result the timetables of history desynchronized. A wiser intelligence might now truthfully say of us at this point: here is a chimera, a new and very odd species come shambling into our universe, a mix of Stone Age emotion, medieval self-image, and godlike technology. The combination makes the species unresponsive to the forces that count most for its own long-term survival.

There seems no better way to explain why so many smart people remain passive while the precious remnants of the natural world dis-

---

E. O. WILSON is a biologist, theorist, naturalist, and author. He is Pellegrino University Research Professor Emeritus in Entomology for the Department of Organismic and Evolutionary Biology at Harvard University, a Fellow of the Committee for Skeptical Inquiry, and Humanist Laureate of the International Academy of Humanism. A two-time winner of the Pulitzer Prize for General Non-Fiction, Wilson coauthored his most recent book, *The Superorganism: The Beauty, Elegance, and Strangeness of Insect Societies*, with Bert Hölldobler.

appear. They are evidently unaware that ecological services provided scot-free by wild environments, by Eden, are approximately equal in dollar value to the gross world product. They choose to remain innocent of the historical principle that civilizations collapse when their environments are ruined. Most troubling of all, our leaders, including those of the great religions, have done little to protect the living world in the midst of its sharp decline. They have ignored the command of the Abrahamic God on the fourth day of the world's birth to "let the waters teem with countless living creatures, and let birds fly over the land across the vault of heaven."

I hesitate to introduce a beautiful subject with an animadversion. Few will deny, however, that the human impact on the natural environment is accelerating and makes a frightening picture.

What are we to do? At the very least, put together an honest history, one on which people of many faiths can in principle agree. If such can be fashioned, it will serve at least as prologue to a safer future.

We can begin with the key discovery of green history: *civilization was purchased by the betrayal of Nature.* The Neolithic revolution, comprising the invention of agriculture and villages, fed on Nature's bounty. The forward leap was a blessing for humanity. Yes, it was: Those who have lived among hunter-gatherers will tell you they are not at all to be envied. But the revolution encouraged the false assumption that a tiny selection of domesticated plants and animals can support human expansion indefinitely. The pauperization of Earth's fauna and flora was an acceptable price until recent centuries, when Nature seemed all but infinite, and an enemy to explorers and pioneers. The wildernesses and the aboriginals surviving in them were there to be pushed back and eventually replaced, in the name of progress and in the name of the gods too, lest we forget.

History now teaches a different lesson, but only to those who will listen. Even if the rest of life is counted of no value beyond the satisfaction of human bodily needs, the obliteration of Nature is a dangerous strategy. For one thing, we have become a species specialized to eat the seeds of four kinds of grass—wheat, rice, corn, and millet. If these fail, from disease or climate change, we too shall fail. Some fifty thousand wild plant species (many of which face extinction) offer alternative food sources. If one insists on being thoroughly practical about the matter, allowing these and the rest of wild species to exist should be considered part of a portfolio of long-term investment. Even the most recalcitrant people must come to view conservation as simple prudence in the management of Earth's natural economy. Yet few have begun to think that way at all.

Meanwhile, the modern technoscientific revolution, including especially the great leap forward of computer-based information technology, has betrayed Nature a second time, by fostering the belief that the cocoons of urban and suburban material life are sufficient for human fulfillment. That is an especially serious mistake. Human Nature is deeper and broader than the artifactual contrivance of any existing culture. The spiritual roots of *Homo sapiens* extend deep into the natural world through still mostly hidden channels of mental development. We will not reach our full potential without understanding the origin and hence meaning of the aesthetic and religious qualities that make us ineffably human.

Granted, many people seem content to live entirely within the synthetic ecosystems. But so are domestic animals content, even in the grotesquely abnormal habitats in which we rear them. This in my mind is a perversion. It is not the Nature of human beings to be cattle in glorified feedlots. Every person deserves the option to travel easily in and out of the complex and primal world that gave us birth. We need freedom to roam across land owned by no one but protected by all, whose unchanging horizon is the same that bounded the world of our millennial ancestors. Only in what remains of Eden, teeming with life-forms independent of us, is it possible to experience the kind of wonder that shaped the human psyche at its birth.

Scientific knowledge, humanized and well taught, is the key to achieving a lasting balance in our lives. The more biologists learn about the biosphere in its full richness, the more rewarding the image. Similarly, the more psychologists learn of the development of the human mind, the more they understand the gravitational pull of the natural world on our spirit, and on our souls.

We have a long way to go to make peace with this planet, and with each other. We took a wrong turn when we launched the Neolithic revolution. We have been trying ever since to ascend *from Nature* instead of *to Nature*. It is not too late for us to come around, without losing the quality of life already gained, in order to receive the deeply fulfilling beneficence of humanity's natural heritage. Surely the reach of religious belief is great enough, and its teachers generous and imaginative enough, to encompass this larger truth not adequately expressed in Holy Scripture.

Part of the dilemma is that while most people around the world care about the natural environment, they don't know why they care, or why they should feel responsible for it. By and large they have been unable to articulate what the stewardship of Nature means to them personally. This confusion is a great problem for contemporary society as well as for future generations. It is linked to another great difficulty: the inadequacy of science education everywhere in the world. Both arise in part from the explosive growth and complexity of modern biology. Even the best scientists have trouble keeping up with more than a small part of what has emerged as the most important science for the twenty-first century.

I believe that the solution to all of the three difficulties—ignorance of the environment, inadequate science education, and the bewildering growth of biology—is to refigure them into a single problem. I hope you will agree that every educated person should know something about the core of this unified issue. Teacher and student alike will benefit from a recognition that living Nature has opened a broad pathway to the heart of science itself, that the breath of our life and our spirit depend on its survival. And to grasp and discuss on common ground this principle: because we are part of it, the fate of the Creation is the fate of humanity.

The Inuit Right to Culture Based on Ice and Snow

Sheila Watt-Cloutier

For those of you who don't know, we Inuit live in four countries of Canada, the United States, Greenland, and Russia. And we're about 155,000 Inuit in the entire world, at the top of the world, in the land of the cold, the ice and the snow.

And we lived very, very traditionally not so long ago. Rapid, tumultuous change has come to our homelands in my lifetime. In fact, I traveled only by dog team on the ice and snow the first ten years of my life. So you can appreciate and imagine that change has happened so very quickly in our homelands in the Arctic. And, of course, that kind of tumultuous change, together with historical traumas, has created an incredible breakdown of our society in the Arctic. And among our young men we have one of the highest suicide rates in North America. So this is the backdrop in which all these other new changes on the new wave are happening.

The first wave arrived in my lifetime, in my mother's lifetime, in my grandmother's lifetime. And now the second wave is really coming hard, and that's environmental degradation. We are already experiencing rapid change—human-induced climate change. We have experienced, of course, the poisoning of our country's food as a result of toxins coming from afar and have had to deal with that intensely at the global level. And so environmental issues indeed are not just

---

SHEILA WATT-CLOUTIER is a Canadian Inuit activist. She has been a political representative for Inuit at the regional, national, and international levels, most recently as International Chair for the Inuit Circumpolar Council. She has worked on a range of social and environmental issues affecting Inuit and has recently focused on persistent organic pollutants and global climate change.

about the environment. When it comes to indigenous peoples, they are very much about the health and well-being of not only our bodies, but also our cultural survival.

So we in the Arctic, for generations, Inuit, we have closely been observing the environment, and we have accurately predicted the weather around us that has enabled us to travel safely on the sea ice to hunt our marine mammals, our walrus and our polar bears. And nowhere else in the world really does ice and snow represent transportation, mobility, and life for a people. And ice and snow, in fact, are our highways that bring us out to the supermarkets, which is the environment, and link us to each other, to other communities.

And several communities already, as we speak, are so damaged by global warming and climate change that relocation at the cost of millions of dollars is now the only option. And among the harms that we have suffered both in Canada and Alaska are the eroded landscape, the contaminated drinking water, the coastal losses because of erosion, longer sea-ice-free seasons, and the melting permafrost, which is now causing beach slumping and increased snowfall in some areas, not enough snow in other areas. New species of birds and fish and insects have arrived that we don't even have names for half the time. There are unpredictable sea-ice conditions. Glaciers are melting, creating rivers instead of streams, and now we have more drownings where our hunters thought they could cross safely. There are also large plans now to relocate some of our communities. And it's becoming very stark, and it's becoming a really dangerous reality for many of us up there. And so it is starting to undermine the ecosystem, the very land, ice, and snow that we depend on for our own physical and cultural survival.

And science now has caught up to our hunters, what we have been saying for years. And our hunters are our scientists in their own right. They may not have institutional recognition, but they are scientists in their own right. And they have been observing these changes for decades. And in 2003, the Arctic Climate Impact Assessment released the world's most comprehensive, detailed regional assessment of climate change. It was prepared by more than three hundred scientists from fifteen countries, but it included the traditional knowledge throughout this process.

It's very stark reading for the peoples of the Arctic, and particularly Inuit. It says marine species dependent on sea ice, including polar bears, ice-living seals, walrus, and some marine birds, are very likely to decline, with some facing extinction. For Inuit, warming is likely to disrupt or even destroy their hunting and food-sharing culture, as reduced sea ice causes populations to decline or become extinct. So you see, climate change is not just an environmental issue with unwelcome economic consequences. It really is a matter of livelihood, food, individual and cultural survival. And it is absolutely a human issue affecting our children, our families, and certainly our communities. The Arctic is not a wilderness or a frontier. It is our home. It is our homeland.

Of course, the hunting culture is so misunderstood, and I know that I am in a roomful of my own fellow indigenous peoples who understand that a hunting culture is not just about the killing of animals. It is about teaching our young children, our children, about the opportunities and challenges of life on the land, but also about teaching them very transferable skills, such as patience, courage, how to be bold under pressure, how to withstand stress, how not to be impulsive, how to have sound judgment and, ultimately, wisdom. Those are the very things that traditional knowledge of hunting cultures teaches and passes on to the younger generation. And these skills are so transferable and, in fact, a requirement to surviving a transitioning culture such as ours and many others around the world.

And so it's more than meets the eye when we talk about environmental degradation of the Arctic. This is very much about a people trying to make it in this new world order of globalization, in a way that affords us respect and gives us a place in this world that we have always had.

The question always is, how can we bring some clarity and focus and purpose to a debate that always seems to be caught up in technical arguments and competing short-term economic ideologies? I believed strongly, and I still do, that it would be internationally significant if global climate change were debated and examined in the arena of human rights, an arena that many countries, particularly those in the developed world, say they "take seriously."

So after two years of preparation with really strong people who felt this was the right thing, and with a legal team, I and sixty-two other Inuit from Canada and Alaska filed a legal complaint on December 7, 2005, after concluding that the 1948 American Declaration on the Rights and Duties of Man, supported by the Inter-American Commission on Human Rights, may provide an effective means for us to defend our culture and our way of life. And what we're doing here is not asking necessarily for the world to take a complete economic backward step; what we're saying to the governments is, you must develop your economies using appropriate technologies that limit the pollution, that limit the greenhouse gases that are at the root of what is happening in the Arctic and the melting of the glaciers and the ice and the snow.

And we Inuit and other Northerners, of course, because we're at peril, because governments are taking a short-term view that is favored by many businesses, in fact what we are doing here is we are defending our right to culture, our right to lands traditionally used and occupied, our right to health, our right to physical security, our right to our own means of subsistence, and our rights to residence and movement. And as our culture is based on the cold, the ice and snow, we are in essence defending our right to be cold.

And we do not wish to become a footnote in the history of globalization, because the predictions are stark. We may lose what I had as a child, what has given me the foundation upon which I do this work; my grandson may lose this in his lifetime. So this is very much an issue of children, families, communities, and as we're just coming out the other side of modernization, here comes this second wave. And this second wave, I fear, is going to be even more challenging. So it's very real here.

We have lived in the Arctic for millennia, and our culture and our economy reflects the land and all that it gives, and we are connected to our land, to our ice and to our snow. We want it to be cold. And our understanding of who we are and our age-old knowledge and wisdom comes from the land, and it is that struggle—and you all, I know, relate to this so well—to thrive in that kind of environment that gives us the answers. Always it gives us the answers that we need to survive in the modern world. Our young people, who are making it, are the ones that are spending as much time as they can out there, hunting and fishing and taking in what our elders are giving them, and it is that outlook, that respectful outlook that sees humans' connection to everything, that should be informing the debate on climate change, as these monumental changes absolutely threaten the memory of who we were, who we are, and all that we wish to become. And so we Inuit, and certainly many indigenous peoples from around the world and in the North, remain very connected with each other and with the land.

I always ask the global community, is it not to reestablish that connection that we are all here trying to deal with this issue? Is it not because people have lost that connection between themselves and their neighbors, between their actions and the environment, that we are debating this issue of climate change in the first place?

So I think that by putting the climate change in the arena of human rights, we have moved the focus from being solely that of a political, economic, and technical issue to one of human impacts and consequences that do affect our children, our families, and our communities. We must remain vigilant in keeping climate change as a human and human rights issue.

The Future I Want for My Daughters

Barack Obama

One of the things I draw from the Genesis story is the importance of being good stewards of the land, of this incredible gift. And I think there have been times where we haven't been, and this is one of those times where we've got to take the warning seriously. Part of what my religious faith teaches me is to take an intergenerational view, to recognize that we are borrowing this planet from our children and our grandchildren.

And so we've got this obligation to them, which means that we've got to make some uncomfortable choices. And potentially, religious faith and the science of global warming converge precisely because it's going to be hard to deal with. We have to find resources in ourselves that allow us to make those sacrifices where we say, you know what? We're not going to leave it to the next generation. We're not going to wait.[1]

We cannot afford more of the same timid politics when the future of our planet is at stake. Global warming is not a someday problem; it is now. In a state like New Hampshire, the ski industry is facing

---

BARACK OBAMA is the forty-fourth president of the United States. His years of public service are based on his unwavering belief in the ability to unite people around a politics of purpose. He is a graduate of Columbia University and Harvard Law School, where he was the first African American president of the *Harvard Law Review*. Obama worked as a community organizer and practiced as a civil rights attorney before serving three terms in the Illinois Senate from 1997 to 2004.

---

1. "Democratic Candidates Compassion Forum," 13 April 2008, http://transcripts.cnn.com/TRANSCRIPTS/0804/13/se.01.html.

shorter seasons and losing jobs. We are already breaking records with the intensity of our storms, the number of forest fires, the periods of drought. By 2050 famine could force more than 250 million from their homes, and famine will increase the chances of war and strife in many of the world's weakest states. The polar ice caps are now melting faster than science had ever predicted. And if we do nothing, sea levels will rise high enough to swallow large portions of every coastal city and town.

This is not the future I want for my daughters. It's not the future any of us want for our children. And if we act now and we act boldly, it doesn't have to be. But if we wait, if we let campaign promises and State of the Union pledges go unanswered for yet another year, if we let the same broken politics that have held us back for decades win one more time, we will lose another chance to save our planet. And we might not get many more.

I want our children and our children's children to point to this generation and this moment as the time when America found its way again. As the time when America overcame the division and the politics and the pettiness of an earlier era so that a new generation could come together and take on the most urgent challenge of this era.[2]

Let's be the generation that finally frees America from the tyranny of oil. We can harness homegrown, alternative fuels like ethanol and spur the production of more fuel-efficient cars. We can set up a system for capping greenhouse gases. We can turn this crisis of global warming into a moment of opportunity for innovation, and job creation, and an incentive for businesses that will serve as a model for the world. Let's be the generation that makes future generations proud of what we did here. The time is now to shake off our slumber, and slough off our fear, and make good on the debt we owe past and future generations.[3]

---

2. Barack Obama, "Remarks of Senator Barack Obama: Real Leadership for a Clean Energy Future," 8 October 2007, Portsmouth, NH, www.barackobama.com/2007/10/08/remarks_of_senator_barack_obam_28.php.

3. Barack Obama, "Full Text of Senator Barack Obama's Announcement for President," 10 February 2007, Springfield, IL, www.barackobama.com/2007/02/10/remarks_of_senator_barack_obam_11.php.

Obligation to Posterity?

Alan Weisman

Here's a conundrum for you: If the answer to this question—Do we have a moral duty to leave a world as intact as our own to future generations?—turns out to be *yes,* then future generations may be doomed.

If it turns out to be *no,* however, there may be a chance for future generations to have a beautiful and functional Earth, even an Earth recognizably similar to our own.

Certain changes that we've already set in motion, of course, will have to play themselves out. Fifty years from now, the world will inevitably be somewhat altered. Shorelines will have shrunk some— or a lot, in fact, if the best current advice isn't heeded. Vegetation will have moved around: temperate forests are already edging pole-ward in either direction, and the tropics may dry out for a spell.

But this won't be the first time that nature has responded to dramatic global reconfiguring. Witness five previous extinctions, some far more devastating than even the one we're currently perpetrating, due to cataclysmic events like encounters with asteroids. Yet eventually the Earth always comes out looking gorgeous. Gorgeous, but different: an Age of Reptiles gets replaced by an Age of Mammals,

---

ALAN WEISMAN's reports from around the world have appeared in *Harper's, Atlantic Monthly,* the *New York Times Magazine,* the *Los Angeles Times Magazine, Orion, Mother Jones,* and *Discover* and on National Public Radio and Public Radio International. His most recent book, *The World Without Us* (2007), has been translated into thirty-three languages. A senior radio producer for Homelands Productions, he is also Laureate Professor of Journalism and Latin American Studies at the University of Arizona.

et cetera. But if we act soon, a fair amount of what we see around us may remain.

Speaking of mammals, here's why a moral imperative to save the planet as we know it is likely a deterrent to actually doing so:

First, we're mammals ourselves. We arguably have no more—or less—right to be here than any other species. But do bears, birds, beetles, and the rest also have an obligation to posterity?

I know, that sounds dumb. But why should it? What's the difference?

In his posthumous collection, *Pensées,* Blaise Pascal likened humans to "thinking reeds" suspended, somewhat miserably, somewhere between angels and animals. (There's an implicit assumption here suggesting that compared to us, animals don't think. We really have no way of knowing; an alien appearing on Earth might be far more impressed by the acumen, ingenuity, and equanimity of several species other than our own, ranging from ants to crows to cetaceans.)

But back to the point: when we describe ourselves as thinking creatures, what we mean is that we can imagine the consequences of our actions. Again, I'm not sure that makes us unique—a squirrel storing nuts for the winter seems to know what he's doing, and why. But fine; I'll grant that we can do that. Nevertheless, as we're increasingly reminded, being able to forecast likely results doesn't necessarily mean that we'll act on that knowledge. These days, we're practically swimming in knowledge, bobbing around in a great swamp of the stuff called the Information Age, with more gushing in continually via electronic pipes that are constantly being fed.

Doesn't seem to help solve anything, though.

It would be nice to kick the Information Age up a notch, into an Age of Wisdom. The problem is, philosophers, sages, and moral leaders—religious and secular—have been trying to do that for several thousand years. Yet as a whole, we don't pay attention. Lip service, maybe.

Our libraries are filled with literature and histories of great enlightened men and women. But they and their moral descendants have always been in the minority. In the big picture, that's never really mattered until now: it was a big world, with room for all kinds of opinions, including horrific ones, because even Holocausts eventually pass, and we've just gone on repeating cycles of history, as if, like the rising sun, there'd always be another tomorrow and another chance to try to get it right.

But for the first time in history, something is actually new under the sun, and although the sun itself will go on rising for quite some time to come, suddenly there are grounds for wondering whether our own chances will keep coming around, as they always have until now.

The first time this really dawned on me was in 1991, when I placed a call to John Frederick, an atmospheric physicist at the University of Chicago. At the time I was in Costa Rica, traveling through Latin America as part of a reporting team on assignment to National Public Radio for a series titled *Vanishing Homelands*. My colleagues and I were documenting how entire human cultures were becoming endangered, as their home ground was literally being ripped from beneath them by various kinds of resource exploitation. The night before I called Frederick, I'd heard a shortwave broadcast from Punta Arenas, Chile, the world's southernmost city, describing a new kind of menace. Residents there were increasingly worried that the expanding ozone hole, lately creeping closer overhead, might someday make their beautiful homeland—the islands, fjords, and snow-capped mountains of the Chilean archipelago—unlivable. People were reporting things they'd never seen before, from sunburned children to blind livestock to shriveling plants in their gardens.

Frederick was the scientist in charge of NASA ground monitors measuring ultraviolet radiation reaching Antarctica and Tierra del Fuego, just across the Strait of Magellan from Punta Arenas. "Are these people just being hysterical, or is there something to this?" I asked him. "Does this justify us going all the way down there to report a story? Or is it too soon to know?"

He hesitated before replying. "Had you called last week," he finally said, "I would have told you that it's too soon to know what all this means—that there hasn't been enough time, or research, to correlate ozone loss with effects on humans."

"But now?"

"But this week I've been looking at data we just collected. It looks like the amount of ultraviolet light hitting the Antarctic Peninsula has doubled over the last ten years. We still don't know what this means. But if you wait to do your story until we can say conclusively, it may already be too late."

We ended up making two trips, to southern Chile and to a U.S. Antarctic research station, and what we learned vindicated our decision: scientists showed us alarming evidence that ozone depletion was a threat to living systems. We also saw signs of yet another looming concern: glaciers had receded a kilometer up Chilean fjords over the previous ten years, and grasses were sprouting from newly bare ground in coastal Antarctica.

Today, nearly two decades later, we all know that we humans have rejiggered the atmosphere. Our CFC refrigerants and methyl bromide agricultural fungicides have dug a still deepening and broadening ozone hole, which we can only hope begins to shrink as we gradually replace culprit chemicals with more benign substitutes. However, the solution to chunks the size of Connecticut calving off Antarctic ice shelves, freshwater flooding seaward from beneath Greenland glaciers, and methane bubbling up from thawed permafrost all along the coast of Siberia and Alaska may be far trickier. All the more reason to stop splitting hairs over whether we know enough to act already.

But I suspect the motivation to do so has to be more compelling than morality. Because if history shows us anything, it's that baser instincts get more of us moving quicker. Rather than trying to appeal to reason or higher values—things that most people either don't grasp or care about, or too often simply ignore—I suggest we appeal to greed and selfishness. It simply takes too long to change enough people's minds and hearts to make them act out of obligation to a good greater than themselves. Not that it isn't worth trying. But if we put all our best efforts into raising the masses' consciousness regarding the eco-errors of their ways, the planet will likely be destroyed long before we've swayed a majority.

Where the environment is concerned, majority rules. Wise leaders throughout history may have been able to steer policies or economies without having to enlighten every citizen, because most citizens had little or no power in such matters. But environmental impact is different. Every person adds his garbage to the pile or her exhaust to the atmosphere. Collectively, the sheer numbers of populations and their demands are the deciding factors in the environment's fate.

In fact, attempts to impose environmental awareness, even by the gentlest persuasion—such as touching images of children, accompanied by a narration asking what kind of planet we want them to inherit—have regularly undermined themselves by igniting a backlash. The decade following the first Earth Day will be remembered not so much by the raising of global environmental awareness as by the rise of a great gobbling globalized marketplace on a wave of world trade pacts. The unintended response to the message that the planet's natural resources were limited was a rush by free marketeers to go out and grab all they could while the resources were still around.

So my vote is to stop trying to tap everyone's highest moral self to persuade people of their obligations to posterity or eternity. Let's aim for the lowest common denominator that anyone can grasp: self-interest. For purely selfish reasons, everyone alive has a stake in fixing what ails the planet right now, because the future's already upon us. Forget what our children may face years down the road: already it's hotter and stormier, so our own personal survival is at play here. That's a warning that every organism on Earth is hardwired to hear. And in the same breath, let's suggest that the surest way to prosperity from here on is by designing and selling new systems, tools, dwellings, and services that mimic nature by turning every waste product into something useful, over and over again.

Let's show everyone why it's in our own selfish interest to limit population: not just because it will leave more room on the planet for other species, reduce demand for resources, and lessen the amount of CO_2 expelled skyward, but because in every family there'll be more to go around if there are fewer mouths to feed and bodies to clothe. Let's tempt them with examples like Italy, a country so charming we all want to visit it, yet also a Catholic country whose growth rate surprisingly vies with Catholic Spain's for the lowest on the planet, simply because they educate their women. Italy has one of if not the highest per capita number of female PhDs. An educated woman defers her child-bearing until her studies are through, and then doesn't have so many kids that she can't exercise the interesting and useful profession she's trained for. And because she inevitably contributes something meaningful to her society, everyone benefits from her self-interest.

Everyone, but especially posterity: a world with fewer people in it will be that much easier for our descendants to manage and will present that many more opportunities for them. And will be that much more beautiful.

There's some wisdom in old farmers' adages and Buddhist koans that equally suggest that the future takes care of itself if we take care of today. Be in the moment. The *now*.

I wrote that—and you read it—in the past. In a second, it'll already be the future. Because we're still part of this little slice of time's continuum, we selfishly—come on, let's admit it—dearly wish it to be the best future possible. If each of us *Homo sapiens* starts to do something about fixing the future for our own selfish pleasure and sustenance, there may be more future to come, a future with others of our kind still around to enjoy and exult in it.

ETHICAL ACTION

How can we save the Earth, and so save ourselves?

What if aliens invaded the planet, Derrick Jensen asks, and set about destroying it? Say they poisoned the water, skinned off mountaintops and dumped them on top of valley towns, killed off 10 percent of the plants and animals on Earth, engineered infertility into crops, and so on—we know the list only too well. What would we do? We would mobilize. We would suspend all other activities, set aside petty animosities, and spend what we needed to spend, and we would save the planet.

Who are these aliens? Pogo says, "We have met the enemy and he is us." That may be true, but only in a paradoxical and partial way. Yes, our decisions about where to invest our money and our working lives empower global corporate powers to act like Derrick's aliens. And yes, there's a process of self-reflection that's important here: Who benefits and who is harmed by the purchases I make? Who benefits and who is harmed by the investments I make? In my daily work, have I become an agent of destruction? There are great and powerful forces in the world trying to influence our decisions for their own benefit and to the expense of much else, including ourselves. Being cognizant and independently thoughtful about this is a necessary first step. We *should* be making responsible decisions as consumers. We *should* refuse to invest in companies that profit from death or damage to the Earth. We *should not* choose careers in destructive industries.

So Pogo is right; but he's wrong, too. Of course industrial corporations would like us to blame ourselves for the global mess. But in fact, corporate actions have gone far beyond what we have authorized as consumers, investors, or employees, and it is a mistake (or perhaps a deception) to think that we still control, and thus have moral responsibility for, the acts of corporate powers. We need to wrest away unlimited corporate license, which wrecks democracy as thoroughly as it is wrecking the Earth.

Here's an idea, from Onandaga Wisdomkeeper Oren Lyons, about how we might collectively do so. Imagine a global Council of Elders—maybe Jimmy Carter, Mikhail Gorbachev, Nelson Mandela, Wangari Maathai, or others. Imagine that they meet and name the corporation that is doing the most damage to the Earth. Imagine that they call a global boycott of that corporation and everyone associated with it. Let every social justice organization, every church group, every environmental nonprofit, every student group, labor union, and human rights organization join the boycott, which will last until the corporation agrees to end its aggression against the Earth and against effective democracy and allies its considerable creative powers with life-sustaining work. Then the council will meet and name the second most destructive corporation on Earth. And so it will go, until corporations around the world change their direction toward what is life-enhancing and world-sustaining work—or go out of business.

We could do that.

Or, if we get tired of waiting for the Council of Elders, we could do it ourselves—individually, or as neighborhoods or organizations in concert.

Is it too radical a step to try to reduce or redirect the power of corporations? We are talking, folks, about saving the possibility of ongoing life on Earth.

2

> *Do we have a moral obligation to take action to protect the future of a planet in peril?*
>
> **Yes, for the sake of the children.**

What is it about little children, that we grant them unconditional moral worth?

It may be their innocence. A small child can never deserve to suffer, because she can never do the wrong that might merit suffering in return. The suffering of any child is unjust. *It may be their promise.* Children contain the future of the whole human race in their little bodies. If we value the future of humanity, we must do right by children. *It may be our love for them.* Responsibilities grow from the relation of parent to child. As we acknowledge the responsibility to care for our own children, we acknowledge the responsibility to care for all children, who can't be distinguished one from another on moral grounds. *Or maybe it's just the driving force of evolution.* The protectiveness we feel toward children may be the manifestation of our urge to leave something of ourselves in those who will come next. Regardless of what ethical theory we turn to, we find a moral imperative to prevent harm to children.

But we are harming children, even as (especially as) we believe we are acting to provide for them. Think of what we decide to do for the privileged children—the poison in the plastic car seat, the disease in the pesticide-treated fruit, the disastrous coal company in the college investment portfolio, the mall where there had been frogs, the carbon load of the soccer tournament. Even as they harm the children, these decisions harm their futures, stealing from them a world as rich and delightful as the world we were born to. It's a tragic irony that the amassing of material wealth in the name of our children's futures hurts them.

But what our decisions will do to the children who are not privileged is not just an irony; it's a moral abomination. These children, who will never know even the short-term benefits of misusing fossil fuels, are the ones who will suffer as seas rise, as fires scorch cropland, as diseases spread north, as famine returns to lands that had been abundant. The damage to their future is a deliberate theft, a preventable child abuse.

If we have a moral obligation to protect the children, and if environmental harms and climate change are manifestly harmful to them, then we have a moral obligation to expend extraordinary effort to immediately stop such harms and to redress the harms that we have already done in their name.

Keepers of Life

Oren Lyons

> In our way of life, in our government, with every decision we make, we always keep in mind the seventh generation to come. It's our job to see that the people coming ahead, the generations still unborn, have a world no worse than ours—and hopefully better. When we walk upon Mother Earth, we always plant our feet carefully because we know the faces of our future generations are looking up at us from beneath the ground. We never forget them.
>
> OREN LYONS
in Michael J. Caduto and Joseph Bruchac, *Keepers of Life*

Our societies are based on great democratic principles of the authority of the people and equal responsibilities for the men and the women. This was a great way of life across this Great Turtle Island, and freedom with respect was everywhere. Our leaders were instructed to be men of vision and to make every decision on behalf of the seventh generation to come; to have compassion and love for those generations yet unborn. We were instructed to give thanks for All That Sustains Us.

OREN LYONS is the Faithkeeper of the Turtle Clan of the Onondaga Nation of the Haudenosaunee. He is Distinguished Service Professor Emeritus of American Studies at the University of Buffalo and cofounder of the national American Indian quarterly news magazine, *Daybreak*. He is also a board member of the Harvard Project on American Economic Development and chairman of the Board Honoring Contributions in the Governance of American Indian Nations.

Thus, we created great ceremonies of thanksgiving for the life-giving forces of the Natural World; as long as we carried out our ceremonies, life would continue. We were told that "The Seed is The Law." Indeed, it is The Law of Life. It is The Law of Regeneration. Within the seed is the mysterious force of life and creation. Our mothers nurture and guard that seed, and we respect and love them for that, just as we love *I hi do' hah,* our Mother Earth, for the same spiritual work and mystery.

We were instructed to be generous and to share equally with our brothers and sisters so that all may be content. We were instructed to respect and love our Elders, to serve them in their declining years, to cherish one another. We were instructed to love our children, indeed, to love ALL children. We were told that there would come a time when parents would fail this obligation and that we could judge the decline of humanity by how we treat our children.

We were told that there would come a time when the world would be covered with smoke, and that it would take our Elders and our children. It was difficult to comprehend at the time, but now all we have to do is walk outside to experience that truth. We were told that there would come a time when we could not find clean water to wash ourselves, to cook our foods, to make our medicines, and to drink. And there would be disease and great suffering. Today we can see this and we peer into the future with great apprehension. We were told there would come a time when, tending our gardens, we would pull up our plants and the vines would be empty. Our precious seed would begin to disappear. We were instructed that we would see a time when young men would pace back and forth in front of their chiefs and leaders in defiance and confusion.

So, then, what is the message I bring to you today? Is it our common future? It seems to me that we are living in a time of prophecy, a time of definitions and decisions. We are the generation with the responsibilities and the option to choose the Path of Life for the future of our children. Or the life and path that defies the Laws of Regeneration.

Even though you and I are in different boats, you in your boat and we in our canoe, we share the same River of Life. What befalls me, befalls you. And downstream, downstream in this River of Life, our children will pay for our selfishness, for our greed, and for our lack of vision.

Five hundred years ago, you came to our pristine lands of great forests, rolling plains, crystal-clear lakes and streams and rivers. And we have suffered in your quest for God, for Glory, for Gold. But we have survived. Can we survive another five hundred years of "sustainable development"? I don't think so. Not with today's definitions of "sustainable." I don't think so.

So reality and the Natural Law will prevail, the Law of the Seed and Regeneration. We can still alter our course. It is NOT too late. We still have options. We need the courage to change our values to the regeneration of our families, the life that surrounds us. Given this opportunity, we can raise ourselves. We must join hands with the rest of Creation and speak of Common Sense, Responsibility, Brotherhood, and Peace. We must understand that the law *is* the seed and that only as true partners can we survive.

On behalf of the Indigenous People of the Great Turtle Island, I give my appreciation and thanks. *Dah ney' to.* Now I am finished.

We Bear You in Mind

Scott Russell Sanders

You are still curled in the future, like seeds biding your time. Even though you are not yet born, I think of you often. I feel the promise of your coming the way I feel the surge of spring before it rises out of the frozen ground. What marvels await you on this wild Earth! When you do rise into the light of this world, you'll be glad of your fresh eyes and ears, your noses and tongues, your sensitive fingers, for they will bring you news of a planet more wonderful and mysterious than anything I can tell you about in mere words.

Mere words are all I have, though, to speak of what I've treasured during my days, and to say what I hope you'll find when you take your turn under the sun. So I write this letter. As I write, I'm leaning against the trunk of a fat old maple in the backyard of our house here in the southern Indiana hills. It's early one April morning, and the birds are loudly courting. I'm surrounded by the pink blossoms of wild geraniums, the yellow of celandine poppies, the blue of phlox. A thunderstorm is building in the western sky, and a brisk wind is rocking the just-opened leaves. My pleasure from wind and rain, from cloud drift and birdsong, from the sound of creeks tossing in their stony beds, from the company of animals and the steady presence

---

SCOTT RUSSELL SANDERS is Distinguished Professor of English at Indiana University. He is a writer of fiction and essays whose most recent books are *A Private History of Awe* (2007) and *A Conservationist Manifesto* (2009). Among other honors, he received the Associated Writing Programs Award in Creative Nonfiction, the Lannan Literary Award, the Mark Twain Award, and Indiana University's highest teaching award.

of trees—all of that immense delight is doubled when I think of you taking pleasure one day from these same glories.

Even here in a tame backyard, Earth's energy seems prodigious. The grasses and ferns are stiff with juice. The green pushing out of every twig and stem, the song pulsing out of every throat, the light gleaming on the needles of white pines and in the bright cups of flowers, the thunderclouds massing, the wind rising—all speak of an inexhaustible power. You will feel that power in your day, surely, for nothing we do could quench it. But everything we do may affect the way that power moves and the living shapes it takes on. Will there be whales for you to watch from a bluff on the Oregon coast, as I watched with my own children? Will there be ancient redwoods and cedars and white oaks and sycamores for you to press your cheeks against? In your day, will there be monarch butterflies sipping nectar in gardens, bluebirds nesting in meadows, crayfish scuttling in creeks, spring peepers calling from ponds?

Because of the way my generation and those that preceded us have acted, Earth has already suffered worrisome losses—forests cut down, swamps drained, topsoil washed away, animals and plants driven to extinction, clean rivers turned foul, the very atmosphere unsettled. I can't write you this letter without acknowledging these losses, for I wish to be honest with you about my fears as well as my hopes. But I must also tell you that I believe we can change our ways, we can choose to do less harm, we can take better care of the soils and waters and air, we can make more room for all the creatures who breathe. And we are far more likely to do so if we think about the many children who will come after us, as I think about you.

I think of you as lightning cracks the horizon and thunder comes rolling in like the distant rumble of trains. When I was little, thunderstorms frightened me, so when the rumbling began, my father would wrap me in a blanket and carry me onto the porch and hold me close as the sky flashed and the air shook and the rain poured down. Safe in his arms, I soon came to love the boom and crash. So when my own two children were little, first Eva and then Jesse, I wrapped them in blankets and carried them onto the porch to watch the lightning and hear the thunder and feel the mist and smell the rain. Now I am doing the same with my grandchildren. Maybe one day a parent or grandparent will hold you during a storm and then you will not only read what I'm saying but feel it in your bones.

The smell of rain reaches me now on a wind from the west, and my skin tingles. The stout maple thrums against my back, like a thick string plucked. This old tree is tougher than I am, more supple, more durable, for it stands here in all weathers, wrapped in bark against the heat and cold, deeply rooted, drawing all it needs from dirt and air and sun. Often, when I come home feeling frazzled from the demands of the day, before I go into the house I stop here in the yard and press my hands against this maple, and I grow calm.

I hope you will find companion trees of your own, where you can hear the birds hurling their lusty cries and watch the flowers toss their bright blooms. May you climb into the branches to feel the huge body swaying beneath you and the wind brushing your face like the wings of angels.

I hope you'll be able to live in one place while you're growing up, so you'll know where home is, so you'll have a standard to measure other places by. If you live in a city or suburb, as chances are you will, I hope you'll visit parks, poke around in overgrown lots, keep an eye on the sky, and watch for the tough creatures that survive amid the pavement and fumes. If you live where it never snows, I hope you'll be able to visit places where the snow lies deep in winter. I want you to see the world clarified by that coating of white, hear the stillness, bear the weight and cold of it, and then relish warmth all the more when you go indoors. Wherever you live, I hope you'll travel into country where the land obeys laws that people didn't make. May you visit deep forests, where you can walk all day and never hear a sound except the scurry and calls of animals and the rustle of leaves and the silken stroke of your own heart.

When I think of all the wild pleasures I wish for you, the list grows long. I want you to be able to chase fireflies as they glimmer in long grass, watch tadpoles turn into frogs in muddy pools, hear loons calling on clear lakes, glimpse deer grazing and foxes ambling, lay your fingers in the pawprints of grizzlies and wolves. I want there to be rivers you can raft down without running into dams, the water pure and filled with the colors of sky. I want you to thrill in spring and fall to the ringing calls of geese and cranes as they fly overhead. I want you to see herds of caribou following the seasons to green pastures, turtles clambering onshore to lay their eggs, alewives and salmon fighting their way upstream to spawn. And I want you to feel in these movements Earth's great age and distances, and to sense how the whole planet is bound together by a web of breath.

As I sit here in this shaggy yard writing to you, I remember a favorite spot from the woods behind my childhood house in Ohio, a meadow encircled by trees and filled with long grass that turned the color of bright pennies in the fall. I loved to lie there and watch the clouds, as I'm watching the high, surly stormclouds rolling over me now. I want you to be able to lie in the grass without worrying that the kiss of the sun will poison your skin. I want you to be able to drink water from faucets and creeks, to eat fruits and vegetables straight from the soil. I want you to be safe from lightning and loneliness, from accidents and disease. I would spare you all harm if I could. But I also want you to know there are powers much older and grander than our own—earthquakes, volcanoes, tornados, thunderstorms, glaciers, floods. I pray that you will never be hurt by any of these powers, but I also pray that you will never forget them. And remember that nature is a lot bigger than our planet: it's the shaping energy that drives the whole universe, the wheeling galaxies as well as water striders, the shimmering pulsars as well as your beating heart.

Thoughts of you make me reflect soberly on how I lead my life. When I spend money, when I turn the key in my car, when I vote or refrain from voting, when I fill my head or belly with whatever's for sale, when I teach students or write books, ripples from my actions spread into the future, and sooner or later they will reach you. So I bear you in mind. I try to imagine what sort of world you will inherit. And when I forget, when I serve only my own appetite, more often than not I do something wasteful. By using up more than I need—of gas, food, wood, electricity, space—I add to the flames that are burning up the blessings I wish to preserve for you.

I worry that the choices all of us make today, in our homes and workplaces, in offices and legislatures, will leave fewer choices for you and your own children and grandchildren, fifty or a hundred years from now. By indulging our taste for luxuries, we may deprive you of necessities. Our laziness may cause you heavy labor. Our comfort may cause you pain. I worry that the world you find will be diminished from the one we enjoy.

If Earth remains a blessed place in the coming century, you'll hear crickets and locusts chirring away on summer nights. You'll hear owls hoot and whippoorwills lament. You'll smell wet rock, lilacs, new-mown hay, peppermint, lemon balm, split cedar, piles of autumn leaves. On damp mornings you'll find spiderwebs draped like handkerchiefs on the grass. You'll watch dragonflies zip and hover, then flash away, so fast, their wings thinner than whispers. You'll watch beavers nosing across the still waters of ponds, wild turkeys browsing in the stubble of cornfields, and snakes wriggling out of their old skins.

If we take good care in our lifetime, you'll be able to sit by the sea and watch the waves roll in, knowing that a seal or an otter may poke a sleek brown head out of the water and gaze back at you. The skies will be clear and dark enough for you to see the moon waxing and waning, the constellations gliding overhead, the Milky Way arching from horizon to horizon. The breeze will be sweet in your lungs and the rain will be innocent.

The rain has reached me now, rare drops at first, rattling the maple leaves over my head. There's scarcely a pause between lightning and thunder, and every loud crack makes me jump. It's time for me to get out from under this big old tree, and go inside to keep this paper dry. So a few more words, my darlings, and then goodbye for now.

Thinking about you draws my heart into the future. I want you to look back on those of us who lived at the beginning of the twenty-first century and know that we bore you in mind, we cared for you, and we cared for our fellow tribes—those cloaked in feathers or scales or chitin or fur, those covered in leaves and bark. One day it will be your turn to bear in mind the coming children, your turn to care for all the living tribes. The list of wild marvels I would save for you is endless. I want you to feel wonder and gratitude for the glories of Earth. I hope you'll come to feel, as I do, that we're already in paradise, right here and now.

For the Children

Gary Snyder

The rising hills, the slopes,
of statistics
lie before us.
The steep climb
of everything, going up,
up, as we all
go down.

In the next century
or the one beyond that,
they say,
are valley, pastures,
we can meet there in peace
if we make it.

To climb these coming crests
one word to you, to
you and your children:

stay together
learn the flowers
go light

GARY SNYDER is a Zen poet, environmental activist, and professor emeritus at the University of California, Davis. He has published more than eighteen books of poetry and prose, including the Pulitzer Prize–winning *Turtle Island* (1975) and his most recent book, *Back on the Fire: Essays* (2008). He has received an American Academy of Arts and Letters Award, the Bollingen Prize, a Guggenheim Foundation Fellowship, the Bess Hokin Prize, the Levinson Prize from *Poetry*, the Robert Kirsch Lifetime Achievement Award, and the Shelley Memorial Award.

Steering the Earth Toward Our Children's Future

John Paul II and the Ecumenical Patriarch His Holiness Bartholomew I

Following a June 2002 symposium on religion, science, and the environment sponsored by the Patriarch on board the ship Festos Palace, *the two leaders signed this joint declaration, which articulates a code of environmental ethics.*

At this moment in history, at the beginning of the third millennium, we are saddened to see the daily suffering of a great number of people from violence, starvation, poverty, and disease. We are also concerned about the negative consequences for humanity and for all creation resulting from the degradation of some basic natural resources such as water, air, and land, brought about by an economic and technological progress which does not recognize and take into account its limits.

ECUMENICAL PATRIARCH BARTHOLOMEW I has been the Patriarch of Constantinople, and thus "first among equals" in the Eastern Orthodox Communion, since 1991. He is the spiritual leader of 300 million Orthodox Christians around the world and has earned the title "Green Patriarch" for his efforts to raise environmental awareness. In 2008 he published *Encountering the Mystery: Understanding Orthodox Christianity Today.*

POPE JOHN PAUL II (1920–2005) was born Karol Józef Wojtyła in 1920 and served as Pope of the Catholic Church and Sovereign of Vatican City from October 1978 until his death almost twenty-seven years later. When he became Pope, John Paul II was already an avid sportsman, jogging in the Vatican gardens, weightlifting, swimming, and hiking in the mountains. John Paul II has been widely acclaimed as one of the most influential leaders of the twentieth century.

Almighty God envisioned a world of beauty and harmony, and He created it, making every part an expression of His freedom, wisdom, and love (cf. Genesis 1:1–25). At the center of the whole of creation, He placed us, human beings, with our inalienable human dignity. Although we share many features with the rest of the living beings, Almighty God went further with us and gave us an immortal soul, the source of self-awareness and freedom, endowments that make us in His image and likeness (cf. Genesis 1:26–31; 2:7). Marked with that resemblance, we have been placed by God in the world in order to cooperate with Him in realizing more and more fully the divine purpose for creation.

At the beginning of history, man and woman sinned by disobeying God and rejecting His design for creation. Among the results of this first sin was the destruction of the original harmony of creation. If we examine carefully the social and environmental crisis which the world community is facing, we must conclude that we are still betraying the mandate God has given us: to be stewards called to collaborate with God in watching over creation in holiness and wisdom.

God has not abandoned the world. It is His will that His design and our hope for it will be realized through our cooperation in restoring its original harmony. In our own time we are witnessing a growth of an ecological awareness which needs to be encouraged, so that it will lead to practical programs and initiatives. An awareness of the relationship between God and humankind brings a fuller sense of the importance of the relationship between human beings and the natural environment, which is God's creation and which God entrusted to us to guard with wisdom and love (cf. Genesis 1:28).

Respect for creation stems from respect for human life and dignity. It is on the basis of our recognition that the world is created by God that we can discern an objective moral order within which to articulate a code of environmental ethics. In this perspective, Christians and all other believers have a specific role to play in proclaiming moral values and in educating people in ecological awareness, which is none other than responsibility toward self, toward others, toward creation.

What is required is an act of repentance on our part and a renewed attempt to view ourselves, one another, and the world around us within the perspective of the divine design for creation. The problem is not simply economic and technological; it is moral and spiritual. A solution at the economic and technological level can be found only if we undergo, in the most radical way, an inner change of heart, which can lead to a change in lifestyle and of unsustainable patterns of consumption and production. A genuine conversion in Christ will enable us to change the way we think and act.

First, we must regain humility and recognize the limits of our powers and, most importantly, the limits of our knowledge and judgment. We have been making decisions, taking actions, and assigning values that are leading us away from the world as it should be, away from the design of God for creation, away from all that is essential for a healthy planet and a healthy commonwealth of people. A new approach and a new culture are needed, based on the centrality of the human person within creation and inspired by environmentally ethical behavior stemming from our triple relationship to God, to self, and to creation. Such an ethics fosters interdependence and stresses the principles of universal solidarity, social justice, and responsibility, in order to promote a true culture of life.

Secondly, we must frankly admit that humankind is entitled to something better than what we see around us. We and, much more, our children and future generations are entitled to a better world, a world free from degradation, violence, and bloodshed, a world of generosity and love.

Thirdly, aware of the value of prayer, we must implore God the Creator to enlighten people everywhere regarding the duty to respect and carefully guard creation.

We therefore invite all men and women of good will to ponder the importance of the following ethical goals:

1. To think of the world's children when we reflect on and evaluate our options for action.
2. To be open to study the true values based on the natural law that sustain every human culture.
3. To use science and technology in a full and constructive way, while recognizing that the findings of science have always to be evaluated in the light of the centrality of the human person, of

the common good, and of the inner purpose of creation. Science may help us to correct the mistakes of the past, in order to enhance the spiritual and material well-being of the present and future generations. It is love for our children that will show us the path that we must follow into the future.

4. To be humble regarding the idea of ownership and to be open to the demands of solidarity. Our mortality and our weakness of judgment together warn us not to take irreversible actions with what we choose to regard as our property during our brief stay on this Earth. We have not been entrusted with unlimited power over creation; we are only stewards of the common heritage.

5. To acknowledge the diversity of situations and responsibilities in the work for a better world environment. We do not expect every person and every institution to assume the same burden. Everyone has a part to play, but for the demands of justice and charity to be respected the most affluent societies must carry the greater burden, and from them is demanded a sacrifice greater than can be offered by the poor. Religions, governments, and institutions are faced by many different situations, but on the basis of the principle of subsidiarity all of them can take on some tasks, some part of the shared effort.

6. To promote a peaceful approach to disagreement about how to live on this Earth, about how to share it and use it, about what to change and what to leave unchanged. It is not our desire to evade controversy about the environment, for we trust in the capacity of human reason and the path of dialogue to reach agreement. We commit ourselves to respect the views of all who disagree with us, seeking solutions through open exchange, without resorting to oppression and domination.

It is not too late. God's world has incredible healing powers. Within a single generation, we could steer the Earth toward our children's future. Let that generation start now, with God's help and blessing.

A Letter to My Boys

Hylton Murray-Philipson

At the beginning of January, I climbed to the top of Turtle Mountain in the heart of the Iwokrama forests of central Guyana. A rocky outcrop gave a magnificent view, and as I scanned the forest stretching as far as the eye could see, swifts flew overhead and a troupe of spider monkeys swung from branch to branch. A shower of rain traversed the landscape in front of us as if from a giant watering can. We were enveloped by stillness and a silence broken only by the shriek of a scarlet macaw.

In the language of the Makushi Indians, Iwokrama means "place of refuge." Little did the Indians realize when they named it how in the twenty-first century it would become rare to find a place where the sounds of the wild are not drowned by the sounds of engines or where a mobile phone is never going to ring. I sat on the rock surrounded by the glory of Creation—reflecting on my life half lived, your lives just beginning, and the future of human life on Earth. Maybe the view from that rock helped me see things more clearly— "Super hanc petram, aedificabo meam ecclesiam" (*on this stone I will build my church*).

There are limits beyond which we cannot go without breaking the covenant between man and God, between humanity and the natural

---

HYLTON MURRAY-PHILIPSON is a former investment banker who tries to harness the power of markets to address environmental issues across the world. He is director of a number of recycling and renewable energy companies in the United Kingdom, adviser to the rainforest project of HRH the Prince of Wales, trustee of the Global Canopy Programme, and honorary chief of the Yawanawa people in Brazil.

world—and those limits are close at hand. We stand at the fork in the road when we must choose between sustainability and catastrophe. On current trends we may have ten years before we cross the point of no return on climate change. Our generation bears a unique responsibility to those who come after—unique, because this set of circumstances has never existed before.

The pace of change in one hundred years has been extraordinary. In 1900, the Wright brothers had not yet flown and the world's population was 1.5 billion. The impact of man on the natural world was limited, and it still made sense to convert natural capital into manufacturing and financial capital to improve the human condition.

Half a century later, in the ruins of World War II, the Bretton Woods Summit laid the foundations of the modern global economy. At that time, there was no mention of "sustainability" or "the environment." The assumption was that nature could be taken for granted. There were plenty of warning signals, even then, that unrestricted exploitation of nature had unfortunate side effects—the near-extinction of the North American bison, the pollution of the Great Lakes, the smogs in London—but the notion of climate change was unheard-of, and the bounty of the planet was still considered unlimited.

As the Beatles said, "Give me more . . . it's all I want." Growth became the yardstick by which politicians and policies were judged—except in Bhutan, where an enlightened monarchy measured Gross National Happiness instead of Gross National Product.

Global population quadrupled to 6 billion in the twentieth century, and the relationship between man and nature has been turned upside down in the process. Economic growth was achieved through burning fossil fuels, releasing in a few decades the energy of the sun trapped over millions of years in coal, oil, or natural gas. As a result, atmospheric levels of CO_2 have risen from preindustrial levels of 280 parts per million (ppm) to 387 ppm. Despite increasing scientific alarm, they continue to rise by 3 ppm per year.

Another unfortunate side effect of the dash for growth is the loss of half of the world's tropical forests. However rich we are on paper, we will have nothing but fool's gold if we lose the other half. Until we alter our perception of value, the world will continue to lose some

5.5 million hectares of tropical forest each year—approximately fifteen thousand hectares a day, or fifty football fields a minute.

Tropical forests produce oxygen, maintain biodiversity, moderate temperature, produce rainfall, and store billions of tons of carbon—and yet the economic model of the world says that this is all "off balance sheet" and without value. An accountant would say that if we cut down everything in sight, sold the timber, burned the rest, and replaced the forest with herds of cattle, we'd all be millionaires. And yet, my boys, we'd be converting an oxygen plant into a methane factory, drying out the sponge that feeds the rains, replacing the diversity of life with the monoculture of production, and emitting up to one thousand tons of CO_2 per hectare as we did it.

Looking down on the forest, I was reminded of Jesus and the temptations in the wilderness. "All the kingdoms of the world, all their power and glory, I will give you," the Devil said, "if you will bow down before me." I could make millions, but the world would be impoverished and your inheritance would be a wasteland.

Our situation triggers a fundamental question of values: Do we value only commodities that we can consume, or do we also value the services of the forest on which life itself depends? Do we consider ourselves above and separate from nature, or do we acknowledge that we are part of the web of life around us? Do we think just of ourselves, or can we consider other forms of life and the generations to come?

The conflict between personal gain and the common good lies at the heart of the problem. Although we collectively struggle with global warming, individually we hack away at the Earth's air-conditioning unit. As we fight to stay within tolerable limits of atmospheric CO_2, the destruction of tropical forests is the suicide note of humankind.

Of one thing I am certain. Halting the destruction of tropical forests is imperative if we are to have any hope of achieving climatic stability by the middle of the century. It is the best way to buy time. With forests we have a chance to halt global climate change; without them, we do not.

The world is crying out for change—not for tinkering at the margins, but for profound change in the way we lead our lives. There on that rock in the middle of Guyana, my spirit soared with the swifts and I recalled the ancient wisdoms of indigenous people. "We need to tread gently on the Earth, for the Earth is our home and our mother. We inherit the Earth from our parents, and we borrow it from our children. If we hurt the Earth, we hurt ourselves."

The economic model off which the world is working came from an Age of Innocence and is not fit for our purpose in the twenty-first century—the Age of Consequences. The rules for 1 billion people cannot be the same for 6 billion, going on 9 billion by 2050. This is a defining moment, full of danger but also full of hope. "Business as usual" would end in tears, but if we can dare to think differently, your lives can be more wonderful than any that have gone before.

The current financial crisis creates the opportunity to do things differently. In 2008, stock markets around the world lost US$32 trillion in "value" and plunged the world into recession. Yes, that's $32 thousand billion—one thousand times the $30 billion that the Stern Report called for to halve the destruction of tropical forests.

The financial meltdown is redefining boundaries between public and private sectors as billions upon billions of dollars, pounds, and euros are poured into bailouts, restructurings, and stimulus packages. A fundamental reappraisal is required of the regulatory framework that allowed the excesses of "subprime" in the pursuit of private profit. But if that reappraisal merely tinkers with the bonus culture and does not question the viability of the model itself, a truly historic opportunity will be lost. Now is the moment to step back, reflect where this is all heading, and begin to value the natural services on which life itself depends.

Why would we pay US$100 a ton for industrial carbon capture and storage (CCS), while ignoring the CCS function of the standing forest? The forests are reacting to higher levels of atmospheric CO_2 by bulking up to the tune of one additional ton of CO_2 per hectare per year, trying to help us even as we hack away at our life-support system. That ton of natural sequestration needs to be valued the same as any other ton of CO_2 around the world. After all, the environment does not notice the difference; it is just the limitations of our accounting treatment that get us into trouble. We are still seeing through the glass darkly, not looking at the matter face-to-face.

Globally, we spend US$3 trillion per year on insurance, but we pay nothing to insure the future of human life itself. Can we not raise a 1 percent levy to finance the conservation of forests around the world on which we all depend?

We can, and we must.

Individual nations cannot solve these issues on their own. There has to be a global agreement on global sustainability, encompassing rich and poor, black and white, Muslim, Christian, Jew. We are all in the same business, the business of survival. There is no point pursuing wealth for its own sake—we need to recall the wisdom of indigenous people who assess the consequences of their actions not on quarterly earnings, but on the prospects of generations to come.

Now is the last chance to seize the challenge of climate change. If we put it off for ten more years, there is no realistic chance of staying within reasonable climatic limits this century, defined as limiting average temperature increase to 2 degrees or staying within 350 parts per million of CO_2 in the atmosphere. Unlike previous generations, we do not have the excuse of ignorance. The reports cascade onto my desk in London—the Stern Review, the Inter-Governmental Panel on Climate Change (IPCC), the Eliasch Review, Mckinsey . . . The reports are in; the jury is out. Although the truth they contain is "inconvenient," we cannot bury our heads in the sand for a moment longer.

Empires come and go. We can reduce interest rates, repair balance sheets, or print money. But once the glaciers and the forests have gone, there is no way to get them back. We may have billions in the bank, but human life on Earth will end.

As I looked out on the teeming life in the forest, I recalled the creation story in the book of Genesis—how God created a mist to water the Earth before He created man. How, in other words, the rainfall of the forests is a precondition to human life on Earth.

At this defining moment, I pray for wisdom to remind ourselves that we are part of nature and not above it, and I pray that we recognize the things that are truly valuable before it is too late. For your sakes, as we search for a viable way forward, I pray that we turn to the Tree of Life as well as to the Tree of Knowledge.

You Choose

Derrick Jensen

Do you believe that this culture will undergo a voluntary transformation to a sane and sustainable way of living?

I've asked that question of thousands of people, and almost no one says yes. The answers range from emphatic no's to derisive laughter.

My next question: For those of us who care about life on this planet, how will this understanding—that this culture won't voluntarily stop destroying the natural world, eliminating indigenous cultures, exploiting the poor, and killing those who resist—shift our strategy and tactics?

The answer? We don't know, because we don't talk about it, and we don't talk about it because we're all so busy pretending that, against all evidence, there will be some miraculous transformation.

But with all the world at stake, it's foolishly and reprehensibly irresponsible to rely on some miraculous transformation we all know won't happen. We need to act, and we need to act decisively. The miracle we're waiting for is us.

In his extraordinary book *The Nazi Doctors,* Robert Jay Lifton explored how men who had taken the Hippocratic Oath could par-

---

DERRICK JENSEN is an environmental activist, author, small farmer, beekeeper, teacher, and philosopher. Jensen's speaking engagements have packed university auditoriums, conferences, and bookstores across the nation. His writing has appeared in the *New York Times Magazine, Audubon, Orion,* and *The Sun,* among many other publications. He has written more than a dozen books, most recently *How Shall I Live My Life? On Liberating the Earth from Civilization* (2008), *What We Leave Behind* (2009), and *Songs of the Dead* (2009).

ticipate in prisons where inmates were worked to death or killed in assembly lines. He found that many of the doctors honestly cared for their charges and did everything within their power—which means pathetically little—to make life better for inmates. If an inmate got sick, they might give the inmate an aspirin to lick. They might put the inmate to bed for a day or two. If the patient had a contagious disease, they might kill the patient to keep the disease from spreading. All of this made sense within the confines of Auschwitz. The doctors did everything they could, except for the most important thing of all: they never questioned the existence of Auschwitz itself. They never questioned working inmates to death. They never questioned starving them to death. They never questioned imprisoning them, torturing them. They never questioned the existence of a culture that would lead to these atrocities. They never questioned the logic that leads inevitably to electrified fences, gas chambers, bullets in the brain.

We as environmentalists do the same. We work as hard as we can to protect the places we love, using the tools of the system the best we can. Yet we do not do the most important thing of all: we do not question the existence of this current death culture. We do not question the existence of an economic and social system that is working the world to death, that is starving it to death, that is imprisoning it, that is torturing it. We never question a culture that leads to these atrocities. We never question the logic that leads inevitably to clearcuts, murdered oceans, loss of topsoil, dammed rivers, poisoned aquifers, global warming. And we certainly don't act to bring it down.

Think about the prominent "solutions" suggested to help curb the worst of global warming. What do they have in common? I'm talking about every major "solution," from those proposed by Al Gore (compact fluorescents, inflating tires, reducing packaging); to James Lovelock (nuclear energy); to Newt Gingrich (giving polluters tax credits to lean them toward voluntarily reducing carbon emissions); to Barack Obama (so-called "clean" coal); to scientists suggesting schemes such as dumping tons of iron into the ocean in the hope that this will cause algae to flourish, absorbing CO_2 into the algae's bodies and, by the way, doing god knows how much damage to the already-being-murdered oceans, or injecting sulfur particles high into the atmosphere to reflect sunlight back into space; and so on.

What all of these "solutions" have in common is that they all take industrial capitalism as a given, as that which must be saved; and they take the real, physical world—filled with real physical beings who live, die, make the world more diverse—as secondary, as something which must conform to industrial capitalism. Even someone as smart and dedicated as Peter Montague, who until recently ran the indispensable *Rachel's Newsletter,* can say, about an insane plan to "solve" global warming by burying carbon underground (which of course is where it was before some genius pumped it up and burned it), "If even a tiny proportion of it leaks back out into the atmosphere, the planet could heat rapidly and civilization as we know it could be disrupted." No, Peter, it's not civilization we should worry about. Far more problematical is the very real possibility that the planet could die. Or take Lester Brown's latest book: *Plan B 3.0: Mobilizing to Save Civilization.* To save civilization. Not to save the planet *from* civilization.

Industrial capitalism always destroys the land on which it depends for raw materials, and it always will. Until there is no land (or water, or air) for it to exploit. Or until, and this is obviously the far better option, there is no industrial capitalism.

How do you stop or at least curb global warming? Easy. Stop pumping carbon dioxide into the atmosphere. How do you do that? Easy. Stop burning oil, natural gas, and coal. How do you do that? Easy. Stop industrial capitalism.

When most people in this culture ask, "How can we stop global warming?" that's not really what they're asking. They're asking, "How can we stop global warming, without significantly changing this lifestyle [or deathstyle, as some call it] that is causing global warming in the first place?"

The answer is that you can't.

To ask how we can stop global warming while still allowing that which structurally, necessarily causes global warming—industrial civilization—to continue in its functioning is like asking how we can stop mass deaths at Auschwitz while allowing it to continue as a death camp. Destroying the world is what this culture does. It's what it has done from its beginning.

Any solution that does not take into account—or, rather, count as primary—polar bears, walruses, whippoorwills, bobwhites, chickadees, salmon, and the land and air and water that support them all—is no solution, because it doesn't count the real world as primary and social constructs as secondary. Any such solution is in the most real sense neither realistic nor practical. Any solution that does not place the well-being of nonhumans—and indeed the natural world, which is the real world—at the center of its moral, practical, and "realistic" considerations is neither moral, practical, nor realistic. Nor will it solve global warming or any other ecological problem.

Do we want a living real world, or do we want a social structure that is killing the real world? Do we want a living real world, or do we want a dead real world, with a former social structure forgotten by everyone, because there is no one left alive to remember?

You choose.

To help clarify how utterly insufficient are mainstream responses to global warming (and more broadly the murder of the planet), let's put all this a different way. Pretend that instead of industrial capitalists destroying the planet, it is space aliens. These space aliens are changing the Earth's climate. They are murdering the oceans; 90 percent of the large fish are gone, the oceans are rapidly acidifying, there is six to ten times as much plastic as phytoplankton in the oceans (which would be the equivalent of nine out of every ten bites you take being Styrofoam instead of food), and so on. They are decapitating mountains, they are putting dioxins into every stream, into every mother's breast milk, into the flesh of your mother, father, sister, lover, child. Into your own flesh. They are damming every river. What would you do?

We all know what we would do. We would fight like hell using every tool at our disposal. We would, using any means necessary, destroy their capacity to steal from us, and we would destroy their capacity to murder the world. This is my definition of bringing down civilization: denying the rich their ability to steal from the poor, and denying the powerful their ability to destroy the planet.

Those who inherit whatever's left of the world once this culture has been stopped—whether through peak oil, economic collapse, ecological collapse, or the efforts of brave women and men fighting in alliance with the natural world—are going to judge us by the health of the land base, because that's what's going to support them, or not. They're not going to care how we lived our lives. They're not going to care how hard we tried. They're not going to care whether we were nice. They're not going to care whether we were nonviolent or violent. They're not going to care whether we grieved the murder of the planet. They're not going to care whether we were enlightened. They're not going to care what sort of excuses we had to not act (e.g., "I'm too stressed to think about it," or "It's too big and scary," or "I'm too busy," or "But those in power will kill us if we effectively act against them," or "If we fight back we run the risk of becoming like they are," or "But I recycled," or any of a thousand other excuses we've all heard too many times). They're not going to care how simply we lived. They're not going to care how pure we were in thought or action. They're not going to care if we became the change we wished to see. They're not going to care whether we voted Democrat, Republican, Green, Libertarian, or not at all. They're not going to care if we wrote really big books about it. They're not going to care whether we had "compassion" for the CEOs and politicians running this deathly economy. They're going to care whether they can breathe the air and drink the water. They're going to care whether the land can support them. We can fantasize all we want about some great turning, and if the people (including the nonhuman people) can't breathe, it doesn't matter. Nothing matters but that we stop this culture from killing the planet.

It's embarrassing even to have to say this. The land is the source of everything. If you have no planet, you have no economic system, you have no spirituality, you can't even ask how to stop global warming. If you have no planet, nobody can ask any questions at all.

Those who come after—presuming anyone survives—are going to wonder what the fuck was wrong with us that we didn't do whatever it takes—and I mean whatever it takes—to stop industrial capitalism from killing the planet. It is long past time for brave women and men to do whatever it takes to protect this planet—our one and only home—from this culture's final solution. It is long past time we brought the industrial infrastructure down before it kills any more of the planet. It is long past time for us to be the miracle we've all been waiting for.

ETHICAL ACTION

How can we protect the children?

Okay. It's time for a frank talk with grandparents.

This is given: We love our children and grandchildren more than life itself. We would, in literal fact, do *anything* for them.

This also is given: Grandparents are in a powerful position to protect the children and grandchildren. The first asset we bring to the work is a set of skills, experience, and knowledge gained over a lifetime of productive work. The second asset is political clout. We vote, and there are a lot of us. We donate to political campaigns, and we have a lot to give. The third asset is something that we have in abundance that no other demographic has enough of: we have time. Put these assets together, and grandparents command the power to shape the new world. We can make sure that our children and grandchildren inherit a planet that will sustain their health and nourish their freedom to make a good life.

How? By organizing our huge political power to elect officials who will get down to the most important business of protecting the life-sustaining systems of the planet. Knock on doors for these politicians, then hold them to account. If politicians dither or jabber or make excuses, send them home. Organize for truly fast and effective climate change legislation. Organize for solar power. Organize to keep poisons from the air, the water, the agricultural fields.

Start small. Madeline's Grandparents for Clean Electricity. Fierce Grandmothers for Safe Food. Retired People for Redwing Swamps. The Council of Elders for Responsible City Government. The Grandparents' Coal Boycott.

Redirect already-existing resources. Tell AARP it's time to stop worrying about the health of our colons and golf games and start worrying instead about our legacy. Focus the church education program on environmental toxins instead of the nature of heaven. Get the alumni association off the cruise ships and into the streets with banners.

Get started. At this stage in our lives, it mocks death to waste time. We work all our lives to provide for the future of our children and grandchildren. We cannot let it all slip away in our last decades. A life-sustaining planet, not an MP3 player or a plate of cookies, will be our last and greatest gift to the ones we love the most. Without that gift, all other gifts are meaningless, and the hugs of a grandparent become cynical jokes on the beautiful little ones, who do not deserve the struggles they will face.

P.S. Let us hear no more talk about the extra entitlements of the elders—entitlements to year-round perfect weather, an annual trip to Las Vegas, low taxes, easy Sunday crosswords, reduced greens fees and the world be damned. It may be true that because we have worked hard, we deserve to sit around and enjoy the fruits of our labor; but we owe those fruits also to a stable climate, temperate weather, abundant food, cheap fuel, and a sturdy government—all unearned advantages that our children will not have if we don't act. It's tragic and culturally dysfunctional if our lives culminate in a radical selfishness that makes us angry and bitter (because selfish desires can never be satisfied) rather than in the respect that we truly earn as people of wisdom and responsibility.

3

> *Do we have a moral obligation to take action to protect the future of a planet in peril?*
>
> **Yes, for the sake of the Earth itself.**

The failure to act on behalf of the Earth is, of course, a great imprudence—a cosmic cutting-off-the-limb-you're-sitting-on stupidity. But it is also a moral failure. That is because the Earth (this swirling blue sphere) is not only *instrumentally* valuable. That is, it's not just valuable because it is supportive of human life (which is valuable in itself). Rather, the Earth, like a human being, has value in and of itself. It has what philosophers call *intrinsic value*. We have responsibilities to honor and protect what is of value. So we have the responsibility to honor and protect the Earth as we find it, a rare green jewel in the solar system. That is the moral position of the essays in this section.

Even if there were no humans to love it, to depend on it, to admire it—even if it were of no use to anything at all—would it be better that the Earth exist than not? If you think so, then you believe that the Earth has intrinsic value. Or try this: Say humans all decided to leave the Earth, for whatever reason. Once the last rocket departed, the Earth would have no human use at all. Would it be wrong for the last person leaving the Earth to light a fuse that would blow the planet to dust? If that would be an abomination in your eyes, then again, you acknowledge the intrinsic worth of the world.

Given the intrinsic value of the Earth, we have an obligation—

even beyond our own interests—to protect the Earth as something of inestimable and unique worth.

It's possible that this argument is even more powerful. Some people argue that the Earth itself is a living, sensate being, much as humans are living, sensate beings. Then, for all the reasons that we accord respect and protection to humans, we owe respect and protection to the Earth. The Gaia hypothesis, for example, suggests that the Earth itself is an autopoeic, self-correcting entity; that the Earth itself meets the criteria by which we judge other things as living entities, as alive and not dead. When scientists demonstrate that we are causing harm to the living Earth's systems (atmospheric, hydrologic, meteorologic), they are speaking literally, not metaphorically. Thus our obligation to save lives extends beyond human beings, beyond species, to the very Earth itself.

Sky

Brian Turner

If the sky knew half
of what we're doing
down here

it would be stricken,
inconsolable,
and we would have

nothing but rain

BRIAN TURNER's ongoing love affair with his New Zealand home lies at the heart of his poetry. Turner is one of New Zealand's leading poets and one of its most significant writers on landscape and environmentalism. His first poetry collection, *Ladders of Rain*, won the Commonwealth Poetry Prize, and his sixth, *Beyond,* won the new Zealand Book Award for Poetry. He was New Zealand's Poet Laureate from 2003 to 2005. His other collections are *Ancestors, Bones,* and *All That Blue Can Be.*

A Hinge Point of History

Holmes Rolston III

We live at a change of epochs. We are witness to the end of nature as we enter a new era: the Anthropocene.[1] From this point on, culture more than nature is the principal determinant of Earth's future. We are passing into a century when this will be increasingly obvious, and this fact puts us indeed at a hinge point of history.

Especially in the West, we have lived with a deep-seated belief that life will get better, that one should hope for abundance and work toward obtaining it. We have even built that belief into our concept of human rights: a right to self-development, to self-realization. Such an egalitarian ethic scales everybody up and, at the same time, drives an unsustainable world. When everybody seeks their own good, there is escalating consumption. When everybody seeks everybody else's good, there is, again, escalating consumption. When we have technological powers to produce these goods, we enter the Anthropocene era.

For some this is cause for congratulation, the fulfillment of our destiny as a species. In a *Scientific American* special issue from the late 1980s, *Managing Planet Earth*, the editors claim that the two

---

HOLMES ROLSTON III is Distinguished Professor of Philosophy at Colorado State University. He is a founder and associate editor of *Environmental Ethics*, a scholarly journal in its thirtieth year. His work has been acclaimed in both the academic and the mainstream press. He received the Templeton Prize in 2003 and the Mendel Medal in 2005. The second edition of his book *Science and Religion: A Critical Survey* was published in 2006.

---

1. Paul J. Creutzen, "The Anthropocene," in *Earth System Science in the Anthropocene*, ed. Eckart Ehlers and Thomas Draft (Berlin: Springer, 2006).

central questions today are "What kind of planet do we want?" and "What kind of planet can we get?"[2]

For others this is cause for concern. We worried throughout much of the past century that humans would destroy themselves in interhuman conflict. That fear—at least of global nuclear disaster—has subsided somewhat, only to be replaced by a new one. We wonder, will these Earth managers produce a sustainable development or a sustainable biosphere? The worry for the next century is that if our present course is uncorrected, humans may ruin their planet and themselves along with it.

There are paradoxes and challenges that confront and confound us in this new era. Although we congratulate ourselves on our powers, perhaps humans are not well equipped to manage the sorts of global-level problems we face. The classical institutions—family, village, tribe, nation, agriculture, industry, law, medicine, even school and church—have shorter horizons. Far-off descendants and distant races do not have much "biological hold" on us. Across the millennia of human evolution, little in our behavior affected those remote from us in time or in space, and natural selection shaped only our conduct toward those who were closer. Global threats, however, require us to act in massive concert—of which we may be incapable. If so, humans may bear within themselves the seeds of their own destruction. To put it more bluntly, more scientifically: our genes once enabled our adaptive fit but may in the next millennium prove maladaptive and destroy us.

This wonderland Earth is a planet with promise. But if we are to realize the abundant life for all time, both policy and ethics must enlarge the scope of concern. Humans are attracted to appeals to a better life, to a higher quality of life; if environmental ethics can persuade large numbers of persons that a sustainable biosphere takes priority over sustainable development, that an environment with biodiversity and wildness is a better world to live in than one without these, then some progress is possible. We can still use an appeal to an even more enlightened self-interest, or, perhaps better, to a more inclusive and

---

2. William C. Clark, "Managing Planet Earth," *Scientific American* 261, no. 3 (September 1989), 46–54.

comprehensive concept of human welfare. That will get us clear air, clean water, soil conservation, national parks, recreational wildlife reserves, and bird sanctuaries. Environmental ethics cannot succeed without these things. This is not simply pragmatic; it is quite true.

We have seen this moral transcendence before. The European Union has transcended national interests with surprising consensus about environmental issues. Kofi Annan, former secretary-general of the United Nations, praised the Montreal Protocol, with its five revisions, widely adopted (by 191 nations) and implemented, as the most successful international agreement yet. Every developed nation except the United States and Australia signed the Kyoto Protocol. The Convention on International Trade in Endangered Species of Wild Fauna and Flora (CITES) has been signed by 112 nations. There are more than 150 international agreements (conventions, treaties, protocols, etc.) registered with the United Nations that deal directly with environmental problems.

Humans are a paradox on Earth, both a part of nature and apart from nature. Humans evolved out of nature. But in important senses, they did just that; they evolved into culture, contrasted with nature. Humans are nurtured into an inherited culture. This cultural genius makes possible the deliberate and cumulative, and therefore the extensive, technological rebuilding of nature. Rather than being themselves morphologically and genetically reshaped to fit their changing environments, humans reshape those environments.

Robert Boyd and Peter Richerson explain that humans have a "dual inheritance system"—genetic nature and cultural nurture. Boyd and Peterson find that the existence of human culture is a deep evolutionary mystery on a par with the origins of life itself: "Human societies are a spectacular anomaly in the animal world."[3] The human transition into culture is exponential, nonlinear, reaching extraordinary epistemic powers. To borrow a term from the geologists, humans have crossed an unconformity. In that sense, it is true that Earth is now in a postevolutionary phase.

3. Robert Boyd and Peter J. Richerson, *Culture and the Evolutionary Process* (Chicago: University of Chicago Press, 1985), and Peter J. Richerson and Robert Boyd, *Not by Genes Alone: How Culture Transformed Human Evolution* (Chicago: University of Chicago Press, 2005).

But at this hinge point of history, isn't it still an open question whether we want the future of Earth to turn entirely on humans? Perhaps we are postevolutionary, but do we wish to be postecological? What kind of planet do we want? What kind of planet can we get? We also ought to ask: What kind of planet do we have? What kind of planet ought we to want? We may be entering the Anthropocene era, but we ought to choose not to enter the Anthropocentric era—and the latter is not a necessary implication of the former.

Nature as it once was, nature as an end in itself, is no longer the whole story. Nature as contrasted with culture is not the whole story either. An environmental ethic is not just about wildlands, but also about humans at home on their landscapes, humans in their culture residing also in nature. This will involve resource use, sustainable development, managed landscapes, and the urban and rural environments, of course. Further, it can and ought to involve, now and in the future, the thought of nature as an end in itself, a sustainable biosphere worthy of care and respect for its own sake.

We already see examples of just such a moral gesture. In the defense of life on Earth since time immemorial, organisms have set up territorial boundaries. If they do not defend their places and their resources, they cannot survive and reproduce. But now there is something new, never seen on Earth throughout its billions of years of evolving species. Humans have begun to set conservation of the biodiversity on Earth as a moral and social goal. We set up boundaries (in biodiversity reserves, wilderness areas, national parks), and we set ourselves apart in this setting. Roger DiSilvestro exclaims: "This is something truly new under the sun, and every protected wild place is a monument to humanity's uniqueness. . . . We not only *can* do, but we can choose *not to do*. Thus, what is unique about the boundaries we place around parks and other sanctuaries is that these boundaries are created to protect a region from our own actions. . . . No longer can we think of ourselves as masters of the natural world. Rather, we are partners with it."[4]

We need to become wiser than Socrates. Certainly "the unexam-

---

4. Roger L. DiSilvestro, *Reclaiming the Last Wild Places: A New Agenda for Biodiversity* (New York: John Wiley & Sons, 1993), xiv–xv.

ined life is not worth living," and certainly we should strive to "know thyself." And while the classic search in philosophy has been to figure out what it means to be human, Socrates was sometimes wrong in his search for the good life. Socrates loved Athens, which is well enough. After all, a human is, as Aristotle put it, a "political animal." We live in towns (Greek: *polis*), in social communities, and we cannot know who we are without an examination of the cultures that shape our humanity. But Socrates avoided nature, thinking it profitless: "You see, I am fond of learning. Now the country places and trees won't teach me anything, and the people in the city do." We need to become more inclusive than Socrates: life in an unexamined world is not worth living either.

This is the answer to the would-be planetary managers' questions about what kind of life we want on what kind of planet: *We do not want a denatured life on a denatured planet. That would rupture history, that would dehumanize us all, that would deny the future their abundant life.*

The Planet Is Shouting but Nobody Listens

F. Stuart Chapin III

When I first came to Alaska as an ecologist forty years ago, I was curious to know how the Great Land worked. I took my science toys into the woods and the tundra to learn how plants and animals interact with one another to shape the ecosystems of which they are a part. This was fun and intellectually stimulating. It gave me something to talk about with my science friends. This was the scope of science that I had been taught to pursue. I also came to love Alaska as a place to hike, raise a family, build a house. I have become deeply immersed in its ecosystems and mystique. This is my home, and I care a great deal about what will become of it.

Over the past two decades scientists have seen many changes in ecosystems around the world. The best explanation is that our planet is on a trajectory of change. Change is nothing new; the world has always changed. What is different is that these changes are rapid and directional—moving us along a trajectory to places the planet has not seen for at least several hundred thousand years—and human actions are the largest cause of these changes. Everyone remembers changes that have occurred during their lifetimes. These become part

---

F. STUART "TERRY" CHAPIN III is a professor at the Institute of Arctic Biology and the Department of Biology and Wildlife at the University of Alaska Fairbanks. He is also the principal investigator of the Bonanza Creek Long-Term Ecological Research program, which aims to improve understanding of the long-term consequences of changing climate and disturbance regimes in the Alaskan boreal forest. He has authored over five hundred academic and popular articles and ten books, including the textbook *Principles of Natural Resource Stewardship: Resilience-Based Management in a Changing World* (2009).

of our personal history, but we seldom place these changes in a larger context or consider our causal role. "I remember the summer that was really hot when I was growing up. Today doesn't seem that different," people say. "Besides, I'm in a hurry, and I need to take my kids to soccer practice."

The planet is whispering, but nobody listens.

Alaska is different. Alaska is warming more rapidly than almost any place on the planet, so scientists have a unique opportunity and responsibility to convey to the rest of the world what rapid change means to the people and other organisms that call this place home. Most Alaskans notice this warming. The car mechanic tells me about the weeks of −40 degrees F that he remembers from his childhood but are now reduced to a few days of minor inconvenience. My permafrost-scientist neighbor talks about how quickly the valley-bottom roads break up and need repair as the permafrost thaws: a road repair that used to last five years or more now lasts a matter of weeks. In the Native Athabascan villages of interior Alaska, it is so warm during the autumn hunting season that the meat spoils before it can be processed. People are now prohibited by hunting regulations from synchronizing their hunting patterns to the changing climate as they had done for previous millennia. In addition, these bush communities are not connected to a road system, so rivers are their highways to access the land. The ice no longer thickens early enough for safe winter travel to trap and hunt, and many people have died when their snow machines broke through the ice. These changes affect almost everyone, in many cases with deep personal and cultural pain.

Because of the warming climate, fires in Alaska have become larger and more frequent. Alaskans remember the dense pall of smoke from the 2004 wildfires. The brown haze limited visibility to a hundred yards or less, and I could still taste the smoke when I coughed the following October. Small children and the elderly were encouraged to stay indoors, and many people with asthma and other breathing problems left Fairbanks for the summer.

Climate-driven changes in wildfire have even larger impacts in rural Alaska. Athabascan residents now live in permanent villages that are locked in place by infrastructure. In earlier times, people lived in temporary camps and moved around the land in small family groups. If a fire occurred in one place, people made small adjustments in their traditional seasonal round of subsistence harvest and still accessed the necessary range of ecosystems. With schools, airports, and a modern village lifestyle, this is no longer a viable option. In this modern village context, implemented with the best humanitarian intents, climate-induced increases in wildfire are having immense cultural impacts. Now, when large fires occur, it is a generation before moose return; it is four generations before there are enough lichens to attract caribou. "How can I sustain my cultural ties to the land after a fire, if I can't teach my kids to hunt?"

The planet is calling out, but nobody listens.

Alaska and other places that are experiencing rapid climate change have a special role to play in redirecting the course of our changing planet. In these places we see abundant examples of impacts that are increasingly felt, in one guise or another, throughout the world. The scientific community has been relatively ineffective in conveying this message of planetary change to our society, whose collective choices propel us along this path. As scientists, we are trained to avoid speaking in ways that touch people's souls. New forms and venues of dialogue must emerge through words, images, and tones that enable society to see and hear more clearly the changes that are occurring, to feel deeply and personally their importance, and to recognize the connections between our personal and collective choices and the trajectory of life on this planet.

The urgency of this new dialogue about our changing planet is greater than many around us might suggest. The momentum of a freight train is minuscule compared to that of the global climate system. Recent emissions of carbon dioxide already commit the planet to at least another half-century of rapid warming, which will strongly shape our lives and the ecosystems on which we depend. The choices that we make in the next decade or two preempt the choices that our children and grandchildren will wish they could make about the planet they inherit.

The planet is shouting, but no one is listening.

Society is well poised to take actions that can substantially reduce the pressures that are currently accelerating the rate of planetary change. The science and technology required to reduce rates of climate change are readily available and maturing rapidly. The missing pieces are the dialogues necessary to connect the increasingly obvious planetary changes with the deepest motivations of every person as a steward of planet Earth. Only if we deeply and personally understand our role in this rapidly changing planet will we make the choices that best preserve for our grandchildren the option to experience personally the human connections to nature that have been and will continue to be the cultural wellspring of humanity.

The Bells of Mindfulness

Thich Nhat Hanh

> Every one of us can do something to protect and care for our planet. We have to live in such a way that a future will be possible for our children and our grandchildren. Our own life has to be our message.

The bells of mindfulness are sounding. All over the Earth, we are experiencing floods, drought, and massive wildfires. Sea ice is melting in the Arctic, and hurricanes and heat waves are killing thousands. The forests are fast disappearing, the deserts are growing, species are becoming extinct every day, and yet we continue to consume, ignoring the ringing bells.

All of us know that our beautiful green planet is in danger. Our way of walking on the Earth has a great influence on animals and plants. Yet we act as if our daily lives have nothing to do with the condition of the world. We are like sleepwalkers, not knowing what we are doing or where we are heading. Whether we can wake up or not depends on whether we can walk mindfully on our Mother Earth. The future of all life, including our own, depends on our mindful

---

THICH NHAT HANH is a Vietnamese Zen master, writer, and peace and human rights activist. In 1982 he cofounded Plum Village, a monastery and practice center in the Dordogne in the south of France. The Village works to alleviate the suffering of refugees, boat people, political prisoners, and hungry families in Vietnam and throughout the Third World. He has published more than one hundred books, including *The World We Have: A Buddhist Approach to Peace and Ecology* (2008), which he coauthored with Alan Weisman.

steps. We have to hear the bells of mindfulness that are sounding all across our planet. We have to start learning how to live in such a way that a future will be possible for our children and our grandchildren.

I have sat with the Buddha for a long time and consulted him about the issue of global warming, and the teaching of the Buddha is very clear. If we continue to live as we have been living, consuming without a thought of the future, destroying our forests and emitting dangerous amounts of carbon dioxide, then devastating climate change is inevitable. Much of our ecosystem will be destroyed. Sea levels will rise and coastal cities will be inundated, forcing hundreds of millions of refugees from their homes, creating wars and outbreaks of infectious disease.

We need a kind of collective awakening. There are among us men and women who are awakened, but it's not enough; most people are still sleeping. We have constructed a system we can't control. It imposes itself on us, and we become its slaves and victims. For most of us who want to have a house, a car, a refrigerator, a television, and so on, we must sacrifice our time and our lives in exchange. We are constantly under the pressure of time. In former times, we could afford three hours to drink one cup of tea, enjoying the company of our friends in a serene and spiritual atmosphere. We could organize a party to celebrate the blossoming of one orchid in our garden. But today we can no longer afford these things. We say that time is money. We have created a society in which the rich become richer and the poor become poorer, and in which we are so caught up in our own immediate problems that we cannot afford to be aware of what is going on with the rest of the human family or our planet Earth. In my mind I see a group of chickens in a cage disputing over a few seeds of grain, unaware that in a few hours they will all be killed.

People in China, India, Vietnam, and other developing countries are still dreaming the "American dream," as if that dream were the ultimate goal of mankind—everyone has to have a car, a bank account, a cell phone, a television set of their own. In twenty-five years the population of China will be 1.5 billion people, and if each of them wants to drive their own private car, China will need 99 million barrels of oil every day. But world production today is only 84 million barrels per day. So the American dream is not possible for the people of China, India, or Vietnam. The American dream is no longer possible even for the Americans. We can't continue to live like this. It's not a sustainable economy.

We have to have another dream: the dream of brotherhood and sisterhood, of loving-kindness and compassion. That dream is possible right here and now. We have the Dharma, we have the means, and we have enough wisdom to be able to live this dream. Mindfulness is at the heart of awakening, of enlightenment. We practice breathing to be able to be here in the present moment so that we can recognize what is happening in us and around us. If what's happening inside us is despair, we have to recognize that and act right away. We may not want to confront that mental formation, but it's a reality, and we have to recognize it in order to transform it.

We don't have to sink into despair about global warming; we can act. If we just sign a petition and forget about it, it won't help much. Urgent action must be taken at the individual and collective levels. We all have a great desire to be able to live in peace and to have environmental sustainability. What most of us don't yet have are concrete ways of making our commitment to sustainable living a reality in our daily lives. We haven't organized ourselves. We can't simply blame our governments and corporations for the chemicals that pollute our drinking water, for the violence in our neighborhoods, for the wars that destroy so many lives. It's time for each of us to wake up and take action in our own lives.

We witness violence, corruption, and destruction all around us. We all know that the laws we have in place aren't strong enough to control the superstition, cruelty, and abuses of power that we see daily. Only faith and determination can keep us from falling into deep despair.

Buddhism is the strongest form of humanism we have. It can help us learn to live with responsibility, compassion, and loving-kindness. Every Buddhist practitioner should be a protector of the environment. We have the power to decide the destiny of our planet. If we awaken to our true situation, there will be a change in our collective consciousness. We have to do something to wake people up. We have to help the Buddha to wake up the people who are living in a dream.

Restoration and Redemption

Robin Morris Collin

To the children of my family, and the children of my friends; to the children in my neighborhood and the children in my country; to my kith and kin, my tribe and clan; to the living beings with whom we share this planet, and its living systems with whom we are joined.

My Dears,

 I have not been in such close communication with you as I should have been, so I am employing this oldest of devices to speak to you about our futures. We are so interconnected—by transportation, communication, and viruses (real and virtual), by the cycles of life that define us—that to imagine we are somehow separated is delusional and ultimately leads to illness. Our future depends on this recognition.

ROBIN MORRIS COLLIN is professor of law at Willamette University College of Law in Oregon. Her latest publication, coauthored with Robert W. Collin, is the three-volume *Encyclopedia of Sustainability: Environment and Ecology, Business and Economics, Equity and Fairness* (2009). She is the first U.S. law professor to teach sustainability courses in a U.S. law school. She has been awarded the David Brower Lifetime Achievement Award from the Public Interest Environmental Law Conference as well as the Campus Compact Faculty Award for Civic Engagement in Sustainability. Morris Collin also served as the founding chair of the legislatively created Oregon Environmental Justice Taskforce and a founding board member of the Environmental Justice Action Group of Portland, Lawyers for a Sustainable Future. She cofounded the Conference Against Environmental Racism (CAER) and the Sustainable Business Symposium.

I thank my husband, Robert William (Will) Collin, for his support of this effort, and his perceptive comments.

In the past, we have made decisions that control how we, and you, live. We imagined that our choices were in our best interests, in the best interests of the majority of us living. We understood that there would be sacrifices—winners and losers—but we winners were not the sacrifice zones. Others, outsiders, poor people, people of color, children, might be incidental casualties or collateral damage in the competition we imagined life to be. We believed that we could survive and not consider the sacrifices of the silent and the voiceless. We humans are rationalizing creatures. We are the only creatures that lie. We invest our desires and impulses with reason—the fig leaf for desire. Our powers of reason can betray us all when we want to believe what is not true or good. There is one great truth-teller, and she cannot lie: nature, the womb into which we were born. She is the repository of all, our bones and the true history of our communities.

Not all of us believed that the greatest good for the greatest number was the course of justice or even truth. Even those who started out accepting that theory could not maintain the fiction of indifference to poverty and degradation. We had a future to think of, in common. That made all the difference for some of us. Perhaps it will make the difference for all of us. You be the judge.

If I were you, I would be angry. We wasted so much, and we polluted so much. You do not have the option of profligacy.

We deprived you of the glories of polar bears, and whales, and elephants, who were far older and perhaps wiser than we were. They were a sight to lift the spirit. As I reflect on the way we have lived, the arguments, our differences, I think our greatest failing was our unwillingness to face the most radical uncertainty of human life, now or in the past or in the future: What will it mean to be human without the nature we knew, without the cycles supporting our lives? I confess I do not know and cannot imagine.

You may note only the harm we did. Consequences are one way to assess our acts and lives. But counting and characterizing consequences is only one kind of argument. Those are the arguments that we had in the absence of certainty about consequences. We knew that we had sped the degradation of the cycles of life on which all lives depend. We still have passionate arguments about issues like the rate of destruction and optimal levels of loss. That is all they are—arguments and delay. You may, in fact, care more about our intentions and reasons. They are a different way of judging us.

The kinds of problems we inherited (and made worse by not changing) were huge in scale—global. They overwhelmed the ability of science to predict or manage. Climate change was so connected to economic growth that we were afraid to uncouple them; that showed what the powerful and privileged valued most. Our systems have been exquisitely sensitive to greed, not nature. We often chose against nature and contrary to our innate sense of reverence for it. We chose materialism, conformity, comfort. Our escapist materialism allowed us to evade questions about destroying our atmosphere and our oceans, while making the consequences worse. We shoved that off our shoulders onto yours. I am sorry.

Words the nineteenth-century English poet Percy Bysshe Shelley wrote in *Prometheus Unbound* still describe too many of us:

> In each human heart terror survives
> The raven it has gorged: the loftiest fear
> All that they would disdain to think were true:
> Hypocrisy and custom make their minds
> The fanes of many a worship, now outworn.
> They dare not devise good for man's estate,
> And yet they know not that they do not dare.

The systems we inherited were the products of empires: political, military, and economic empires with an agenda of conquest. In them, greed acquired the royal clothing of formal religion and the power of a gun. Armed and winged, greed marched mercilessly over the Earth and her people. Religion clad this army with vindication, not redemption. Winged like angels and armed like demons, we began organizing our lives around greed. Nature and poor people were trampled underfoot like the grass when elephants fight. We made systems that carved up people and land into commodities. We inherited systems from another age that poisoned the air, destroyed the Earth's temperature-control cycles, and sickened our children and elders. From transportation to agriculture, from housing to employment, we created systems that venerated greed without qualifications or limits. You will learn that greed and the systems modeled by greed will never restore what we have lost.

Still, I can hear the voices of my contemporaries rejecting blame. They insist that they have not actually done anything evil. They did not enslave anyone, they did not kill anyone and take their way of life, they did not choose petroleum, and they did not choose cars instead of public transportation systems. I say to all of us that the only way forward lies in abandoning a time-bound sense of right and wrong. We must make amends for wrongs to the Earth and to the people who abide there for wrongs that degrade them and us, even if the one terrible wrong we committed was to enjoy the fruits of wrongdoing.

The work of redemption begins with us and will fall to you in your turn. Fearlessly and selflessly, heroes must take on the burdens left to them by past generations. Let us be heroes. Then it will be your time. We all, in our own lives, must struggle to restore our sense of reverence for nature and the Earth. We must reconnect in whatever ways restore us. There are a thousand paths worth taking for the joy they offer—not to obtain salvation, not to win our redemption, just for simple joy and the restoration of reverence. Psalm 51 describes this sort of path:

> Create in me a clean heart, O God,
> and renew a right spirit within me.
> Cast me not away from your presence
> and take not your holy Spirit from me.
> Give me the joy of your saving help again
> and sustain me with your bountiful Spirit.

Humans are unique creatures of this Earth, because we have the capacity to choose. Our ability to transcend immediate conditions and to make choices based on other values is a special power very like the creative power of God. When we do not exercise this power, we are behaving without conscience. We are more than the sum of our individual wants and needs. In that, we resemble the deity. We must consider the full range of our powers of choice, and exercise them as if we valued nature and our connection to it.

Our power to choose is both small and vast. It ranges from choices about what to eat to decisions to design whole systems free of waste and pollution. Perhaps the most powerful exercise of our powers lies in our abilities to create stories to frame our sense of reverence and to tell those stories to you, our children.

We find reverence for the Earth in many different ways. Some of us, gardeners and farmers, feel it in our hands when we dig in the dirt. Some of us resonate to the stories of our faiths, stories about the Garden of Eden and its bounty. Some of us feel the responsibility of human awareness of our blessings. Some see the hand of God most clearly in an experience of wilderness. Some connect to that reverence through ceremony. Some experience the embrace of Earth-love in the ways of healthy communities—places where we greet and care for one another, not use each other mercilessly.

I remember communities that lived by the rhythms of nature, punctuating the year with celebrations of natural and supernatural events. I remember the sense of nobility and richness in these communities without indoor plumbing or electricity. They had a sense of right living, a rhythm and pace that matched the nature around them. They, too, had problems; their people were saddled with an experience of multigenerational trauma and rage that continued to kill them when the perpetrators had long since passed. The sense of reverence for nature, the gifts and creatures of the Earth, was still part of the fabric of their daily lives and formal ceremonies.

This reverence for Earth and nature unites us far more than our differences about what to call it separate us. We are the people of the Earth. We do not own her; we belong to her. What she experiences, we experience. What we experience, she experiences. We dream her dreams. We cannot do things to her without doing them to ourselves. That is the truth about who we humans are as a species, and that will be what is remembered about us, not the reasons for our wars, not our warlords.

What would it mean to live without the gifts and blessings of nature that we now enjoy? It would be like dreamless sleep, like nothingness or absence. This nature we have in common defines us, keeps us healthy and sane. The connection we form with the cycles of nature keeps us rich in spirit and growing. Chief Seattle described the connection this way: "Whatever befalls the Earth befalls the sons of the Earth. Man did not weave the web of life; he is merely a strand in it. Whatever he does to the web he does to himself."[1]

Our redemption lies in a multigenerational effort that we must start and you must carry. We must demand reconciliation of our faiths, and amends-making to the Earth from our nations. Our faiths clothed conquest, murder, slavery, and rape with a powerful entitlement. For sins done in the name of faith, faith must atone. Our nations have spent your inheritance on arms. They must give back to the Earth accordingly, in ways that care for the systems that support life, not by using force to take life.

May your spirits be strong and willing. I will dream of you.

1. Chief Seattle speech, version 3, Washington State Library, www.synaptic.bc.ca/ejournal/wslibrry.htm (accessed 16 December 2009).

A Copernican Revolution in Ethics

Kate Rawles

I write this piece as a howl from the heart. It is a howl of anguish but also a howl of hope and exuberant elation. Above all I want to howl an invitation, a plea, a recommendation to anyone who has not yet looked the future fully in the face and who is still dozing unaware: "Wake up. Wake up!"

For all its seriousness, climate change is just a symptom. It is a symptom of the commitment to infinite growth on a finite planet. At a deeper level, it is a symptom of the dominant worldview of industrialized societies, a worldview that is profoundly flawed and that advocates and legitimizes ways of life and social structures that are, in a very literal sense, unsustainable.

A critical flaw is the essentially materialistic, consumption-oriented understanding of "progress," "success," and "development" this worldview offers us. To be "developed" as a country and "successful" as an individual—to achieve what is commonly called "a high standard of living"—requires, on the current materialistic model, ever-increasing quantities of resources. Global energy consumption, to take just one example, has increased by *80 percent* over the last thirty years. Because consumption requires resources

---

KATE RAWLES is a lecturer in the School of Outdoor Studies at the University of Cumbria. In 2002 she was awarded a National Endowment for Science, Technology and the Arts Fellowship to develop Outdoor Philosophy, short courses that combine inspirational experiences of wild places with critical thinking about our relationship with the environment. In 2006 she was awarded a Winston Churchill Travelling Fellowship and cycled 4,500 miles, from El Paso to Anchorage, exploring North American attitudes and beliefs about climate change.

and because both resources and the Earth's capacity to absorb pollution have limits, high standards of living as we currently define them cause serious environmental problems, even though only a minority of Earth's human population enjoy those high standards of living. And the majority of people across the world, who do not share this standard of living, have been convinced to aspire to it.

The fundamental problem this causes has been powerfully summarized by the World Wildlife Fund: if everyone across the world lived like an average Western European, we would need three planet Earths. Substitute North American for Western European and we would require seven planets. The symptoms of our multiple planet living already include mass species extinction, the appropriation of 25 percent of the Earth's land surface for agriculture, and the alteration of the composition of the atmosphere. Yet this is the way of life that is promoted across the world as "developed"—as something to aim for, as the direction of progress. Add to this that the human population is still increasing, and the utter unsustainability of our current trajectory becomes starkly clear.

So I would first howl "wake up"—wake up to the sheer insanity of stoking the engines when the train is already hurtling toward the buffers; and wake up to the awful impacts of "progress" and modern industrialized, high-consumption ways of life on the billions of lifeforms with whom we have coevolved and with whom we share the planet. We are already wiping out roughly one species a minute, a rate that is between one hundred and one thousand times faster than the natural "background" rate and that has been described as the "sixth great extinction." Depending on how far we travel up the scale of predicted temperature rises, we can expect this rate to increase massively. Predictions of this kind can never have absolute precision, but there is a highly unnerving convergence of international scientific opinion that a rise of 2 degrees C in the average global temperature, combined with existing pressure on ecosystems and habitats, is likely to cause the extinction of about 30 percent of all wild species. If the Earth's average temperature rises by 4 degrees, which scientists at the climate conference in Copenhagen have announced is now more likely than not, then about half—half!—of all Earth's current species will be lost forever: an absolutely catastrophic collapse of ecological systems across the world.

Yet I regularly encounter people, including young people just out of school, who are simply unaware that any of this is happening. This seems astonishing. How can our educational system have failed these young people so spectacularly? Even more astonishing, perhaps, is that among those who *are* aware, many hold the view that climate change and massive biodiversity loss will not really impinge on human well-being. This is not just about tendentious climate change skepticism. (Biodiversity loss skepticism is, interestingly, much rarer.) It is firmly rooted in another critical flaw in the industrialized worldview: the portrayal of human beings as somehow separate from the rest of the natural world. We humans study biodiversity and ecological systems, including the climate, but we do not really feel or act like a part of them. It follows from this often unconscious but deeply held belief that while the loss of other species and the profound changes to the natural world may be sad, they are not really threatening.

Even from a human-centric perspective, this is a tragic misunderstanding. This is not about losing a luxury some of us enjoy at weekends. A recent report by the United Nations confirms—again—that current levels of species extinction and habitat loss are already seriously compromising our ability to meet human needs, making the Millennium Development Goals of eradicating hunger and poverty extremely hard—if not impossible—to achieve. And if the well-being of people matters now, then it surely matters tomorrow and next year as well; it matters for all time. If we are right to care that as I write this 900 million people still—in the twenty-first century—lack clean water, adequate nutrition, and basic sanitation, then the needs of their children—everyone's children—and their children's children must matter too. And the outlook for these future people could not be clearer. According the United Nations, the way we currently meet (some of) the human population's needs quite literally threatens our own survival as a species.

Of course, the ethical implications of our impact on the climate and other life-forms arguably extend far beyond self-interest. The suffering and death of individual sentient animals in their billions; the widespread extinction of other species; the degradation of extraordinarily complex ecosystems—these are all ethically significant in their own right as well as in relation to their adverse impacts on people.

But that is not what the dominant worldview tells us. Another critical flaw in this worldview is its presentation of ethics as a purely interhuman affair. This is a staggeringly anthropocentric view. It puts our own species alone at the center of the ethical universe and allocates value to all other species solely in relation to our own needs and interests, the ethical equivalent of the belief that the sun revolves around the Earth.

Its remedy would amount to a Copernican revolution in ethics: the recognition that we are one species among many, and that the way we treat other forms of life matters in its own right, not just because bad treatment of them in the end rebounds on ourselves. The recognition of intrinsic as well as instrumental value in the living world in turn shifts our understanding of ourselves from rightful masters of everything that exists on Earth—only a set of resources for the benefit of people—to members of a vastly complex and interrelated ecological community. From this revised position, compassion for other beings and justice in our relationships with them are as telling, ethically, as our compassion and justice toward other people.

This is all much better expressed in the context of a sea kayak journey. Ideally, I would invite everyone who reads this—and everyone who doesn't—to join me on a trip around the Arisaig Islands, off the western coast of Scotland. As oystercatchers startle up from a rock, vivid orange, black, and white and crying a tad histrionically, the concept of intrinsic value becomes a whole lot less abstract. Neat black guillemots dive as you approach, leaving, on a calm day, perfect ripples. Look down as the ripples glide back into surface stillness and you see a whole other world, a million lives playing out in a forest of seaweed just below you. Look behind and you realize that you've been tailed by a seal. As you catch its eye, it snorts in alarm and dives with a dramatic splash, but sit still on the water for a while, and a whole circle of seals, whiskery faces, curious eyes, emerge and surround you. Their at-homeness in the cold water is astonishing. The idea that these animals' only value resides in their usefulness to people no longer makes any sense at all. Suddenly, a movement at the corner of your eye snatches your attention. A small black fin. A basking shark! The fin alerts some primeval sense of alarm despite your certain knowledge that its owner eats only plankton. The huge fish glides beneath your boat, its massive bulk shadowed in the water. It circles once and then is gone, leaving, despite the fin, a powerful sense of peace.

These are the creatures—the seals, the birds, the seaweed forest inhabitants, the basking shark—that industrialized, consumption-based lifestyles threaten on a daily basis. This is what climate change will destroy. It hurts, really hurts, to know this. But it also opens up a different perspective, and this is our way forward. In fact, to sit in a small boat surrounded by seabirds and seals and feel, even on a calm day, the immense power of the ocean as it moves below you is to open the door to multiple perspective shifts, to seeing many aspects of the dominant worldview in a very different light. Our industrialized societies tell us not only that these astonishing creatures are really only a set of resources but also that we humans can control their lives, can control the interconnections between them, can control nature. Climate change, of course, is telling us more and more loudly how profoundly wrong we are about this; but the ocean can tell us this very loudly indeed, and with an immediacy that climate change, for many, does not yet have. There is nothing abstract or indirect about the power of the ocean when you are in a small boat on even a lazy swell. The fantasy of power over nature is experienced as just that—a fantasy. But it is a fantasy in both senses of the word. And here is my howl of exuberant exultation. It is fantastic to be alive in this world, on this wave, surrounded by these astonishing beings! How unbelievably wonderful to be alive, on such an extraordinarily beautiful, diverse planet and in such amazing company!

Industrialized ways of life are blindly committing us to global climate change and other forms of environmental impact that, in the end, our own species—and millions of other species—may not even survive, at least not without great conflict, unprecedented loss of life, and profoundly destabilizing change. But I also want to howl, "Wake up!" in relation to what industrialized ways of life offer us in exchange for this devastation. I think it was Peter Singer who wrote that we are destroying the world for beef burgers. This captures it for me. I am not denying that there is much to celebrate in industrialized lifestyles and much that we would want to keep. But there is so much dross too. Throwaway consumption, egotistical individualism, crass materialism. Bad food, obesity, ill health. Not to mention increasing levels of inequality, disenfranchisement, insecurity, and crime. And stress, even for the privileged. How many of us are, or have been, or will be, trapped in a way of life in which we spend the best years of our life working nine to five (at least), five or six days a week (at least), earning money to buy stuff that we don't really need? For how many of us does the expression "time poverty" resonate? Even those favored by the current system are often trapped by it—trapped in ways of life that, for all their material wealth, are too busy, too stressful, too fast, offering little real quality of life. And for the less privileged, current global economic systems endorsed by the industrialized world do not just coexist with poverty and violence; they create and exacerbate it.

So here is the howl of hope. I want to live with a lighter footprint and a clearer conscience, and I want more time in my life! There is a win-win here. We could have a higher quality of life, with much more equity, *and* a greatly decreased impact on the environment. The aim of meeting people's basic needs across the world should of course remain—or, rather, become—an absolute priority. But "meeting basic needs" doesn't have to mean "living like a Western European." And study after study has shown that after a certain point, increase in material wealth does *not* lead to an increase in either quality of life or happiness. In fact, quality of life and happiness may even decrease, as greater material wealth—especially when inequitably distributed—typically coincides with increases in mental health issues, violence, crime, eroded social cohesion, and environmental impact.

This, then, is the core challenge and immense opportunity that climate change and the "nature crunch" more generally (as well as the objectively much less threatening but vastly more acted-upon credit crunch) present us with. On one level, it is a challenge that issues very clear and immediate practical prescriptions. We have a diminishing window of opportunity to make the changes required to keep global climate change below "dangerous" levels. Anything we can do to reduce our personal and professional carbon footprint—in relation to food, transport, travel, heating, electricity consumption, purchases—is of critical importance. Education and communication are also key, as, of course, is making politicians, who inevitably tend toward timidity, aware that the electorate is ready to support them in making bold decisions—aware that we *want* them to redesign existing legal, fiscal, and regulatory systems in ways that will facilitate low-carbon, low-impact living and working.

But there is also a deeper challenge, a challenge that is absolutely critical if we are to succeed. This is the challenge that plays out at the level of worldviews. The challenge is to radically reassess what is currently meant by "success," "development," and "progress." It is to completely overhaul the values that underpin these concepts and to craft instead an understanding of "quality of life" that everyone on Earth could aspire to—without requiring multiple planets. And it is to reject those aspects of the industrialized worldview that tell us we are separate from nature and can survive environmental degradation, however profound, pretty much unscathed. This is not about going back to the caves or even about giving up things that we deeply value. We can keep the best of the industrialized world's education, communication, medical advances, time-saving appliances (I personally have no desire to wash my clothes by hand), music, literature, painting, low-impact technology, and even transportation systems. But we can surely let go of throwaway consumption and the "hedonic treadmill" of ceaselessly needing more to achieve an ever-elusive state of happiness; and we can let go of a value system that tells us we are someone only when we buy the right kind of stuff. There is a positive, life-enhancing vision to embrace instead: the pursuit of quality of life through quality of time, quality of experience, good relationships, cohesive communities, creativity, meaningful employment. And time to hang out with other species—those who are singing on street corners as well as those who are dancing on waves.

Climate change offers us the opportunity to make radical, positive change. It is a chance to make human life, now and in the future, better and fairer in ways that are compatible with ecological community. One-planet living, after all, is not really optional, and I can't be alone in not wanting to live in a way that inescapably implicates me in environmental destruction and social inequity; in threatening the sparrows that live in the ivy on the front of my house, the basking sharks, the oystercatchers. The way I get to work, the way I heat my house, the food I eat—all inexorably exact an unacceptable toll. Imagine the sheer relief of being free to celebrate our coexistence with other creatures and other people with a clear conscience, in the context of a wider human society working to dispense with poverty rather than collude with it, aiming to live alongside other species rather than to dominate them. Imagine living in a society motivated by compassion and justice rather than short-term profit and material acquisition. Imagine time to live life to the full. Work a bit less, consume a bit less, play a bit harder: we can truly celebrate a low-carbon lifestyle and everything it entails.

Wake up! We can have it all, on this one extraordinarily wonderful Earth.

ETHICAL ACTION

How can we fulfill our obligations to protect the Earth—for its own sake?

It's easy to get behind policies that benefit us directly, and relatively easy to rally support for policies that help human beings in general. We are asked now for something different—to support policies that protect the Earth, even if the benefit to us is negligible or hard to understand. This changes the name of the game—from self-interest to sacrifice.

Of course, it is possible to argue that when it comes to environmental actions, what we call "sacrifice" is not really sacrifice. Because giving something up is often a way of getting something of equal or greater value, sacrificing might be a species of self-interest. We can do what's right and still be selfish—a win-win game! Trade complex travel plans for peace of mind. Walk to the store instead of drive, and get a healthier, longer life. All this is smart. All this is to the good.

But it misses the point.

To "sacrifice" is to *make sacred*. In the long history of devotion, people made sacrifices to honor the Most Worthy, giving up something of great value—a goat, a rooster, sometimes a son. It is exactly this renunciation of self-interest that brings a person into closer relation to the divine.

So a better way to think about sacrifice is to ask, What ways of acting will reduce my fixation on self-gratification, and so bring me into closer relation with what is Good-in-itself, the Earth, the unfolding of life? Try these three:

1. Live with the seasons. There is a time for every purpose under heaven. A time for strawberries and melons, a time for acorn squash.

A time for basking in the sun, a time for playing in the snow. A time for rain, a time for drought. The seasons bring their ripeness and their joy. Live with them, plant for them, celebrate them, anticipate what each will bring.

2. Stay close to the ground. Walk where you are going; go where you can walk. Refuse to fly. Move close to your family. Sit on the steps or the curb or the fallen log. Dig potatoes. Pick up wind-thrown apples. Gather your neighbors and plant food in the vacant lots. Sled. Lie on your back and watch the stars or the reflection of the city on the clouds. It's connection with the Earth that gives us life and grounds our joys.

3. Honor the Earth. Have you forgotten how? Then think of how you honor your grandmother, the wizened woman in the nursing home—by visiting her, spending time holding her hand, singing her songs of the season, telling her stories of how she cared for you, asking her to remember how you were, bringing her small gifts, stroking her arm, listening to her, making sure her hair is clean and people are kind to her. The Earth is equally beloved and holy: honor it the same way.

(Full disclosure: You will sacrifice out-of-season fruit shipped across the equator, winter vacations in Florida, spring skiing in Vail, irrigated crops and lawns, rapid travel, airport security lines, seatbacks in an upright position, long commutes, poisonous potatoes, perfect apples, constant TV, isolation from neighbors, loneliness and guilt—and the psychic and carbon costs of these.)

4

> *Do we have a moral obligation to take action to protect the future of a planet in peril?*
>
> **Yes, for the sake of all forms of life on the planet.**

The history of the moral development of the human race, Aldo Leopold wrote, is the expansion of the sphere of our moral concern. As we develop (as individuals and as cultures), we acknowledge duties to our selves, then our families, our tribes, people like us. With increasing sophistication, we acknowledge duties to those of other races, other nationalities, other sexes, then all creatures that feel pain. Each progressive step is based on the dawning recognition that the qualities that make us worthy of moral concern are qualities that we find also in others, who are not so different from us.

The next step in moral development is to expand the sphere of moral concern to include other forms of life. We recognize that the sparrow and whale are not so unlike us. They have interests; that is, they seek some things and avoid others. They thrive or fail to thrive. Like us, they are the product of millions of years of evolution, shaped by the lives and deaths of literally trillions and trillions of plants and animals, the intricate interrelationships between living things and the world in which they live. How could we acknowledge duties to anything without also acknowledging duties toward this array of lives? If there is anything in the universe worthy of respect, worthy of marvel and wonder and awe, and ultimately worthy of protecting and defending, then so is the rich variety of life on planet Earth.

And there is yet another step to achieve in our moral development. That is to expand the sphere of our moral concern from current lives to those in the future. There is no morally significant difference between a present polar bear, say, and her daughter yet to be born. If one is the subject of our moral concern, then so must be the other.

This is the grounds for our obligations to future generations—not just of humans, but of all species, all equally worthy products of time and chance.

Our current actions, however, are greatly diminishing the biological diversity of the Earth. Even conservative estimates suggest that within only the past few decades, anthropogenic actions have caused a sixth great extinction. The unique and morally outrageous thing about this sixth extinction event is that it was caused by a single species, a single species who knew what it was doing, a single species who could have chosen otherwise but did not.

Wild Things for Their Own Sakes

Dave Foreman

The time of Man is but an eyeblink in the great span of Earth's being, yet humans of all kinds find it hard to think of an Earth-time when we were not here or of an Earth-time to come when we will not be here. So we think Earth is ours.

And what of the other Earthlings?

We see wild things as good only for what good they are to us.

We see ourselves as the meaning of life and Earth as the acknowledged end of all that has come before. All this is self-evident truth. Or so we believe.

This self-centeredness is at the heart of what is wrong with our thinking and our believing, whether natural or supernatural.

I am not talking about fleeting high-minded words sometimes found in the human cultures swarming Earth. Rather, I am talking about what is done—not in what we pretend, but in how we behave.

And we—nearly all of us—behave as though all besides us is dross. It is stuff. Stuff for us to use and throw away. Wild things are food. Pests in our way. Threats to be tamed or killed.

DAVE FOREMAN is a wilderness conservationist and the founding director of The Rewilding Institute, a conservation think tank advancing ideas of continental conservation. He also cofounded Earth First! and The Wildlands Project. Foreman has served on the board of directors for the Sierra Club and New Mexico Wilderness Alliance, on the board of trustees for the New Mexico Chapter of The Nature Conservancy, and as editor of the *Earth First! Journal* and *Wild Earth*. He speaks widely on conservation issues and has authored a number of books, including *Rewilding North America: A Vision for Conservation in the 21st Century* (2004).

Before Charles Darwin, our thinking was built on the sand of imagination and flights of fancy, not on the hard bedrock of things as they are. Ernst Mayr, perhaps the greatest biologist of the twentieth century, wrote that Darwin overthrew all philosophy before his time, that all must be rethought in the bright light of his *natural* world.

Darwin saw that all living things—all Earthlings, if you will—were descended from a common ancestor.

He saw that there were no living things frozen in their *types* or *essences*, that wild things change through time and place.

He saw that in Life's descent with modification (evolution) there is no foretold end or goal.

He saw that there was no guiding hand but rather natural and sexual selection, chance and accident.

But since Darwin's time, both those who think and those who only believe have been as though asleep to learning from Darwin's view of life.

It is time to ask at long last, "Where do Darwin's insights take us when we struggle with values and ethics?"

Aldo Leopold, the great conservationist of twentieth-century America, wrote that we could no longer be the conquerors and lords of the land community; instead we must become plain citizens and members of it. Standing on Darwin's insights, Leopold crafted his Land Ethic:

> *A thing is right when it tends to preserve the integrity, stability, and beauty of the biotic community. It is wrong when it tends otherwise.*

In the sweep of nearly four billion years of earthly life instead of our teeny, puny glimpse of five thousand years of human empire, Leopold's Land Ethic is the rock on which we must build. His is the ethic that comes from a Darwinian worldview.

In this light, which is the light of 3 or 4 billion trips around the sun, the key statement of values for us today is this:

> *Wild things exist for their own sake.*

This means that they have worth inborn in their being and are not just stuff for us to do with as we please.

The heart of the relationship between Man and other Earthlings winds and whips about the meaning of *wild*. Historian Roderick Frazier Nash wrote in *Wilderness and the American Mind* that in Old English, a wild thing is a *wildeor*, a self-willed animal; wilderness is the home of wildeors. Wilderness is self-willed land, writes philosopher Jay Hansford Vest. Self-willed things, deors or land, are not under the willful thumb of humans, nor should they be.

But the fifty-thousand-year-long tale of humanity is the growing reach of Man's will over wild things, of taking wild things and turning them into more and more human things. It is the taming, the domestication, the *homo*genization of the world—and of Earthlings of all kinds. It is the steady, now overwhelming wave from wild things to humans and human-willed things.

The ethical challenge for humanity—as a species and as individuals—is to step back and let the being of wild things be, to go widdershins so that wild things and their self-willed Earth grows and man-willed parts become smaller.

So, this book asks, do we have an obligation to future generations of all living things? The question is wrong-headed. It begins with the thought that we have the right to trash, ransack, plunder, and shrink the world and its wild things to mere stuff and gobble them up—unless someone can show us otherwise. It puts the burden of proof on those wild things, on that tomorrow, not on us today. That is unjust.

Our asking should instead begin on the bedrock that we of course have an obligation to wild things of all species, today and tomorrow, to honor their intrinsic value and thus to act only in ways that keep whole the beauty, integrity, and stability of Earth. Those who want out of that obligation will have to fully show how it is okay to snuff life for short-term, selfish ends. This shifts the burden of proof in a strong and mindful way. Those who would shatter life will have to show in a deep, wide way why this careless, carefree, uncaring behavior is good. That will be hard—so hard—to do.

We'll know we have come further toward looking full-on at the wreckage of Earth when the next book asks, Do we have a right to bring to naught the wild beings of Earth?

Spray Glue Goes. Maggots Stay.

Carly Lettero

My mother touched a drop of holy oil to my grandfather's forehead as he took his last breath. She leaned over him to whisper the Lord's Prayer. My sister held his hand, which stopped shaking for the first time in nearly a decade. My aunt stepped away from his bed. *Is it over yet?* she asked. *Is it over?* I watched the color drain from his face, caught my sister's glance, listened to my mother pray in a shaky whisper. Yes. It was over.

My grandfather's death marked the end of a generation in our family. His was a generation of devout Ukrainian Orthodox who wore suits on Sundays and danced the polka in the smoky church basement. His was a generation that worked hard in Chicago's factories and commuted home to the suburbs. His was a generation of acquired things—of big houses full of knickknacks and overflowing closets and stockpiled basements.

I did not know what to do in the moment of death. The hospice nurse announced the time—ten p.m. exactly. An early winter snow flurry pushed across Lake Michigan. In a few minutes we would call my cousins and uncle, and they'd be on their way. Then the under-

---

CARLY LETTERO is a writer, interdisciplinary researcher, and environmental activist. She founded the Campus Carbon Challenge, a grassroots initiative encouraging people to reduce their carbon dioxide emissions by making changes to their daily routines. She is also a former resident of Penn State University's Renewable Energy Homestead at the Center for Sustainability, where she lived within her ecological footprint by growing her own food, generating her own renewable energy, and processing her own waste.

taker would come. But the still moments after death were uncharted territory for me, so I followed my mother's lead.

She stood next to the bed, calmly watching as the hospice nurse began to wash my grandfather's body. He lost nearly one hundred pounds in the last few years. Only his face and hands were recognizable now. The nurse pulled on a pair of rubber gloves. I wondered if there was an afterlife. I wondered what my grandfather would have said if he had been able to speak these last few years. The nurse tore open a package of hermetically sealed cotton swabs. I wondered if my grandfather was reuniting with my grandmother in some joyous swirl of energy above us. I wondered if Parkinson's disease was lurking in my genes too. As the nurse ripped into a package of individually wrapped wipes and plastic padded sheets, my wonder turned to irritation. Why on Earth was he using so much disposable crap to give my grandfather his last bath?

Moments after my grandfather's death and I was thinking about garbage. What was wrong with me? I deliberately thought of things I was grateful for instead. Thank you for my mother, for my blue eyes, for the pocket full of candy you always carried. Thanks for teaching me to spit watermelon seeds, hook a worm, and appreciate garage sales. But the nurse stuffed my grandfather's hospital gown into the garbage bag, and I lost my train of thought. Why shouldn't I think about garbage? Why should I think about it only when it's convenient? Shouldn't garbage be a consideration in every moment, especially the ones that are sacred?

The nurse finished, the undertaker came and went, and I drove home long after midnight. As I tried to fall asleep, I remembered finding the body of a sea lion on the Oregon coast. Waves had rolled it from the saltwater suspension of the ocean onto the windswept beach. Sand pooled around its thick black skin and buried its flippers. Turkey vultures pecked into its stomach and chest, and black fluid flooded the holes. Closer still and the holes were crawling with bugs. Maggots slithered through the black liquid and burrowed into the soft tissue. Flies swarmed above the body, landed, got swept down the beach by a gust of wind, and swarmed again. So much life spiraled out from this one dead animal. In stark contrast to my grandfather's death, nothing was wasted.

I woke the next morning, ironed my black suit, and drove to the funeral home. My family gathered on the mauve and gray couches to meet with the funeral director. My aunt, the executor of the will, laid out the necessary paperwork and updated us on the state of my grandfather's affairs. Everything was in order. After all the hospital bills and funeral expenses, we would each receive a modest inheritance. He always hoped to leave us something, and I was thankful he could.

But there was no mention of the other things he'd left us. My generation learned how to pronounce "Chernobyl" in middle school. In high school, we watched birds suffocate in the black oil of the *Exxon Valdez*. We are the heirs to garbage piles that leach toxic chemicals into our groundwater and soil, an atmosphere choked with unprecedented levels of carbon dioxide, mountains that have been destroyed for coal and metal. This list, as we all know, goes on and on. I am not thankful for these things. I am heartbroken.

Sometimes I am confused too. I do not understand how previous generations could have let things go so terribly wrong. I am furious that I cannot drink water from streams, that I will never hear the call of an Atitlan grebe, and that Montana's glaciers will disappear before my unborn kids get to see them. Sometimes I am devastated. I have fantasized about snuggling into a bed of old-growth moss and forgetting about all this.

But I cannot forget. I feel obligated to make things better for future generations. Not just to leave things as I found them but to do everything I can to help water run cleaner and wilder and to help dwindling populations of owls thrive and to defuse the time bomb of climate change. I feel obligated to do something all the time. As burdensome as the obligation feels some days, it also empowers and guides me. I am grateful to be alive when my studies and career and votes and everyday decisions can help the natural world. I am hopeful that my generation is increasingly more conscientious about how our actions reach into the future and shape the planet our grandchildren and great-grandchildren will be born into. I am curious about how and where and when we will find solutions.

My sense of obligation to the future, and the slurry of emotions that goes along with it, arises out of my reverence for life. All of life.

For the life of my grandfather, the seal, the turkey vultures, and the maggots. For the generations I carry in my genes and will carry in my womb. For the life of the churning glaciers, the silenced Atitlan grebe, and the rushing water. And for the life of the soil, where our garbage is buried. These other expressions of life nourish my body and mind and spirit. I want to reciprocate.

We drove through the wrought iron gates of the cemetery in a funeral procession. The road wound past the graves of my great-grandparents and great-aunts and -uncles and ended in front of the mausoleum. The cemetery was in the direct flight path of Chicago's O'Hare airport, and as the priest began the funeral, the white belly of a jumbo jet began its climb over us. The priest chanted in Ukrainian, then English; swung a ball of incense; drew crosses in the air with his open hands; then placed a spoonful of dry dirt on my grandfather's casket. *From the Earth we came,* he said, *and to the Earth we must all return.*

But my grandfather will not be returning to the Earth anytime soon. It seems he took every possible precaution to avoid it. He was laid to rest in a metal casket. When the priest finished the ceremony, two men who smelled like cigar smoke and fresh dirt filed into the room. They pulled a white bag of impermeable plastic over the casket and sealed the edges with spray glue. They spoke to each other in rushed Ukrainian and managed to carry the casket outside and hoist it onto a hydraulic lift. They raised the whole package precariously toward the cold afternoon sky, slid it into a cement wall, and closed it in with a sheet of Styrofoam, a thin piece of marble, and four screws. Were bugs able to squeeze through the fortress, they would die of formaldehyde poisoning from the embalming fluid. This body entombed in a wall was no longer my grandfather. It had become toxic waste.

My mother, sister, and I huddled together, our pockets stuffed with crumpled Kleenex. Family and friends joined us and said what they could. *Eighty-seven years of a wonderful life. I'm glad his suffering is over. He left us with so many memories.* And they were right.

On the drive to the funeral luncheon at a local Polish restaurant, my thoughts turned to garbage again. Not the garbage of that moment, but the rotting trash of all the moments leading up to that one. A lifetime of cereal boxes and plastic wrap, televisions and transistor radios, ties and dress shoes, flocked synthetic Christmas trees and wrapping paper, Tupperware and microwaves, cars and tires and barrels of gasoline.

If I am to leave future generations a world that is better than the one I am inheriting, I am going to have to change almost everything, from the way I live to the way I die. It is not a matter of simply handing down a policy or new technology. To leave things better for the future, I need to hand down traditions, like polka dancing and garage sale hunting and knowing what to do in the moment of death. Traditions take time and repetition and witness and careful attention to details. It is not easy. But my grandfather, and the generations before him, did lots of things that were not easy. They fought two world wars, survived the Great Depression, and sent a man to the moon. Can't my generation stop throwing stuff away, curb greenhouse gas emissions, and ban pollution? Thomas Jefferson once said that every generation needs a new revolution. I believe this is ours.

My grandfather handed down many of the values I'll need in these revolutionary times—hard work, steadfast determination, persistence, patience, faith, and even humor. He smiled easily, and I imagine he would smile at this too. It is a time of moral spring cleaning. It is hard to get started, but once it gets going, it feels so good. It is time to reinvent the way we live in the world and get rid of everything that is not working. The spray glue goes. The maggots stay.

When I die, wash my body with a cotton cloth. Bury me in a split-wood coffin crafted from trees that died a natural death. Lay me to rest in clothes I have already worn thin. Do not seal out the water and bugs and burrowing critters. Let me be absorbed back into the Earth. Let my body turn to soil. Even when I'm dead, let me nourish the future.

Ornithophilia

Shepard Krech III

Decline weighs heavily on environmental history. Seldom cataclysmic, and often unnoticed until it cumulates over the course of a lifetime, it frequently provokes sharp, poignant, and personal retrospective stories about "the good old days." Take mine. It begins with the Chesapeake Bay, whose eastern shore I have known for sixty years. My earliest memory of the bay dates from the late 1940s, when the water was clear, the eelgrass abundant, and the crabs, eels, and fish countless. Today the bay, a watershed that flushes and dilutes poorly, is in large parts murky, grassless, and barren. Oysters are all but gone, blue crabs are in free fall, and marine life throughout is in peril. Fish kills and toxic blooms are common, hypoxia is rampant, and NOAA has given the bay its worst possible grade on eutrophication. In short, the bay ranks as one of America's sickest estuaries. The culprits? Population pressure. Nitrogen, phosphorus, and potassium from insufficiently regulated agricultural, industrial, and residential activity. Lack of political will. Too many unwilling to grant the bay

---

SHEPARD KRECH III is professor of anthropology and director of the Haffenreffer Museum of Anthropology at Brown University. He has conducted ethnographic and historical research among the Gwich'in in the Canadian Arctic and black Americans in Tidewater Maryland. The recipient of numerous fellowships and a trustee of the National Humanities Center, he is author or editor of ten books, including *The Ecological Indian: Myth and History* (2000), *Spirits of the Air: Birds and American Indians in the South* (2009), and *Encyclopedia of World Environmental History* (2003, edited with J. R. McNeill and Carolyn Merchant).

standing in their calculus of whose rights matter now or in the future. If the bay is a commons, then it looms as another tragedy.[1]

I remember especially the bay's birds. Then—a half century ago—small shorebirds and landbirds blew through in spring and fall on migratory missions that took them far to the north or south. In winter, finches and sparrows came from the north, and we obliged them with seed and suet. Some people called the difficult-to-distinguish little ones "dicky birds," collectively; today birders sometimes refer to them as "little brown jobs." Yet as species, each was different from the others, each in its own way a treasured bit of nature and marvelous product of evolution. And in summer, great blue and green-backed herons and great and snowy egrets lined the shores, ospreys took up residence on navigation markers, laughing gulls and common terns claimed their fill of fish whose schools broke the surface of rivers and bays. Many birds arrived in spring to breed: summer tanagers, blue grosbeaks, yellow-billed cuckoos, indigo buntings, common yellowthroats, yellow-breasted chats, white-eyed vireos, and dozens of others that raised broods above and around such year-round residents as quail (the northern bobwhite) and Carolina wrens. Their coming brought a riot of color and song, creating an aesthetic overload.

One can still see and hear these birds even if their numbers are greatly diminished. For others the story is bleaker and tracks the bay's degraded fate. The bay had always been famed for its waterfowl and still was in the 1950s. The ducks in particular drew many from the north to live there—including my parents. At the top of the order were canvasbacks, redheads, and greater and lesser scaup, or, as they were widely known, cans, redheads, and blackheads, present in astounding numbers and performing in tight knots in swift-winged flight when not resting in dense rafts on the water. When ice denied them access to eelgrass on the flats where the Susquehanna River met the bay, they came in great numbers to our part of the bay. As spec-

---

1. S. Bricker, B. Longstaff, W. Dennison, A. Jones, K. Boicourt, C. Wicks, and J. Woerner, *Effects of Nutrient Enrichment in the Nation's Estuaries: A Decade of Change*, NOAA Coastal Ocean Program Decision Analysis Series No. 26 (Silver Spring, MD: National Centers for Coastal Ocean Science, 2007).

tacular as they were, their equal could be seen in the sky-high flights of mallards, black ducks, American widgeon, northern pintail, and other dabbling ducks, seemingly without end, that awed us into late-afternoon silence. Today, most of these birds are gone. So is much of the habitat on which they depend both winter and summer. For the ecologist, the two are not unrelated.

Many of us hunted these ducks. Hunters in our spare time from being doctors, farmers, lawyers, judges, insurance salesmen, retirees, or students, we shot ducks, geese, and other birds. At the start of the season we hauled decoys nested in bushel baskets from basements and barns, rigged them in the water, and tolled birds in, and our dogs, eyes wide, shivering in anticipation, retrieved them. When not in a duck blind on the edge of the shore or a goose pit in a field of corn stubble, I accompanied my father and a bird dog in search of quail that coveyed up in the hedge- and tree-rows and the fields in weedy fallow, and woodcock that haunted bottomland woods.

We spent a lot of time not just trying to add to the larder—we consumed what we killed—but also observing, talking about, and contemplating these birds. Even if we came home empty-handed, we relished the ritual and social nature of the hunt and of simply being in nature. In his *Meditations on Hunting*, José Ortega y Gasset captured eloquently what I think we felt: the happiness of the hunter in nature, the "mystical union" between hunter and prey, and the "conscious and almost religious" humbling of the hunter, for whom success was at times beside the point in an activity so deeply problematic.[2] Like sportsman-conservationists in our own and previous generations, we supported—out of self-interest, for sure, but also from concern—conservation efforts on breeding grounds that would ensure the return of our quarry.

One spring, I saw an American woodcock dance. In its courtship display a male, hoping to attract a mate, erupted from a clear spot

---

2. José Ortega y Gasset, *Meditations on Hunting* (New York: Charles Scribner and Sons, 1986[?]), 49–50, 97, 123–124.

on the ground in waning light on a spring evening, spiraled up and up into the sky, and every few seconds called a nasalized *peent, peent, peent.* At the top of the corkscrew he reversed direction precipitately, maple-leafing down to the same small clearing faster than he could be kept in focus in binoculars. Lying quietly on the ground nearby, I could attest to this. I was amazed at what I had seen, and after I had thought about it a bit I vowed never again to pull the trigger on woodcock and never have. Later I discovered *A Sand County Almanac,* in which the ecologist Aldo Leopold spoke to the contradiction engendered by hunting something for which one also feels deep affection or passion—love, even. For him too it was about woodcock. His resolve—"since learning of the sky dance I find myself calling one or two birds enough"—became mine except that his one or two became none for me.[3] Today I think of this as my Aldo Leopold moment.

By that time, I knew that I was happy when I could be with and think about birds. I still am. I believe that if only each one of us can latch on to something in nature that allows for reflection, brings self-awareness, and enhances well-being, then one is poised to take the next step and intervene in favor of that thing or the habitats on which it depends. One matter should reassure us: while decline might be common in environmental history, it is neither inevitable nor irreversible. If the view is long, the course of history is often a roller coaster of alternating decline and ascension interrupted by stability.

For me, the latching on had to do with birds. Birds enchanted me. I grew to be very fond of them—to be passionate about, even to love them. I didn't have a label then for what I felt, but "ornithophilia" will do. From the Greek *ornithos* (bird) and *philos* (loving), it is etymologically akin to biophilia, love for life, which, in the sense of focusing on or affiliating with life, the biologist E. O. Wilson proposes is innate.[4] An agnostic on genetic programming, I nevertheless believe that the

3. Aldo Leopold, *A Sand County Almanac and Sketches Here and There* (New York: Oxford University Press, 1968), 34.
4. E. O. Wilson, *Biophilia* (Cambridge, MA: Harvard University Press, 1984).

strongest case for a learned and even intuitive biophilia, though subject to wide cultural variation, can be made for birds.

My fondness for birds has in part to do with awareness of their importance. As pollinators, seed dispersers, nutrient depositors, scavengers, predators, insectivores, and ecosystem engineers, they have played important roles as providers of "ecosystem services."[5] Moreover, they and humankind globally have interacted with each other throughout time. Birds have widely been useful not just for flesh but also for raw materials for myriad objects. The stuff of transactions, birds have frequently become commodities in marketplaces or tribute for the powerful. Around the world, the domestication of birds has linked them to people symbiotically, and here and there the relationship with wild birds has been of mutual benefit, as when greater honeyguides lead people to beehives: when the people break into the beehives for themselves, the honeyguides also gain access. Lastly, through time, people have pursued birds, predictably with opportunism rather than prudence, and they have transformed or destroyed their habitats, causing avian extinctions in ancient and modern times, on islands and continents, in species as diverse as honeycreepers and the dodo, passenger pigeon, great auk, Carolina parakeet, and (unless it rises again) ivory-billed woodpecker.[6]

Birds, the anthropologist Claude Lévi-Strauss suggested, are good not just for narrow utilitarian ends but for contemplation, for which reason they or their representations show up on bodies, in material culture, and in a wide range of social, political, religious, and narrative contexts where they resonate with culturally specific meaning.[7] This seems to be the case everywhere from ancient times through today. In general, the more intense a people's involvement with nature, the deeper the potential for knowledge, discrimination, and naming of birds and for more complex relationships with, and understanding of,

---

5. Cagan H. Sekercioglu, "Increasing Awareness of Avian Ecological Function," *Trends in Ecology and Evolution* 21 (2006), 464–471.

6. Shepard Krech III, "Birds," in *The Encyclopedia of American Environmental History*, ed. Kathleen A. Brosnan (New York: Facts on File, 2010).

7. Claude Lévi-Strauss, *Totemism* (Boston: Beacon Press, 1963).

birds. For many, birds comprise one part of an animate world that can be enlisted to help one heal or harm others or to achieve other ends. For many, certain birds, through augury, divine the future or evoke witchcraft, sorcery, or ghosts. For many, birds figure in narratives about the past. Because they fly, birds often serve as messengers to gods perceived to be also of the air. Birds are truly spirits of the air.[8]

Today, in a world stressed by global exploitation and rapid environmental and climate change, birds also serve as canaries in the coal mine. In the last forty years, the winter ranges of birds typical of those that frequent the Chesapeake Bay have shifted northward, some by two hundred to three hundred miles.[9] Several years ago, American robins showed up in the Arctic, where the Inuit had no name for them. The feeding and reproductive behavior of the ivory gull, black guillemot, and king eider have been adversely affected by the retreat of Arctic ice. The causes can be traced to global warming and climate change. Birds are stressed in other ways as well. The red knot, which stops in Delaware Bay in late May on its epic migration from wintering grounds in Tierra del Fuego to breeding grounds in the high Canadian Arctic in order to gorge on the eggs of breeding horseshoe crabs, is endangered by commercial fishing's unregulated desire for crabs for bait. Communication towers and tall, brightly lit buildings claim an estimated 100 million birds annually in collisions. Habitats and the birds that depend on them under current or imminent stress include eastern forests, which despite increase are fragmented or monocultural; prairies and their wetland "potholes," everywhere gravely threatened; low-lying coastal habitats, either cleared for residential development or in certain danger if, as expected, sea levels rise millimeter by millimeter; and northern boreal forests, stressed by new insects, a warming climate, and the commodification of natural resources, including hydrocarbons. The examples could be multiplied. Each loss—of a bird, of habitat—impoverishes us.

8. Shepard Krech III, *Spirits of the Air: Birds and American Indians in the South* (Athens: University of Georgia Press, 2009).
9. National Audubon Society, *Birds and Climate Change: Ecological Disruption in Motion* (New York, 2009), www.audubon.org/news/pressroom/bacc/pdfs/Birds%20and%20Climate%20Report.pdf (accessed 6 March 2009).

What, if anything, can or should we do? When the causes are of our own human making, as with climate change or unregulated and avoidable practices, they call for an ethical stand and response. For many, one seemingly insurmountable problem when it comes to global climate change is that it is too vast and abstract to grasp. Here, again, Aldo Leopold suggests a way forward. It is inconceivable, Leopold wrote, that "an ethical relation to land can exist without love, respect, and admiration for land, and a high regard for its value."[10] With *philos,* we capture love, passion, respect, admiration, value—and ethics. For land, substitute the Earth if one can get one's mind around it, but if not (which is likely for some), then replace it with something particular and concrete, such as birds. Hinge the particular (e.g., *ornithos*) for which one has fondness (*philos*) to the general, and tackle problems that surely link both. In the specific, as in the framing Earth, there remains vast scope for ethics and practical good.

10. Leopold, *Sand County Almanac,* 223.

Heirloom Chile Peppers and Climate Change

Gary Paul Nabhan

It was my first pilgrimage back to the delta region of the Mississippi since Hurricanes Katrina and Rita. It did not take much to see how the floodwaters had stained and befuddled most everything in sight. Whole windrows of trees had their roots knocked out from under them; houses and cabins had bathtub rings on their walls, and even the ones up on stiltlike pilings slumped in the mud below them. This was a land made weary by recent hurricanes, whose frequency has been increasing as the death march of global climate change increases to a thunderous volume.

But my pilgrimage was not merely one for mourning; I had decided to drive some 120 miles southwest from New Orleans to Avery Island, where one of the rarest but most notorious chile peppers of North America still grows. It is the tabasco pepper, the only heirloom chile variety of *Capsicum frutescens* that has some 140 years of cultivation and culinary use in North America. Its commercial production in the United States is limited to patches within Avery Island's 2,200 acres, which are also a hot spot for migratory birds, reptiles, and subtropical plants along the Gulf Coast.

GARY PAUL NABHAN is a research social scientist based at the Southwest Center of the University of Arizona, founder of the Renewing America's Food Traditions Alliance, and the wild chile–eating champion of the Stinkin' Hot Desert. He is also an Arab American writer, lecturer, food and farming advocate, rural lifeways folklorist, and conservationist. He has been honored with the John Burroughs Medal for Nature Writing and a MacArthur "genius" award. He has authored numerous books, including *Where Our Food Comes From: Retracing Nikolay Vavilov's Quest to End Famine* (2008).

As I stopped at a toll bridge that links Avery Island to New Iberia, Louisiana, I suddenly realized that the core production area of tabascos in North America lies just a few feet above water. Although the island's highest point is 150 feet above the ocean, most of Avery's acres hover just above sea level. Indeed, water not only surrounds the fields and hot sauce factories of the McIlhenny Company; it rages in on them during tropical storms and hurricanes. Although Avery Island narrowly escaped the wrath of Hurricanes Katrina and Gustav, the McIlhenny-Avery clan watched in desperation as Rita came close to obliterating their entire operation: the fields where the peppers grew were completely inundated, and whitecaps ripped across their surface. Their low-lying factory—filled with barrels of fermenting tabasco pepper mash and finished salsas—came within four inches of being fully flooded. Since then, the McIlhennys have reinforced the structure of their hot sauce factory to withstand 140-mile-an-hour winds, and its crews are still at work bulwarking levees to stand up against any 18-foot storm surge that threatens them in the future.

Those are not Chicken Little scenarios: recent assessments suggest that maximum hurricane wind speeds along the Gulf Coast are likely to increase 2 to 13 percent and rainfall around the delta will increase 10 to 31 percent over the coming century. If ocean levels rise as predicted, tabasco peppers may lose their home on Avery Island before many of us are planted in the ground. Tabascos may be able to grow elsewhere in the Americas, but if their fields are permanently lost from the island, a link between Cajun culture, bayou habitats, and the unique terroir of a heritage food will have been severed for good.

It is the endangerment of such relationships between nature, culture, plants, *and the stories and songs which celebrate them* that is making me mourn my way across the delta. As I wander away from a modest lunch of crayfish étouffée, I listen to Cajun and zydeco classics "Hot Chile Mamma" and "Jalapena Lena" on the jukebox. Global climate change will uproot trees and scour out bayous, to be sure, but it will also wreak havoc on the place-based cultures of this Earth. If they survive it all, it is likely that they will find many of their traditional icons, metaphors, and symbols swamped or washed away.

If this were true for only tabasco peppers and their Cajun caretakers, the cynics among you might rightly accuse me of crying wolf. But as I spread a map of North America across my steering wheel, I try to locate where other heirloom chile peppers of limited distribution but exquisite flavor and pungency happen to occur.

Around St. Augustine, Florida, the datil pepper variety of *Capsicum chinense* has been a mainstay in Minorcan-influenced pilaus, paellas, and bouillabaisses since the early 1800s. Kin to the fiery habanero, it is now grown by backyard gardeners in places across the continent, but never tastes quite like it does when grown on its home ground. And yet its core production, too, is dangerously close to sea level. In recent years, Hurricane Wilma and Tropical Storm Tammy have saturated its fields and diminished its yields. If the Floridian prophets of climate change are anywhere near the mark, there will be much more chaos in store for datil pepper growers, picklers, and salsa producers.

Moving northward up the Atlantic seaboard, the Fish Pepper of Baltimore's crab shacks and oyster houses has been custom-grown by African American gardeners along the Chesapeake Bay for nearly two centuries. Most of the seeds of this heirloom that have survived since the 1940s passed through the hands of one Horace Pippin, an African American folk artist who got them from Baltimore cooks and caterers, who made piquant cream sauces with them. Over sixty years ago, Pippin dispersed some of the Fish Pepper seeds beyond the shores of the bay; perhaps it is good that he did, for today 3.5 million Marylanders have built their homes in the coastal zone where these chiles once grew. But the bay may some day reclaim this land. Maryland's hurricane season is now longer and more severe than it has been at any time within recorded history. In 1999, Hurricane Floyd exceeded the storm surge magnitude of a 500-year storm, dumping twenty inches in a matter of hours. The Fish Peppers would have had to mutate into fish to have survived the flooding that season.

Of course, we will lose more than a few pods of pepper here and there should such trends continue. We will lose not only farms, but farmers as well. We will lose not only culinary traditions, but entire American cultures as we know them today. We will have lived through the final rupture between our sense of place and our sense of taste.

Such concerns may seem trivial to climatologists who fully fathom the magnitude of what we are facing. But for most of America, we need to talk about climate change in terms of what our neighbors care about. They care about sharing food, sharing stories and songs and seasonal celebrations. Many silently acknowledge that our lands and waters are linked to our sustenance and survival. If something as bright and as hot as a chile pepper can conjure up such connections for them, so be it. Let it burn in their memories.

Imagining Darwin's Ethics

David Quammen

Drawing ethical conclusions from Darwinian evolutionary theory can be a dangerous exercise, productive of much woozy mischief. This is especially true when the conclusion-drawing is based on willful or careless misperceptions of what Darwinian evolutionary theory actually says. People have been eager to miss the point that it's a descriptive framework explaining what happens and how, not a prescriptive framework stating what *should* happen and how. From the misguided "Social Darwinism" of Herbert Spencer and others in the mid-nineteenth century, to the eugenics of Francis Galton slightly later, to the murderous social engineering of the Nazis in the twentieth, some extrapolations of supposedly "Darwinian" ethics got weird and ugly during the decades after first publication, in 1859, of Darwin's *On the Origin of Species*. That wasn't Charles Darwin's fault. He wrote what he wrote, and from it people saw, heard, and construed what they wanted to. It's happening still, on both the anti-evolution and the pro-evolution sides of the aisle.

The subject of Darwin's own ethics is a separate question. He was a man of rectitude, but not of faith; his rectitude wasn't grounded in piety. He couldn't be called "spiritual" by any stretch of the term. He believed that life is a wondrous phenomenon, yes—a sublime mystery of which the first origin was unknowable to him—but that, once begun, life had proceeded over the ages through a series of transmu-

---

DAVID QUAMMEN is a science, nature, and travel writer whose work appears regularly in *National Geographic* and occasionally in *Harper's*, *Outside*, and other magazines. His books include *The Song of the Dodo* and *The Reluctant Mr. Darwin*.

tations and bifurcations that are explicable by physical laws. In other words, it had evolved. Those physical laws (if I read him correctly) didn't carry moral valence or imply moral result; they simply were.

Still, without embracing any metaphysical system of morals, Darwin was an ethical man: faithful husband, kind father, loyal friend, conscientious citizen, doting grandfather, gentle patron of worthy causes and needy people. His personal ethics existed alongside his scientific thinking, and sometimes challenged it tensely, rather than deriving from it. Among the many things Darwin taught us, by his life as well as his work, is that a stony agnostic evolutionist, seeing a world of material causes and effects, could nevertheless feel an imperative to be good. And he did. Wearing his scientific hat, he came to the conviction that ethics had evolved naturally in *Homo sapiens,* as an adaptation offering social utility—much as the physiological capacity for speaking complex languages had evolved as a parallel adaptation. You can find some of this in chapter 5 of *The Descent of Man,* published in 1871, where he speculates about the evolution of "moral faculties." In his personal life, he behaved ethically (at least most of the time) because . . . well, it just seems to have been his inclination, based on character and social conditioning. He was a man of radical ideas and conservative, scrupulous habits. When he died, in 1882, his uneasy but firm reconciliation of materialism and ethical responsibilities remained intact.

Now to the point: What would Charles Darwin say if he came back today and saw the mess that we're making of planet Earth?

All right, "mess" is a vague word, but please note that I don't ask, *What would he say about climate change?* That's putting the question too narrowly. Although anthropogenic climate change is at last getting its due attention, we need to remember that it isn't the biggest ethical and geo-bio-physical crisis we face; it's only a subcategory of the biggest. The biggest, of course, is the crisis of plummeting biological diversity, presently occurring by way of species extinction, the eradication of localized subspecies, the simplification of ecological processes, and the reduction of genetic variation within populations. Stated plainly, in terms so familiar they've begun to seem dreary: we are perpetrating a mass extinction. And as we've all heard before, this event bodes to descend to a superlative nadir of awfulness, commensurate with the five greatest mass extinctions in Earth's history:

the Ordovician, the Devonian, the Permian, the Triassic, and the Cretaceous. We can call this one the Holocene extinction (as some experts have proposed), since that's our present geological epoch. And we shouldn't forget that, unlike all others, the Holocene extinction is attributable not to asteroid impact or catastrophic vulcanism or some other form of external accident, but to the actions of a single earthly species: us. Those actions and their direct effects fall under six major headings: habitat destruction, habitat fragmentation, overharvest (especially on islands and in the oceans), transfer of invasive species from one ecosystem to another, cascades of extinction that tumble through ecosystems, and finally, climate change, yes, because it exacerbates the effects of habitat loss, habitat fragmentation, and invasive species. The combined result of these six trends is so egregious that eons from now, when paleontologists from the planet Tralfamadore arrive to investigate, they will wonder what the hell happened in the late Holocene to life on Earth.

Charles Darwin was gone before the worst of this cataclysm had developed or the eventual dimensions could be foreseen. We don't know what he would say of it. We can only make guesses, based on who he was and what he *did* say.

Extinction, it must be admitted, was not a subject that worried him. He wrote dispassionately of it in *On the Origin of Species*. For instance: "The extinction of old forms is the almost inevitable consequence of the production of new forms." That's from chapter 10, his discussion of how species and higher groups have succeeded one another throughout time. (He wasn't aware, nor was any other scientist in his day, that the fossil record is punctuated by mass extinctions.) He viewed the extinction of species as parallel to the death of individuals: a natural termination making room for others. It was slow and incremental, he thought, like evolution. "Extinction," he said three chapters later, "has played an important part in defining and widening the intervals between the several groups in each class." He meant that it has set diversification into relief, by pruning away intermediate forms. It has made taxonomic classification possible. We can easily draw a line between reptiles and mammals because their common ancestors have long since died away. And again, forty pages later, in his recapitulation: "The extinction of species and of whole groups of species, which has played so conspicuous a part in the history of the organic world, almost inevitably follows on the principle of natural selection"—that is, *his* cardinal principle—"for old forms will be supplanted by new and improved forms." Extinction in moderation: he saw it as routine, inevitable, necessary, and good.

What Darwin didn't consider was that, though speciation occurs slowly and locally, extinction can be widespread and fast. Nor did he foresee that our own species might so dominate the planet, transform its landscapes and its climate, and appropriate its resources that the very conditions necessary for speciation and evolutionary divergence—and thereby the replenishment of diversity—might no longer exist.

Discussing extinction as a natural process, Darwin spoke with his brain. But he also had a heart and a conscience. He had children and, by the time of his death, one precious grandchild. He cared about the shape of the world, including the world his offspring would inherit. His devotion to his subject was emotional as well as scientific, and we can safely assume he would have been saddened, alarmed, maybe outraged, to see the very matter of that subject leaching away. He was a sometime geologist, a sometime paleontologist, but most essentially a biologist: he studied what lives. In his famous concluding passage of *The Origin*, remember, he wrote: "Thus, from the war of nature, from famine and death, the most exalted object which we are capable of conceiving, namely, the production of the higher animals, directly follows." And then:

> There is grandeur in this view of life, with its several powers, having been originally breathed into a few forms or into one; and that, whilst this planet has gone cycling on according to the fixed law of gravity, from so simple a beginning endless forms most beautiful and most wonderful have been, and are being, evolved.

He wasn't just talking about the grandeur of an idea. The diversity of life, its beauty and wondrousness, had given focus, meaning, and joy to Darwin's own life. Surely he would have cared about bequeathing that joy to the future.

Evening Falls on the Maladaptive Ape

Robert Michael Pyle

If I were a giant clam, instead of an ape, and if I lived on the Great Barrier Reef, and if clams were in charge, my moral choice would be clear: to do something about it. "It" is the fact that the Great Barrier Reef, as we know it, is likely to be history within brief decades, due to warming and acidification of southern ocean waters. Even as other ocean currents slow and the great North Atlantic gyre slugs up into a polluted sump, ice caps melt, and Bangladesh and Bikini slip slowly beneath the water's lap, that country of corals—the Great Barrier Reef—is expected to collapse as a living, functioning system.

So, as that clam, what is this thing I should do? Actually, in order to examine a moral choice, you need two things: a moral outlook and ethical agency. *Tridacna gigas,* you guess, gets by without a moral framework, and, being sessile, it surely lacks agency. So while it would be nice to do something about it, I really can't: I guess I'm off the hook. When crown-of-thorns sea stars occupied our reef, just as when hungry Pacific lionfish (escaped from broken aquariums during a hurricane) recently infiltrated the Atlantic, we clams and such really were left with only one choice: adapt or die.

Biologist ROBERT MICHAEL PYLE writes essays, fiction, and poetry on a tributary of the Lower Columbia River. A Guggenheim Fellow and recipient of the Society for Conservation Biology's Distinguished Service Award, he has worked as a consulting lepidopterist in Papua New Guinea, as the Northwest Land Steward for The Nature Conservancy, and as a visiting writer or professor at many universities. Among his fifteen books are *The Thunder Tree, Where Bigfoot Walks, Chasing Monarchs,* and *Time in Gray's River,* which won the National Outdoor Book Award.

Now, shedding my giant crenellated turquoise-mantled shell and donning my ape-suit once more, things become more complex. We so-called higher primates do possess a moral sensibility, or what passes for it, or so we tell ourselves. At the very least we know what such a trait should look like. Even chimps seem to express a knowledge of what's "right" and "wrong." What's more, they act on it. Yet their agency is limited to interpersonal relationships: they can't pass laws, exercise regulations, levy taxes or credits, extend or deny permits. In the end, for them too, it's adapt or die. And more and more, it's looking more like die, unless Jane Goodall & Co. can help it.

But *Homo sapiens*—named by Linnaeus in an extraordinary act of optimism, especially for a Swede, as "all the same and smart"— has not only his morals and an elaborate ethics derived from them, but also an astonishing degree of active agency. Since the big bamboo scaffolding of human enterprise, like the Barrier Reef, ice cap, and rainforest, is in early to mid-collapse, it is worth asking whether said sapient ape has more than a clam's chance to do something about it, more than a chimp's chance of making it.

Agency cuts both ways. It is our big brains and opposable thumbs, after all, that brought everything—clam, coral, ice, and ape—to this pretty pass. So can our technical and manual powers translate into actual ethical agency that's up to the task—speeding the currents on their way, jump-starting the gyres while cleansing their load, cooling the ice, caring for the corals? It might very likely be too late for all that, and just too hard a task for us; but if it's not, will we? The evidence elsewhere (war and torture on every hand, for example) is not encouraging. In addition to morals and agency, we have several other traits that might belong largely if not exclusively to humans, among them rue, regret, dread, and hindsight. Combined with moral sense, intelligence, and agility, they surely should drive an ethical response to our crumbling scaffold. But can they?

The key question might be, how have we come by our morals and ethics? I can imagine only three sources. First: they were bestowed by a higher power. Many, maybe most, folks believe that we have a moral imperative to "be good"—a mandate from on high, delivered up from some power outside ourselves, mainlining morals straight to our hearts. Others, happy with happenstance as the whole show, require no rule maker or referee for their cosmos. To my mind, any intelligence that bestowed a moral system on just our species among millions, plus the tools to wreck it all, would not be worth the term; and any god remotely satisfied with the results to date would be a risible god. But such a provenance is surely the handiest, when it comes to self-justification: when "God is on our side," anything goes.

Second hypothesis: we bestowed our ethical construct upon ourselves. This seems clearly to be the case, at least in part: our laws are laws of men (mostly *sensu stricto*), though often erected in the name of whatever indulgent or vengeful deity their authors contrive, the better to meet their needs and desires. And just as clearly, this hasn't worked out, except to create, empower, and maintain privilege, beggar the poor, and plunder the common weal. For those ends, such self-published ethics have worked very well.

The third possibility is that our morals and ethics have evolved, biologically and socially. This too seems obvious to a certain extent, since dogs, cats, elephants, cetaceans, great apes, and maybe others all demonstrate elements (I won't say rudiments) of a moral outlook: decency and affection toward others, loyalty, and suchlike. This is hopeful, as it suggests that human behavior could conceivably evolve into something worthy of itself. Aldo Leopold posits such growth in "The Land Ethic," comparing a time when chattels might be killed on whim with ethical impunity, to a modern era of greater restraint or compunction. Today Leopold might question whether we have come much past the chattel-wasting days, what with Mexican drug cartels in Juárez, systematic rape in Uganda, bomb-belts and drone missiles at wedding parties in Afpakiraq, and bulldozer brigades in Palestine, as "Thou shalt not kill" becomes "except when thou needest to."

In calling for a land ethic, Aldo took his vision a step further. He dared hope for a time when we would see, agree, and act on the idea that "a thing is right when it tends to preserve the integrity, stability, and beauty of the biotic community. It is wrong when it tends otherwise." He thought the way to get there would be to "quit thinking about decent land-use as solely an economic problem," to "examine each question in terms of what is ethically and esthetically right, as well as what is economically expedient." Many might regard such a shift as even less likely than mass observance of the Golden

Rule, and so might it be; others would see it as a narrow side-issue of the larger dilemma, even peripheral. Yet it seems to me that such an ethic as Aldo describes would be the only sort of moral resolve that could conceivably give comfort and relief to the current crisis. An ethic based on concern for future human generations alone doesn't strike me as having much heft, since it hasn't stemmed our excesses against one another so far. But an ethic founded on the well-being of the supportive fabric might have a chance.

I think our moral and ethical equipment is largely self-contrived, built upon an evolved capacity, more jury-rigged than refined, and better developed in some than in others. So while the individual will, desire, skill, and devotion for doing so may all be in place, our collective power to repair the reefs and waters, currents and gyres, clouds and caps may be no more effective than the corporate efforts of clams: not only because the task is so large and intractable, but also because the overall ethic is so feeble, and its agency so malleable. And because alongside our clear capacity for honesty, generosity, mercy, and cooperation lurks our apparent imperative for power, greed, and domination. The friction between them is what poet Carolyn Wright calls "our struggle for the preservation of compassion and decency in a perennially fallen world."

And what was this fall? As I see it, it was a two-part tumble. In more than a merely metaphorical way, it was first the very one implied in Genesis: sex. Not doing it, but doing it for eons in the absence of Planned Parenthood. For the very unrestrained "weight of the whole human race" (Waylon Jennings) is that from which we probably cannot return to our prelapsarian grace: the 6 billion were 3 when we last turned around, and will be 12 when we next take a look. Then there was industry, the removal of the people from the land to the factory, and, in particular, our embrace of fossil fuel. Rudyard Kipling already had it right in "The Gods of the Copybook Headings": "In the Carboniferous Epoch we were promised abundance for all / By robbing selective Peter to pay for collective Paul." Or as Fritz Schumacher put it in *Small Is Beautiful*: "As the world's resources of non-renewable fuels—coal, oil, and natural gas—are exceedingly unevenly distributed over the globe and undoubtedly limited in quantity, it is clear that their exploitation at an ever-increasing rate is an act of violence against nature which must almost inevitably lead to violence among men." As it has.

Schumacher didn't know about the coming violence to the climate, but he did write that "in the excitement of the unfolding of his scientific and technical powers, modern man has built a system of production that ravishes nature." One who intuited remarkably early that the ravishment would extend to the very elements was Vladimir Nabokov. In a 1974 interview, he said, "Pollution itself is a lesser enemy of butterfly life than, say, climatic change." In my own travels all across the United States in 2008 in search of butterflies, I saw a number of examples that corroborated his prescient fears. The Uncompahgre fritillary, for one, barely teeters on the roof of the San Juan Mountains.

I am even more angry than sad at how much will be lost before complexity arises again. Extinction events are nothing new, but this one didn't have to happen. The grand evocation of diversity that we know, or call it the "Creation," could have lasted a lot longer, and we could have stuck around to enjoy it. But my anger isn't yet widely shared. It will be, and if it takes longer than I live to get there, I won't mind missing that point. A world without *Tridacna,* or the reefs to hold them? A world without pika geeking from the high rocks? Count me out. I'd rather be a clam.

For us too, it is a matter of adapt or die—there is no way out of that circumstance for any organism. Yet our flimsy ethical condition, regardless of political persuasion or economic system, still thinks perpetual growth and ultimate good are compatible. They are not. As Fritz Schumacher put it, "Greed and envy demand continuous and limitless economic growth of a material kind, without proper regard for conservation, and this type of growth cannot possibly fit into a finite environment." Nor is it merely a matter of greed and envy. As we saw in the recent economic implosion, the entire social system suicidally depends on perpetual growth. Kipling wrote, of the old values and verities, that "we found them lacking in Uplift, Vision, and Breadth of Mind / So we left them to teach the Gorillas, while we followed the march of Mankind"—directly into the lap of "the Gods of the Marketplace," where we engage in carbon trading to pay back Peter. Maybe the gorillas, or some other ape, could make a better job of it next time around.

In spite of the worst we can do, the tatters of nature we leave behind will carry on. The insults will be redressed when, as Robinson Jeffers imagines in "November Surf,"

> The cities long gone down, the people fewer and the hawks more
>   numerous
> The rivers mouth to source pure; when the two-footed
> Mammal, being someways one of the nobler animals, regains
> The dignity of room, the value of rareness.

In time, coyote may give rise to a dozen new canids. Collembola might found whole new orders of invertebrates, with lifeways beyond our imagining. And if some sort of upright ape is among the array, it might not bear much resemblance to those who pulled the plug, pushed the plunger, pulled the trigger on the chamber that, oops! really *was* loaded.

Our moral sense is really the collective genetic and social experience of what's good for us and what isn't. Just like any other species, we are subject to natural selection of the most advantageous traits for the species in the long run. And that can change: love without issue, anathema once, has become heroic and adaptive. Evolution never before encountered self-interest such as ours, empowered by such powerful tools for planetary alteration: the ability of the individual animal or group to frustrate the good of the whole. To my mind, the greatest reason for bringing an ethical response to bear on annealing the damaged climate is to give evolution a fair shake in the next iteration.

Our moral imperative to turn down the heat is merely our mandate from nature to adapt—to survive. But fat chance we've got, now. For this species, and the many others we are dragging down with us, it is late in the game. Cap-and-trade is unlikely to do the job, especially if economic and population growth remain the order of the day. With luck, the smart and humble adherents of the small and the local may survive to have another go after the larger collapse, may even thrive under the uncomfortable and inconvenient conditions to come. But I doubt very much that the culture we know will long persist, absent truly radical changes in the way it works. We are the maladaptive ape, at twilight. Evolution will mock our tardy rage.

ETHICAL ACTION

How can we preserve a future for grass sparrows and sea-grass, blue whales, krill, coral reef fish, lingonberries, grizzly bears—in fact, for all the extraordinary, beautiful, and tragically contingent variations of life on Earth?

We may not be able to preserve a future for all these marvelous creatures, but surely it's a moral minimum to *stop killing them and destroying their habitats*. That is the ecological analogue of the Hippocratic Oath: First, do no harm. Herewith, an Oath for the Animals:

1. I will not buy poisons or introduce them in any way into the world. Nor will I buy food or anything else that was sprayed with poisons.
2. I will not destroy a natural place—not by building a house on unspoiled land or tearing out a meadow for a lawn, or plowing a grassland for crops.
3. I will not step foot in a Kmart or Walmart or a (former) friend's house or any other building that was built on the newly bulldozed remains of what once was a native habitat.
4. I will not refrain from guffawing when someone says it is possible to mitigate the destruction of a natural place by creating a new marsh or wetland.
5. I will not own an outdoor cat or other domestic animal that preys on wild birds. I will not be afraid to badger my neighbors about theirs.
6. I will not be stingy about giving my support, my money, my time, and my vote to those who create natural reserves—marine reserves, wildfowl reserves, wilderness reserves, public land

trusts, urban wilderness parks, and other places where animals can thrive.

7. I will not buy products produced by destroying tropical or temperate rainforests—not coffee, not tea, not fine wood, not toilet paper, not cedar shingles or two-by-fours. Nor will I buy fish or other wild creatures that are not sustainably harvested.

8. I will not oppose subsistence hunting or other respectful hunting for food, but I will howl about trophy hunting, plinking, or any other destruction of animals for "sport," "fun," "father-son bonding," or any other grotesquerie.

9. I will not plant exotic species in my garden—none of the holly and the ivy, no Scotch brooms. The land I control, city or country, will be native habitat for wild creatures, even if I have to plant it myself.

10. I will not take more than I need from nature's bounty, understanding that what I take is taken from some other creature who has an equal right to the conditions for life.

11. I will not pretend that a person can be an upright citizen who kills wild creatures in the course of his business, or causes them to be killed, or profits from the destruction of their habitat.

12. I will neither dither nor fuss nor quibble, delaying action until scientists prove *beyond any doubt* that, say, polar bears can't get along just fine without polar ice, or owls won't die if all the rodents are poisoned. I will take precautions when risks make that a reasonable course of action.

13. I will not worry about being a sanctimonious pain in the butt. I wouldn't worry one minute about trading a friendship for the ongoing existence of meadowlarks. *Someone* has to take a public stand, and if that makes other people feel guilty, maybe it's about time.

5

Do we have a moral obligation to take action to protect the future of a planet in peril?

Yes, to honor our duties of gratitude and reciprocity.

The gifts of the Earth (what we cravenly call "natural resources" or "ecosystem services") are freely given—rain, sun, fresh air, rich soil, all the abundance that nourishes our lives and spirits. Perhaps they are given to us by God or the gods; maybe they are the fruits of a fecund Earth. It doesn't matter to the argument; let that be a mystery, why we are chosen to receive such amazing gifts. What is important is that they are given. We do not earn these gifts. We have no claim on them. If they were taken away, there would be nothing we could do to get them back. At the same time, we are utterly dependent on these gifts. Without them, we quickly die. This unequal relationship, the relationship of giver and receiver of gifts, makes all the moral difference.

We understand the ethics of gift-giving. To receive a gift requires us to be grateful. To dishonor or disregard the gift—to ruin it, or waste it, or grind it into the ground, to turn it against the giver or lay greedy claim to it or sourly complain—all these violate our responsibilities as a recipient. Rather, to be grateful is to honor the gift in our words and our actions, to say, "This is a great gift," and to protect it and use it well. In this way, gratitude calls us to attentiveness, celebration, and careful use.

Furthermore, an important part of gratitude is reciprocity, the responsibility to give in return. We give in return when we use our gifts well for the benefit of the Earth and the inhabitants who depend on its generosity. In this way, gratitude for our abundant gifts is the root of our moral obligation to the future to avert the coming environmental calamities and leave a world as rich in possibilities as the world that has been given to us.

To Commit a Crime Against the Natural World Is a Sin

Ecumenical Patriarch Bartholomew I

The Ecumenical Throne of Orthodoxy, as a preserver and herald of the ancient Patristic tradition and of the rich liturgical experience of the Orthodox Church, today renews its long-standing commitment to healing the environment. We have followed with great interest and sincere concern the efforts to curb the destructive effects that human beings have wrought upon the natural world. We view with alarm the dangerous consequences of humanity's disregard for the survival of God's creation.

We believe that Orthodox liturgy and life hold tangible answers to the ultimate questions concerning salvation from corruptibility and death. The Eucharist is at the very center of our worship. And our sin toward the world, or the spiritual root of all our pollution, lies in our refusal to view life and the world as a sacrament of thanksgiving, and as a gift of constant communion with God on a global scale.

We envision a new awareness that is not mere philosophical posturing, but a tangible experience of a mystical nature. We believe that our first task is to raise the consciousness of adults who most use the resources and gifts of the planet. Ultimately, it is for our children that we must perceive our every action in the world as having a direct effect upon the future of the environment. At the heart of

---

ECUMENICAL PATRIARCH BARTHOLOMEW I has been the Patriarch of Constantinople, and thus "first among equals" in the Eastern Orthodox Communion, since 1991. He is the spiritual leader of 300 million Orthodox Christians around the world and has earned the title "Green Patriarch" for his efforts to raise environmental awareness. In 2008 he published *Encountering the Mystery: Understanding Orthodox Christianity Today*.

the relationship between man and environment is the relationship between human beings. As individuals, we live not only in vertical relationships to God and horizontal relationships to one another, but also in a complex web of relationships that extend throughout our lives, our cultures, and the material world. Human beings and the environment form a seamless garment of existence, a complex fabric that we believe is fashioned by God.

People of all faith traditions praise the Divine, for they seek to understand their relationship to the cosmos. The entire universe participates in a celebration of life, which St. Maximos the Confessor described as a "cosmic liturgy." We see this cosmic liturgy in the symbiosis of life's rich biological complexities. These complex relationships draw attention to themselves in humanity's self-conscious awareness of the cosmos. As human beings, created "in the image and likeness of God" (Genesis 1:26), we are called to recognize this interdependence between our environment and ourselves. In the bread and the wine of the Eucharist, as priests standing before the altar of the world, we offer the Creation back to the Creator in relationship to Him and to each other.

Indeed, in our liturgical life, we realize by anticipation the final state of the cosmos in the Kingdom of Heaven. We celebrate the beauty of creation and consecrate the life of the world, returning it to God with thanks. We share the world in joy as a living mystical communion with the Divine. Thus it is that we celebrate the beauty of creation, and consecrate the life of the world, returning it to God with thanks. We share the world in joy as a living mystical communion with the Divine. Thus it is that we offer the fullness of creation at the Eucharist, and receive it back as a blessing, as the living presence of God.

Moreover, there is also an ascetic element in our responsibility toward God's creation. This asceticism requires from us a voluntary restraint in order for us to live in harmony with our environment. Asceticism offers practical examples of conservation.

By reducing our consumption—in Orthodox theology, *encratia,* or self-control—we come to ensure that resources are also left for others in the world. As we shift our will, we demonstrate a concern for the Third World and developing nations. Our abundance of resources will be extended to include an abundance of equitable concern for others.

We must challenge ourselves to see our personal, spiritual attitudes in continuity with public policy. Encratia frees us of our self-centered neediness, that we may do good works for others. We do this out of a personal love for the natural world around us. We are called to work in humble harmony with creation and not in arrogant supremacy against it. Asceticism provides an example whereby we may live simply.

Asceticism is not a flight from society and the world, but a communal attitude of mind and way of life that leads to the respectful use, and not the abuse, of material goods. Excessive consumption may be understood to issue from a worldview of estrangement from self, from land, from life, and from God. Consuming the fruits of the Earth unrestrained, we become consumed ourselves, by avarice and greed. Excessive consumption leaves us emptied, out of touch with our deepest self. Asceticism is a corrective practice, a vision of repentance. Such a vision will lead us from repentance to return, the return to a world in which we give to as well as take from creation.

We are called to be stewards, and reflections of God's love by example. Therefore, we proclaim the sanctity of all life, the entire creation being God's and reflecting His continuing will that life abound. We must love life so that others may see and know that it belongs to God. We must leave the judgment of our success to our Creator.

We lovingly suggest, to all the people of the Earth, that they seek to help one another to understand the myriad ways in which we are related to the Earth and to one another. In this way, we may begin to repair the dislocation many people experience in relation to creation.

If human beings treated one another's personal property the way they treat their environment, we would view that behavior as anti-social. We would impose the judicial measures necessary to restore wrongly appropriated personal possessions. It is therefore appropriate for us to seek ethical, legal recourse where possible, in matters of ecological crimes.

It follows that to commit a crime against the natural world is a sin. For humans to cause species to become extinct and to destroy the biological diversity of God's creation, for humans to degrade the integrity of Earth by causing changes in its climate, by stripping the Earth of its natural forests or destroying its wetlands, for humans to injure other humans with disease, for humans to contaminate the Earth's waters, its land, its air, and its life with poisonous substances—these are sins.

In prayer, we ask for the forgiveness of sins committed both willingly and unwillingly. And it is certainly God's forgiveness which we must ask, for causing harm to His own creation.

Thus we begin the process of healing our worldly environment, which was blessed with beauty and created by God. Then we may also begin to participate responsibly, as persons making informed choices, both in the integrated whole of creation and within our own souls.

We are urging a different and, we believe, a more satisfactory ecological ethic. This ethic is shared with many of the religious traditions. All of us hold the Earth to be God's creation, where He placed the newly created human "in the Garden of Eden to cultivate it and to guard it" (Genesis 2:15). He imposed on humanity a stewardship role in relationship to the Earth. How we treat the Earth and all of creation defines the relationship that each of us has with God. It is also a barometer of how we view one another. For if we truly value a person, we are careful as to our behavior toward that person. The dominion that God has given humankind over the Earth does not extend to human relationships.

We must be spokespeople for an ecological ethic that reminds the world that it is not ours to use for our own convenience. It is God's gift of love to us, and we must return His love by protecting it and all that is in it.

The Lord suffuses all of creation with His divine presence in one continuous legato from the substance of atoms to the Mind of God. Let us renew the harmony between heaven and Earth and transfigure every detail, every particle of life. Let us love one another, and lovingly learn from one another, for the edification of God's people, for the sanctification of God's creation, and for the glorification of God's most holy Name.

Amen.[1]

1. Ecumenical Patriarch Bartholomew, "To Commit a Crime Against the Natural World Is a Sin," Environmental Symposium, Saint Barbara Greek Orthodox Church, Santa Barbara, California, 8 November 1997.

Sacred Ancestors, Sacred Homes

Nirmal Selvamony

> You wander far from the graves of your ancestors
> and seemingly without regret.
> CHIEF SEATTLE

With the emergence of the industrialist society and its rationalist worldview, even mythology has lost its hold over people. Its place has been usurped by modern-day equivalents such as sustainable development, eco-consciousness, biocentrism, and ecocentrism. Certainly, none of these is as compelling and motivating as one's own ancestors. It is becoming increasingly apparent that addressing issues of climate change and environmental degradation and achieving a maximally harmonious relationship to our planet are impossible without acknowledging the due place of one's ancestral homestead. And if that is the case, a social order based on ancestral sacralization is our best option for a viable future.

NIRMAL SELVAMONY (currently an associate professor) has been teaching English at Madras Christian College, Chennai, India, since 1982. Presently, he is a visiting faculty member and Coordinator of Languages at the Central University of Tamil Nadu at Thiruvarur, Tamil Nadu, India. He introduced ecocriticism in the Indian academy and founded a forum called OSLE-India (Organization for Studies in Literature and the Environment–India; www.osle-india.org) to promote it. With Watson Solomon, he founded the *Indian Journal of Ecocriticism* and has researched extensively in Tamilology, especially *tiNai* theory, *Tolkaappiyam*, and Tamil musicology. He has evolved an ecocritical theory known as "Oikopoetics" and has published twelve books and more than fifty research papers. He is a guitarist, playwright, and director of plays.

In traditional Tamil spirituality, *tiNai* is the word for a family of human and nonhuman members living off their indigenous, ancestral homestead. Both a family and a kind of living, *tiNai* is where you ought to be and what you ought to do. Human inhabitants of *tiNai* regarded plants and animals, too, as members of their own family. For example, in the coastal *tiNai*, a laurel tree was sister to a girl, and the shark was one of the ancestors of several coastal families.

Like other alternative location-based ways of life, such as the bioregion, *tiNai* is marked by smallness of scale, territoriality, and values orientation. But it also has a unique feature: continuity, especially ancestral sacral continuity. A life-way based on ancestral continuity has deep ecological implications that recommend it as our best future option.

The idea of continuity in *tiNai* is apparent from its etymology. *TiNai* derives from language referring to "that which is united and therefore firm, firm conduct, a region where people and other organisms are united with nature and coexist." The same roots refer to the fifth relation in music, where the two tones (the tonic and the fifth) unite perfectly, like spouses who experience oneness rather than simply togetherness. *TiNai* thus creates a continuity among its three members—humans, nature, and the sacred, which is the ancestor.

Continuity in *tiNai* is found at more than one level. Indigeneity may be said to embody the continuity between organisms or objects and a place; tradition, on the other hand, is cultural continuity. But because tradition is closely connected with place, cultural continuity is not possible without continuity between a place and its people.

When *tiNai* society gave way to state society, the different continuities possible in the former society weakened or disappeared entirely. One major form of discontinuity is displacement. State formation involved displacement of different kinds. When common land was appropriated by the state, people who depended on it were displaced. State projects such as irrigation altered the cultural geography as well as the topography. In the ancient Tamil country, monarchs often welcomed immigrants, creating new settlements that caused displacement in the natural and cultural landscapes. War and natural disasters also displaced people.

Organisms were displaced as well. Scientist Charles Elton observes that "we live in a period of the world's history when the mingling of thousands of kinds of organisms from different parts of the world is setting up terrific dislocations in nature."[1] Human agency in such "mingling" is considerable; one of its most trenchant modes is colonialism. After exhausting their own forests, European colonizers "laid claim to vast terrains the world over" for the sake of wheat and cattle and their weeds and diseases.[2] The British introduction of merino sheep to Australia in 1797 and the creation of pasture land out of the Eastern highlands is only one case in point. "Ecological imperialism" is in fact a form of natural displacement that results in cultural and human displacement.

In the context of human displacement, honoring one's ancestors has been the most effective form of continuity; this practice allows peoples to avoid displacement and ensures indigeneity and tradition. This idea has been powerfully expressed by Chief Seattle, who declared that American Indians, unlike the white man, would not abandon their ancestral homesteads for any reason whatsoever.[3] The relationship between colonizers and their dead is discontinuous; this emboldens them to invade foreign lands, violate nature, and ultimately destroy themselves and others. In contrast, an important aspect of *tiNai* is ancestral continuity, for it hallows one's homestead and makes it sacred.

For example, what are commonly known as "sacred groves" or "sacred forests" are homesteads sacralized by the spiritual presence of the deceased ancestor. The sacred forest is named after such ancestors as AyyanappaaTTan (*ayyan*, "leader," + *paaTTan*, "ancestor") and ETTukkaiyamman (*eTTu*, "eight," + *kai*, "hand," + *amman*, "goddess"). The power that an ancestor gains after death is indicated by her eight hands. These ancestors dwell in the homesteads and guard them. In fact, reference to sacred ancestral homestead is found even in the earliest extant Tamil text, which speaks of the

---

1. Charles S. Elton, *The Ecology of Invasions by Animals and Plants* (Reprint, London: English Language Book Society and Methuen & Co., 1966).

2. Madhav Gadgil and Ramachandra Guha, *This Fissured Land: An Ecological History of India* (Delhi: Oxford University Press, 1997).

3. Paul Eschholz and Alfred Rosa, eds., *Outlooks and Insights: A Reader for College Writers*, 3rd ed. (New York: St. Martin's Press, 1991).

scrub jungle as the favorite dwelling of Maayoon, who probably was the prototypical ancestor of all the ancient families that inhabited that region. Some ancestors were identified with an animal, bird, or plant—a practice known as "totemism." The shark was in all likelihood one of the totems of the people of a coastal *tiNai*, even as the salmon is along the North Pacific Coast. Anthropologists refer to ancestral sacralization as "animism."

Long after the alien caste system was overlaid on the indigenous *tiNai* social order in Tamil Nadu, *tiNai* as the sacred ancestral homestead still survives in the form of the social institution known as "the clan shrine." Though many Tamil people do not dwell on their ancestral homesteads today, they visit them on special occasions. These homesteads were the original homesites of the ancestral parents of several present-day families.

Now, as people of the world experience displacement and discontinuity—the shattering of the connections between people, their places, and their ancestors—we should not be surprised to find disharmonious and destructive ways of life. We would do well to recover the spiritual basis of the *tiNai*, understanding that the places of our ancestors are sacred places where we ourselves belong, and thereby recover the continuity of people and places throughout time.

The Giveaway

Robin W. Kimmerer

Red over green, raspberries bead the thicket on a summer afternoon. The blue jay picking on the other side of this patch has a beak as red-stained as my fingers, which go to my mouth as often as to the bowl. I reach under the brambles for a dangling cluster and there in the dappled shade is a grinning turtle, shin deep in fallen fruit, stretching his neck up for more. I'll let his berries be. The Earth has plenty and offers us abundance, spreading her gifts over the green, strawberries, raspberries, blueberries, cherries, and currants, that we might fill our bowls. *Niibin,* we call it in Potawatomi, summer, "the time of plenty," and also time for our tribal gathering, for powwows and ceremony.

Red over green, the blankets spread on the grass beneath the arbor are piled high with gifts. Basketballs and furled umbrellas, peyote-stitched key chains and Ziploc bags of wild rice. Everybody lines up to choose a gift, while the hosts stand by, beaming. The teenagers are dispatched to carry choice items to elders seated in the circle, too frail to navigate the crowd. *Miigwech, miigwech,* the "thank-yous" circle among us. Ahead of me, a toddler, besotted with abundance, grabs a whole armload. Her mother bends and whispers in her ear. She stands indecisive for a moment and lays it all back down, save a neon yellow squirt gun.

ROBIN W. KIMMERER is a professor of botany in the Department of Environmental Science and Forestry at State University of New York College. Her work combines her Potawatomi heritage with her scientific and environmental expertise. She is the author of numerous scientific articles and the book *Gathering Moss: A Natural and Cultural History of Mosses* (2003), which was awarded the 2005 John Burroughs Medal.

And then we dance. The drum begins the giveaway song and everyone joins the circle in regalia of swaying fringe, nodding feathers, and rainbow shawls. And some T-shirts and jeans, too. The ground resonates with the fall of moccasined feet. Each time the song circles around to the honor beats, we dance in place and raise the gifts above our heads, waving necklaces, baskets, and stuffed animals, whooping to honor the gifts and the givers. Amid the laughter and the singing, everyone belongs.

This is our traditional giveaway, the *minidewak,* an old ceremony, well loved by our people and a frequent feature of powwows. In the outside world, someone celebrating a life event can look forward to receiving presents in their honor. In the Potawatomi way, this expectation is turned upside-down. It is the honored one who *gives* the gifts, who piles the blanket high to share good fortune with everyone in the circle.

Often, if the giveaway is small and personal, every gift will be handmade. Sometimes a whole community might work all year long to fashion the presents for guests they do not even know. For a big intertribal gathering with hundreds of people, the blanket is likely to be a blue plastic tarp strewn with gleanings from the discount bins at Wal-Mart. No matter whether the gift is a black ash basket or a potholder, the sentiment is the same. The ceremonial giveaway is an echo of our oldest teachings.

Generosity is simultaneously a moral and a material imperative, especially among people who live close to the land and know its waves of plenty and scarcity. The well-being of one is linked to the well-being of all. Wealth among traditional people is measured by the ability to have enough to give away. Hoarding the gift, we become constipated with wealth, bloated with possessions, too heavy to join the dance.

Sometimes there's someone, maybe a family, who doesn't understand and takes too much. They heap their acquisitions beside their lawn chairs. Maybe they need it. Maybe not. They don't dance, but sit alone, guarding their stuff.

In a culture of gratitude, everyone knows that gifts will follow the circle of reciprocity and flow back to you again. This time you give and then you receive. Both the honor of giving and the humility of receiving are necessary halves of the equation. The grass in the ring is trodden down in a path from gratitude to reciprocity. We dance in a circle, not in a line.

After the dance, I saw a little boy in a grass dance outfit toss his new toy truck down on the grass, already tired of it. His dad made him pick it up and then sat him down. A gift is different from something you buy, possessed of meaning outside its material boundaries. You never dishonor the gift. A gift asks something of you. To take care of it. And something more.

Among our people we say that the plants are our oldest teachers. They were here first and have had a long time to learn how to live in a good way. Where better to look for wisdom than among those who make food of air and light and give it all away?

I don't know the origin of the giveaway, but I think that we learned it from watching the plants, especially the berries, who offer up their gifts all wrapped in red and blue. We may forget, but the language remembers. Our word for the giveaway, *minidewak*, means "they give from the heart." At its center lives the word *min*. *Min* is a root word for "gift," but it is also the word for "berry." In the poetry of our language, might speaking of *minidewak* remind us to be like the berries?

The berries are always present at our ceremonies. They join us in a wooden bowl. One big bowl and one big spoon, which are passed around the circle, so that each person can taste the sweetness, remember the gifts, and say "thank you." They carry the lesson, passed to us by our ancestors, that the generosity of the land comes to us as "one bowl, one spoon." We are all fed from the same bowl that Mother Earth has filled for us. There is but one spoon, the same size for everyone.

It's not just about the berries, but also about the bowl. The gifts of the Earth are to be shared, but gifts are not limitless. The generosity of the Earth is not an invitation to take it all. Every bowl has a bottom. When it's empty, it's empty.

"One Bowl, One Spoon" speaks of our responsibilities in return for the gifts of the Earth. How do we refill the empty bowl? Is gratitude alone enough? I think that berries teach us otherwise. When berries spread out their giveaway blanket, offering their sweetness to birds and bears and boys alike, the transaction does not end there. Something beyond gratitude is asked of us. The berries trust that we will uphold our end of the bargain and disperse their seeds to a new place to grow, which is good for berries and for boys. They remind us that all flourishing is mutual. We need the berries and the berries need us. Their gifts multiply by our care for them, and dwindle from our neglect. We are bound in a covenant of reciprocity, a pact of mutual responsibility to sustain those that sustain us. And so the empty bowl is filled.

Those berries have great wisdom. When they ask us to carry them away, they know that not every seed will make it to a sunny summer meadow. Some will grow, but most will wait. Berry seeds are packages of possibility, a promise of abundance in a plain brown wrapper. They can last in the soil for decades, waiting for the right conditions. They plant their own future, safeguarding the possibility of berries, and therefore the possibility of bears and boys, for everything is connected. Somewhere along the line, people have abandoned berry teachings. Instead of sowing richness, we diminish the possibilities for the future at every turn.

The uncertain path to the future, impoverished or abundant, may be illuminated in language. In Potawatomi, we speak of the land as *emingoyak*, "that which has been given to us," a gift that must be reciprocated with our own. In English, we speak of the land as "natural resources" or "ecosystem services," as if the lives of other beings were our property, our entitlement, which becomes valuable only when consumed. As if the Earth were not a bowl of berries, but an open pit mine, and the spoon a gouging shovel.

Imagine that while our neighbors were holding a giveaway, someone broke open their home to take whatever they wanted. We would be outraged at the moral trespass. So it should be for the Earth. The Earth gives away for free the power of wind and sun and water, but instead we break open the Earth to take fossil fuels. Had we taken only "that which is given to us," had we reciprocated the gift, we would not have to fear our own atmosphere today.

We are all bound by a covenant of reciprocity: plant breath for animal breath, winter and summer, predator and prey, grass and fire, night and day, living and dying. Water knows this, clouds know this.

Soil and rocks know they are dancing in a continuous giveaway of making, unmaking, and making again the Earth.

We live in a moral landscape. The land is reading us law over and over, but we forget to listen. Our elders say that ceremony is the way we can "remember to remember." In the dance of the giveaway, remember that the Earth is a gift that we must pass on, just as it came to us. We forget at our peril. When we forget, the dances we'll need will be for mourning. For the passing of polar bears, the silence of cranes, for the death of rivers and the memory of snow.

When I close my eyes and wait for my heartbeat to match the drum, I envision people recognizing, for perhaps the first time, the dazzling gifts of the world, seeing them with new eyes, just as they teeter on the cusp of undoing. Maybe just in time. Or maybe too late. Spread on the grass, green over brown, they'll honor at last the giveaway from Mother Earth. Blankets of moss, robes of feathers, baskets of corn, and vials of healing herbs. Silver salmon, agate beaches, sand dunes. Thunderheads and snow drifts, cords of wood and herds of elk. Tulips. Potatoes. Luna moths and snow geese. And berries. More than anything, I want to hear a great song of thanks rise on the wind. I think that song might save us. And then, as the drum begins, we will dance, wearing regalia in celebration of the living Earth: a waving fringe of tallgrass prairie, a whirl of butterfly shawls, with nodding plumes of egrets, jeweled with the glitter of a phosphorescent wave. When the song pauses for the honor beats, we'll hold high our gifts and ululate their praises: a shining fish, a branch of blossoms, and a starlit night.

The moral covenant of reciprocity calls us to honor our responsibilities for all we have been given, for all that we have taken. It's our turn now, long overdue. Let us hold a giveaway for Mother Earth. Spread our blankets out for her and pile them high with gifts of our own making. Imagine the books, the paintings, the poems, the clever machines, the compassionate act, the transcendent idea, the perfect tool. The fierce defense of all that has been given. Gifts of mind, hands, heart, voice, and vision, all offered up on behalf of the Earth. Whatever our gift, we are called to give it and to dance for the renewal of the world.

In return for the privilege of breath.

From the Mountain, a Covenant

Courtney S. Campbell

> The Obsidian Trail, August 29. The forest gives way finally before a massive lava flow that comes from the Three Sisters, the snow-covered peaks a vertical mile above me. To the south, I can make out the Obsidian Cliffs, their glassy volcanic remnants sparkling in brilliant sunshine.

Although I affirm that truth can be found in many places, I am a Christian, and I believe that in Christian narrative I will find the greatest truth about moral responsibility. A coherent contemporary Christian ethic will be faithful to moral teachings in scripture and in Christian tradition, but it will also engage in dialogue with philosophical traditions, respond to scientific information and methods, and incorporate understandings of the human experience.

Because a Christian ethic must integrate scientific description, ongoing research that reveals the "inconvenient truth" of global climate change is of momentous importance. New disclosures from scientific research about a planet in peril require a transformation of the central Christian moral paradigms about the Earth, nature, creation, human responsibility, and providential control. Whereas significant

---

COURTNEY S. CAMPBELL is Hundere Professor in Religion and Culture at Oregon State University, where he previously served as chair of the Philosophy Department and as director of the Program for Ethics, Science, and the Environment. Prior to coming to OSU, Courtney was a research associate at The Hastings Center, a think tank for medical ethics. He authored papers for the National Bioethics Advisory Commission on the ethical questions of human cloning and of research on human tissues.

claims in Christian theology and ethical teaching may be partly to blame for climate change, a new ethic committed to incorporating scientific, philosophical, and experiential insights must re-vision how to act ethically in the present so as to promote a viable future. This requires not only trusting in scientific research as a public authority, but also paying close attention to the global human experience of living on a planet in need of healing. In this essay, I will assume the validity of scientific studies on climate change and explore its transformative significance for central themes in a Christian moral ecology.

> I turn away from the cliffs onto a path that leads through a meadow. Full of aster and paintbrush, the meadow is stunning in its simplicity, breathtaking in its majestic framing of the high peaks above. The path narrows, and finally gives out entirely. I begin a steep scramble up rocky scree from volcanic eruptions, the terminus of a glacial moraine. Now above the forest and meadow, I consider the lichen embedded in the rocks, a simpler but no less intricate and fragile life ecosystem than the trees surviving in the lava. A scramble across two ridges brings me onto the Collier Glacier, at one time the largest glacier in the state. Its snow and ice crystals sparkling, at this time of year it is the source of water pure and cold.

Relational Dependency and Gift Ethics

An initial step in the necessary moral transformation is to choose among the primary models of the human relationship with nature. The principal historical model of human beings as authoritative lords and viceroys over the Earth must clearly be discarded, for ethical, theological, and scientific reasons. This model, cultivated by the deadly vices of pride, gluttony, and avarice, has fostered an attitude of separateness from Earth and corresponding practices of attempted domination and mastery over nature. It supports the perspective that resources should be accumulated and consumed in service to the anthropocentric purposes of the human viceroy. This underwrites an assumption that only the experience of the comparatively advantaged persons counts, and therefore marginalizes the experience of the poor and vulnerable. This is directly contrary to an insistent biblical imperative to care for the stranger, the vulnerable, and the voiceless.

Moreover, the viceroy model is biologically inaccurate. Humans are dependent on numerous life-forms for our biological existence and sustenance. It is of no small consequence that the Latin root for Earth and soil—*humus*—is also the linguistic cognate for "human" and for "humility." Our language discloses the profound interrelation of the Earth, persons, and moral sentiment. Awareness of the biological reality that our very existence is contingent on Earth, soil, water, air, and so on should cultivate moral dispositions of gratitude, humility, and solidarity. These dispositions are at the core of an ethic of radical dependency and interdependency that must supplant historical models of anthropocentric authority without accountability.

If we understand our radical dependency on nature and on the sacred, and our interdependency within the human community over place and time, then we will experience life as a gift. I am dependent for my very being on powers beyond my control—biological, divine, and human—as well as for that which sustains my being over time. I must be consciously aware that I am, as one sacred text puts it, a "beggar before God."

I am a recipient of the constitutive physical, social, and spiritual elements of life, conveyed through nature, nurture, and the sacred; the myth of the self-made person is precisely that, a myth. Predecessor generations have bestowed on me a legacy that in my radical dependency I experience as a gift and manifestation of generosity.

This background moral dynamic places me in a form of relationship that my religious tradition has historically referred to as one of "covenant." I find myself in a covenantal relationship with nature, with the Earth; with the sacred, with God; and with other persons, with community. The content of this covenant concerns my respectful and grateful use of the gifts I have received, which requires of me trust, communal benefit, mutual aid and reciprocity, and accountability.

One further implication flows from the transformation of moral paradigm from domination to dependency. My dependency on prior generations, and my interdependency with current generations, situates me in a covenantal relation with future generations. The responsibilities embedded in covenantal relationship require me not only to treat the gifts respectfully and gratefully, but to act so as to extend the availability and accessibility of the gifts into the future. I assume a promise to bestow these gifts throughout all generations of time. In this manner, the ethics embedded in dependency, generosity, gift, and covenantal relation provide a basis for affirming a general responsibility to generations who reside in a future I will not live to see. More specifically, my covenantal commitment is to ensure an open future; I am responsible to act so as not to foreclose or predetermine life options for future generations, but rather to preserve and, when possible, expand on these options.

This transformational paradigm spans and binds together the past, present, and future, "all generations of time." The binding together of generations with each other, of communities with God, of persons with nature, is the *ligio,* the attachment or connection, that is present in the core of our term "religion."

> The beauty of my surroundings is beyond words—a cobalt sky, a blinding white glacier, the red volcanic rocks and nearby summits, thedark, dark green of the forest expanse covering endless miles below. This is a sacred place. It evokes awe, wonder, mystery. But I've taken only a few steps before I realize the Collier Glacier is different than I remember. Five years ago, the glacier was crusted with snow; now the snow has melted, leaving only ice. The snow provided good, solid footing, a relief from the piercing scree. But the ice is slippery and treacherous, and I don't have the right equipment. I've left my crampons and ice axe at home. Crossing a portion of the glacier to reach a rocky island, I fall several times. Glacial water trickles down my legs, fills my boots, and chills my feet. Only when I reach the island and find a spot to rest on the scree do I look down and see that what I thought was water in my boots is blood.

Vision and Awe

Caring for creation is clearly a divine mandate in the creation stories of scripture and Christian tradition. After each creative act in biblical stories of origins, the narrator observes, "And God saw that [the Creation] was good," a pattern repeated until the creation story is complete, which evokes a further elaboration: "And God saw everything he had made, and it was very good." In each case, valuing requires seeing. In order to care for the Earth, we must be able to envision its intrinsic goodness, its fragile and intricate life-supporting ecosystems, the mysteries by which life manifests itself. Perhaps the vision is gained by stepping *back* in a reflexive mode (a form of transcendence). Perhaps it is acquired by stepping *into* a forest, river, or garden (a mode of immanence). Ultimately, however, I think caring vision requires that we understand we are both *part of* nature and *apart from* nature and thus uniquely able to make assessments of value. When we sustain this delicate dialectic, aesthetic assessment can be transformed into the core religious sentiment, an experience of awe and wonder.

The sentiment of awe necessarily issues in restraint and caring to avoid profaning and polluting the sacred, whether in a remote mountain wilderness or a backyard garden. I find a profound moral ecology in the biblical story of Jesus' first discourse to his disciples. In the course of admonishing his followers to serve the cause of God's kingdom, Jesus *first* instructs them to "consider the lilies of the field, how they neither toil nor spin." Living in faith is necessarily connected with attentiveness and awareness—to others and to God, surely, but also to the intricacy of nature. Creation care requires learning from nature about how to live our lives and where to place our priorities. We are apart from and a part of nature, deeply aware of the fragility and transience of life in the moment.

Awe and wonder cultivate dispositions of gratitude, humility, and solidarity as the self opens to a realization of dependency on powers beyond our control for all we are and for all the gifts with which we have been entrusted. And this realization contains ethical content: as I have been a recipient in my dependency, so I am called to care for the vulnerable and poor, including the vulnerable of a fragile future whose very existence may be dependent on my actions in the present.

> At the crest of the glacial ridge, I stop, gasping for breath, gasping too at the splendor of what I see. To the east, the Hayden Glacier drops off nearly three thousand feet to Soap Creek, and then beyond, to the pine forests and high desert of central Oregon. To the north loom the Cascade pinnacles of North Sister, Little Brother, Mt. Washington, Three-Fingered Jack, Mt. Jefferson, and even Mt. Hood. Beyond the smog-filled valley to the west, the Coast Range emerges. I can see roughly thirty thousand square miles—volcanoes, glaciers, crevasses,

opal lakes, dark forest lands and lighter grassy plains, rivulets and streams. Look south, and I'm confronted with a further 1,100-foot climb to the summit. I listen carefully. Rocks fall from the nearby slopes, set free by melting ice and snow.

Eschatology and Prophecy

Creation care is not, however, based only on stories of the past. A full Christian response to climate change and our planetary peril will also attend to profound convictions about the end of time, a future era of peace and justice under the reign of God. The Lord's Prayer, for example, speaks of the coming of God's kingdom, within which divine purpose will be done *on Earth* as in heaven. The earliest Christian communities were formed in anticipation of the imminent return of Jesus as Christ to initiate the era of peace and justice, when all things, including the Earth, will become new. A recent Pew study indicated that nearly 80 percent of contemporary Christians affirm a belief in this "second coming" of the Messiah; nearly one-quarter believe the eschatological era will occur within their lifetimes.

Christian eschatology can be appropriated for absolutely asinine political purposes, of course; on some accounts of a world in its death throes, it makes little sense to contemplate responsibilities to future generations. My claim, by contrast, is that eschatological convictions provide a basis for a radical critique of the existing social order and status quo. In the context of imperial Rome, for example, Christian eschatology challenged social conventions, political systems, and economic patterns. In our context of environmental crisis, Christian eschatology is one ground for radical critique of the social, political, and economic conventions that produce unsustainable patterns of living.

If creation summons the voice of awe, eschatology issues forth in the voice of prophecy. The prophetic voice speaks truth to power, in the memorable phrasing of Reinhold Niebuhr. The truth of greatest environmental significance is the voice of scientists speaking about climate change and planetary peril. But the history of the Hebraic prophetic tradition, as well as the ministry of Jesus of Nazareth, illustrates that a prophetic voice that speaks the truth is most frequently rejected. The prophet has a distressing message, after all, and the powerful have a self-interested stake in the status quo. The powerful manifest, in biblical phrasing, "hearing that does not hear, seeing that does not see," and so the prophetic voice bears a Cassandra-like burden: knowledge of truth to which no one else is attentive.

Knowledge of even inconvenient truths is essential for a sustainable life now and for living in covenantal relationship with future generations. On one account, for example, it would require 5.57 Earths for every person in the world to live as I do. This inequity will clearly not pass any moral test. Because of the quality of life the eschaton anticipates in the future, I need to begin conforming now to the eschatological requirements of personal sufficiency and collective sustainability.

The prophetic voice commonly represents the moral memory of a community. It challenges the community to a task of remembering its formative ideals and the practices that must be enacted for communal life to be sustained into the future. As moral memory, prophetic voices offer re-"minding," mindfulness in the midst of mindless self-delusion and evasion. This voice does not shrink from speaking the truth about the evasion of moral responsibility by diffusion through multiple actors. Most important, the prophetic voice witnesses to the advantaged their responsibility to care for the vulnerable, for the wretched of the Earth, and for the Earth itself.

The prophetic voice is a voice of critique, but also a voice of promise—that change is possible, that a different world, with a renewal of relationships and responsibilities cannot only be imagined but will be realized. The eschatological conviction rules out despair. Christian communities have historically reinterpreted eschatological meaning to generate an ethic that would sustain a community of believers into a future different from what the original faith promised. What sustained the community was the virtue of hope, hope in the eschatological promise of a better future in time, and beyond time. Only an impoverished imagination can assume we live in the best of possible worlds.

> I climb slowly from ridge to narrow ridge, aware that one or two missteps on the steep slopes will initiate a rockslide and possibly a fatal

fall. I think I have crested the small summit, but then a new set of cliffs appears, a higher set of boulders emerges, a steeper traverse is necessary. I find it amazing that while I have no human companion to encourage, to inspire, to push me, I am surrounded by orange and black Cascade butterflies. I imagine Jesus, giving his Sermon on the Mount, directing his disciples to "consider the butterflies of the mountain." How fragile life is here—delicate alpine flowers, these butterflies, grasses blown with the winds, the tenacious lichen, not to mention one solitary and bloodied climber. Be attentive, be aware of this fragility.

Finally, the summit. Elation overtakes me as, four thousand feet below, the dark blue Chambers Lakes glisten in the sun. The South Sister and Cascade peaks farther to the south now come into view. How I wish I could share this sight, this experience, with someone. The cell phone I bring for emergencies gets no signal here. I can talk only to God.

So the Future Can Come Forth from the Ground

Deborah Bird Rose

Jessie Wirrpa often took me by the hand when we were walking in difficult places. Rough country was her homeland: an expanse of rugged savannah woodland in the northwestern corner of the Northern Territory of Australia. Jessie's people had been there "forever": their ancestors were said to have come out of the ground in this country, and she and her descendants live there still. When I went to live among her people in 1980, I was a novice anthropologist, an inexperienced bush woman, a person with so much learning to do even to gain basic competence that I sometimes wondered: Could I do it? Was it worth the effort? Jessie took charge of me, and if she too sometimes pondered these same questions, she was kind enough not to be obvious about it.

She was teaching me country, its complexity and its simplicity. Indigenous country is made up of sets of interdependent homelands, each with its own people, animals and plants, landforms, water, stories, creation sites, and ceremonies. Country is the place of creation. According to another of my teachers, the Earth itself gave birth to the creation ancestors known as Dreamings, and they made or became everything that is, including the shape of the country: "Dreamings

---

DEBORAH BIRD ROSE is a professor of social inclusion at Macquarie University, Sydney. Her work focuses on entwined social and ecological justice and is based on her long-term research with Aboriginal people in Australia. She co-edits the Ecological Humanities section of the *Australian Humanities Review* and has written numerous essays and books, including *Reports from a Wild Country: Ethics for Decolonisation* (2005). Her current book project, *Wild Dog Dreaming*, explores ethical relationships with more-than-human others in this era of loss.

put up all those big hills, rivers. . . . Dreaming. Tree, everything, sugarbag [honey], tucker, goanna, fish, that's not nothing. All from Dreaming. Everything all from Dreaming."

When Jessie went walkabout she called out to her ancestors. "Give us fish," she would call out. "The kids are hungry." Her country included the dead as well as the living, the Dreaming ancestors as well as her own parents and grandparents. The Owlet Nightjar was her Dreaming ancestor, and there were signs of his story everywhere: the place he walked, the place he made a fire and cooked food, the place he encountered a huge snake and decided to go home. Of course, there were owlet nightjar birds, too, lots of them. Indeed, everywhere she went, she encountered signs of life. A discarded stone spear point, some charred sticks from a campfire, a Dreaming tree that got knocked by lightning when her oldest father died. No distinction between history and prehistory for Jessie: in her country the present rolled into the past on waves of generations of living beings who had all worked to keep the place alive.

Westerners like me implicitly understand that a temporal orientation is also a spatial orientation: we face the future; the past is behind us. Indigenous Australians make temporal-spatial links too, but theirs work differently. They face the source; those who come after them are called the "behind mob." Each generation follows along behind their ancestors, and their descendants follow along behind them. I imagine this mode of time as waves of generations; we face the source, which is where we all came from ultimately, and we follow our predecessors back to the source, leaving behind us a "new mob" or "new generation" to take over. Those behind us walk in our footsteps, as we walk in the steps of our old people. From time to time a Western person may experience a dizzying sense of historical inversion—of the past jumping ahead, or of time running backward.

Within this indigenous world of time, space, and generations in motion, the future is delightfully complex. On the one hand, it can be assumed to be following behind us in the form of the next generations of people, plants, animals, and others. More significantly, though, it is in the ground. The future is waiting to come forth, to be born and to live, and then to return into the source, riding the waves of generations that have kept country and all the creatures alive "forever."

To imagine the future in the ground requires attention to what is happening all around us: taking care of country, to use the indigenous phrase, is all about taking care of the future. Inevitably, therefore, to look at collapsing ecosystems is to see terrible loss. Not only the loss of topsoil, riverbanks, and, ultimately, many of the species and habitats that support each other's lives, but also the loss of the future. Who or what can come after us if the source of life is being wrecked? Where will the next generations come from, and how will they live if there is no country to bring them forth and sustain them?

Concern for those who come after is one huge issue. Another issue is that of betrayal. To wreck the ground of creation is to betray all those who came before us. From the beginning right up to today people have bequeathed responsibilities. Failure to sustain those relationships of care lets them start to unravel. In a degrading and dying world, at this very moment creation is coming unstuck, disintegrating right back to the beginning. That dizzying sense of time running all over the place confronts me vividly as I try to think about this great unmaking—an unmaking of the possibilities for the future that is also an unmaking of all the work and care of the generations that precede us.

I have heard talk about a light at the end of the tunnel: Is it the brightness of day out in the world of life, or is it a train bearing down on us? Only time will tell, and our only choice seems to be to keep going toward it, whatever it may be. Perhaps we may imagine that as long as we keep moving, we can continue to hope.

Western metaphors draw on this kind of imagery—the linearity of the tunnel, the necessity of staying on track, the future as a distant point toward which we are moving. Hope conceived in the future tense often taps into this imagery: somewhere farther down the line there will be a better day, a place of light, a flourishing world, a new dawn. Indeed, religious traditions that promise an end of time—either as eternity in otherworldly paradise, or as eternal paradise on Earth—may offer little reason why we should concern ourselves with destruction in the here and now. After all, are we not hastening onward toward a vision of transformed perfection?

Jessie and her people don't hold out such expectations. Rather than hoping that somehow we'll make it through, their way of thinking of the future impels us to take care of the ground right now, right where we are, because we are here, because this is our source, because our purpose in life is to bequeath life, not to unmake it. Jessie expressed such ideas as "true stories"—true accounts of the real world and our responsibilities as humans. True for humans everywhere, and true for other creatures as well. As I consider the accelerating destruction of life on Earth, I wonder: Are we capable of becoming humans who take care of country?

Often, too, I wonder how to enliven our awareness that in the end life is not ours to make or unmake: it is not our place to ask whether it is worth the effort. The profundity and simplicity of Jessie's caring for country is, for me, an ethical claim. The future is not a promised land waiting for us to arrive, nor does it bear down on us. The future is in the ground. It is life, and it wants to come forth and flourish. The future is creation in everyday life, and like all everyday miracles, it is as fragile as it is resilient. We are members of creature communities, and our appropriate work is to honor the bequest by taking care of it so that the future can come forth.

A Conference in Time

Ursula K. Le Guin

Sometime between the beginning and the end,
who was it called, what great lamenting voice?
A word rang out across the inner sky
of jeweled zodiac zone and sailor's star,
across the close-shored seas, the long-known lands,
to come, to meet in council in eternal Rome.
And while the pope sleeps warm in downy satin,
and homeless folk as always huddle cold
on Roman cobbles, those who were summoned here
from counted centuries and chasms of time
gather together.
They are not manifest to mortals,
though sleeping pigeons tuck their heads in tighter,
and children dream strange dreams. Pilots aloft
look down through dark to not quite see
the dome of Peter glimmering with glory,
flicker of myriad wings of candleflame;
a tired night-watchman by the Tiber almost hears
the ruined temples under ground groan welcome.

URSULA K. LE GUIN is the author of seven books of poetry, twenty-two novels, over one hundred short stories, four collections of essays, twelve books for children, and four volumes of translation. Three of her books have been finalists for the American Book Award and the Pulitzer Prize. She has received the National Book Award, five Hugo Awards, five Nebula Awards, the Kafka Award, a Pushcart Prize, and the Margaret A. Edwards Award, among many others. Her recent publications include *Powers* (2007) and *Lavinia* (2008).

They come: those who are called the Living, those
forgotten and called dead, but in their presence
is life itself, and all things are remembered.
The two that walk apart are here, apart,
Allah, Jehovah, each one with his book;
and Jupiter whose mantle is the sky
and all its weathers, comes with his thunderstone;
and Zeus the stern seducer; the one-eyed Wanderer,
two ravens on his shoulders, with his staff;
Jesus healed from wounding, his ghost-winged dove
and blue-robed mother following modestly;
Freya with her beasts; the Many-Breasted One,
and striding, implacable Athene; and then, clear
and sudden as in summer with the wind
of dawn the young light strikes across the sea,
Aphrodite comes, and the stones of Rome
tremble with fierce desire at her coming.
The Lady of the Crossways, dark and bright, is here,
and goddesses of trees and hidden springs,
and shaggy sinuous river-gods, Pan leaping like a goat.
Down from the northlands pace the druid powers,
shadowy lords of the great fallen forests.
The gods of Egypt stalk with head of cat
or hawk or vulture, in white pleated linen,
narrow of waist, with long and narrow feet.
The bright lord of the Farsis blinds with flame
the dull-eyed, cruel idols. And slowly, one
by one, innumerable images
of clay and stone and bronze crowd silently
from storehouse, gateway, kitchen, army camp.
Old, old, crude, fat, headless Venuses of chalk,
out of the barrows and the shallow graves,
come in their ancient dignity of awe. So all the streets
and ways and open places of the city
are filled and shiver with fullness of immortal being.

So it is everywhere: for in the City
of Mexico blood starts from the pyramids:
the Plumed Serpent and the Turquoise One have met.
The lion of great stones that is Cuzco rises, scenting
from the altars of Machu Picchu smoke of copal,
and Pachamama strokes the dark head of the lion.
Who dances on the ashen steps the Ganges washes,
where a million forms of the endless dreams of Vishnu
gather in conference? Who comes to the drum-call
through jungles, over Africa's long plains,
and who to Coyote's quavering cry across the desert?
Corn Woman rises, and with White Shell Woman
walks in beauty, and the tall kachinas
stride through the streets in silence to the kiva.
The gods are gathering in all the hallows,
Ayers Rock, Kuala Lumpur, and Stonehenge,
all deities in all their names and aspects,
a flood that follows darkness round the world,
luminous, intangible, and countless;
and as it is in Rome it is in all sacred places.

The first to come, who is also the last,
opener and closer of the door, speaks from
his two mouths: "What is to be done?" he says.

                        The voice
is like a great wind blowing in the darkness,
like a warm breath. The gods are that breath.
God is that wind. It answers: "Mourn."
The anger of God gathers in the night,
a stormcloud deploying: "Have I not destroyed them
in their iniquity, before?"
The power of God bursts forth in a scream
of steel, a burst of blinding fire: "War!"
The mercy of God murmurs, "Forgive them."

But it is Aphrodite born of ocean foam,
Corn Woman risen from the furrowed dirt,

Persephone daughter of the bread men eat,
with her dark-crowned husband, and the shadows
of all the lives that lie in Earth, who answer
without a word. It is the mortal Earth
herself, who breathes the being of the gods,
whose long warm sigh blows all the words away
into the silence that was always there.

            The city falls. It takes a day
to fall, or centuries. The seven hills
stand for a while longer.
            Prayers are spoken,
sung, breathed out; was there once an answer?
The conference is over. Dust blows across the mesas,
lies on the altar, fills the bowl of stone.
The poles grow blue and dark.
Low shores surrender to the sea. Islands go under.
Over drowned Mumbai and Bengal
the sun sets with a red grin like Kali's.
Insatiable Tlaloc has had his fill at last.
The web Anansi wove hangs loose, a cobweb.
The roofs of the world collapse, Himalayas, Andes
flow down in torrents of barren rock and water
as thick as Shiva's hair, as bright as Inti was.
White horses charge from the despoiled ocean,
salt manes blowing, hoofs trampling, destroying.
No hand is on their reins to guide them.

There were places and their names and gods.
There were those who inhabited this house.
Across the sands of Libya a hawk flew hunting,
past the red mesas a lean coyote trotted,
a hummingbird in Nicaragua flared like a thrown ruby.
Over the broken Yangtze dam did dragons soar,
did phoenix or raven nest in rubble of cement
and rusted steel beneath the Golden Gate?
These names are named no longer. Voices
are few and far between, they are strange voices,

the mouths that speak have changed their shape,
and who is there to tell us the meaning
if a word is spoken?
    If it is not in the end,
it was not in the beginning.

                  The poles whiten, spray of iron oceans
grows stiff as Styrofoam; islands arise again with steep
cliff-coasts of stone; ice groans in the great silence.

Among the many ways and other worlds,
galaxies of galaxies, immense vibrations,
quiver of atoms, quanta, passage of kalpas,
the Earth goes on her changing circling way,
dances her dance and does her chancy business,
sweeping her shadow down the shores of light,
maker of darkness, mother of the night,
whose children too cast shadows where they go.

ETHICAL ACTION

How can we express gratitude to the Earth for all its gifts?

Write a song of praise and sing it in the street.

Say thank you before morning coffee, which is a gift of grace from the water and the soil, which owe you nothing.

Celebrate the season of harvest with feasting, the season of scarcity with fasting, the season of new life with dancing, and the season of ripeness with listening.

For every gift you are given, give something in return: a planted seed, a suet scrap, a moment to notice the moon.

Be glad for ponds.

Take nothing without noticing. A deep breath, a carrot salad, a drink of water or wine.

Write thank-you notes, which is what your mother taught you. Write to the soil, "This is a great gift and you are kind to give it and I hope you are well in the new year." Bury the note in the garden.

Hold each gift in your hands—fresh snow, a tomato, a child's crayon drawing; examine it closely to understand how beautiful it is, and astounding. This is how a gift becomes sacred.

Make something of every gift you are given.

Use it, but use it wisely and well.

Imagine, when you awake each morning, what you will make of the new day, that greatest of all astonishing gifts.

Listen closely when the gift is music. Return it abundantly when the gift is love. Touch it gently when the gift is fragile. Protect it fiercely when the gift is vulnerable. Laugh aloud when the gift is joyous. Share it when the gift is truth. Use it bravely when the gift is freedom. When the gift is money, give it away.

Above all, do not pretend to understand why you have been chosen to receive these gifts. This is the mystery of life.

6

> *Do we have a moral obligation to take action to protect the future of a planet in peril?*
>
> **Yes, for the full expression of human virtue.**

Ethical thought going back to Aristotle has assumed that right and wrong actions ought to be judged not by the sort of results that the actions produce, but by whether the actions express the appropriate virtues. The moral person is not necessarily the one who *accomplishes* good, but the one who *is* good. So what is it to be good, to be fully human, the best example of our kind? The answer to this question is important, because we have a choice about what kind of person we will be. In character and action, we can choose to embody love or hatred, peace or violence, humility or arrogance, courage or timidity, wisdom or stupidity, care or indifference, empathy or callousness. We are virtuous people to the extent that our characters exhibit these virtues and turn away from what is not worthy of us as human beings.

So a virtue ethic does not ask, Do we have obligations to future generations to avert climate catastrophe? Rather, it asks, What kind of person would put his immediate interests ahead of the interests of the multitudes to come? What kind of person could imagine that the suffering or possible extinction of other forms of life asked nothing of her? What kind of person, with the power to help in at least a small way to avert harm on a scale never before seen on Earth, would turn away from the chance—indifferent, unknowing, or self-deceptively in denial?

To recognize oneself as part of the world, created by and creative of a rich and beautiful network of lives, is part of what it means to be fully human. The virtues that express that humanity have to do, at the very least, with compassion—an empathy that extends beyond one's own time and species—and with an equally broad-minded sense of justice. This is the strength of character of the person who will bravely, stubbornly, lovingly act in defense of a world he will never know.

A Newt Note

Brian Doyle

One time, years ago, I was shuffling with my children through the vast wet moist dripping enormous thicketed webbed muddy epic forest on the Oregon coast, which is a forest from a million years ago, the forest that hatched the biggest creatures that ever lived on this bruised blessed Earth, all due respect to California and its redwood trees but our cedars and firs made them redwoods look like toothpicks, and my kids and I were in a biggest-creature mood, because we had found slugs waaay longer than bananas, and footprints of elk that must have been gobbling steroids, and a friend had just told us of finding a bear print the size of a dinner plate, and all of us had seen whales in the sea that very morning, and all of us had seen pelicans too which look like flying pup tents, and how *do* they know to all hit cruise control at the same time, does the leader give a hand signal? as my son said, and one of us had seen the two ginormous young eagles who lived somewhere in this forest, so when we found the biggest stump in the history of the world, as my daughter called it, we were not exactly surprised, it was basically totally understandable that suddenly there would be a stump so enormous that it was like someone had dropped a dance floor into the forest, that's the sort of thing that *happens* in this forest, and my kids of course immediately leaped up on it and started shaking their groove thangs, and danc-

---

BRIAN DOYLE is the editor of the award-winning *Portland Magazine* at the University of Portland, Oregon. He is also the author of nine books of essays, nonfiction, and "proems," most recently *Thirsty for the Joy: Australian & American Voices* (2008). His writing has appeared in the *Atlantic Monthly*, *Harper's*, *American Scholar*, *Orion*, and the *Best American Essays* anthologies.

ing themselves silly, and I was snorting with laughter until one kid, the goofiest, why we did not name this kid Goofy when we had the chance in those first few dewy minutes of life I will never know, well, this kid of course shimmed over to the edge and fell off head over teakettle, vanishing into a mat of fern nearly as tall as me, but the reason I tell you this story is that while we were all down in the moist velvet dark of the roots of the ferns, trying to be solicitous about Goofy and see if he was busted anywhere serious but also trying not to laugh and whisper the word "doofus," one of us found a newt! O my god! dad! check it *out!*

Of course the newt, rattled at the attention, peed on the kid who held it, and of course that led to screeching and hilarity, and of course on the way home we saw damselflies mating, which also led to screeching and hilarity, but the point of this story isn't pee or lust, however excellent a story about pee or lust would be. It's that one day when my kids and I were shuffling through the vast wet moist forest we saw so many wonders and miracles that not one of us ever forgot any of the wonders and miracles we saw, even though we saw only tiny shreds and shards of the ones that are there, and what kind of greedy criminal thug thieves would we be as a people and a species if we didn't spend every iota of our cash and creativity to protect and preserve a world in which kids wander around gaping in wonder and hoping nothing else rubbery and astonishing will pee on them? You know what I mean?

Worship the Earth

John Perry

Religion has come in for hard knocks lately.[1] But at least one aspect of religion is well worth saving—the practice of worship, and the associated institution of taboos. Worship is a natural human tendency that helps people to rise above their own concerns; to take account of things that are deep, important, and worthy of respect and adoration; and to give thanks for the contribution those things make to the fact and quality of existence. Associated with worship is the idea that certain things are holy; offenses against the holy are not only bad but truly vile; they are not only to be avoided but ought not even to be considered; that is, they are taboo. The problem with the major religions is that the object of worship is ill-chosen, and most of the associated taboos that have not been long forgotten are of no practical use in dealing with the big problems that confront us. I suggest that the object with the most credentials to be worshipped is the Earth, and that elevating offenses against the Earth to the level of unthinkable taboos would be a great idea.

JOHN PERRY is Distinguished Professor of Philosophy at the University of California, Riverside, and Professor Emeritus at Stanford University. His books and articles deal with personal identity, the philosophy of language, consciousness, and other philosophical topics. He has received a Nicod Prize, a Humboldt Award, and a Guggenheim Fellowship and is a member of the American Academy of Arts and Sciences and the Norwegian Academy of Arts and Sciences. He is cohost, with Ken Taylor, of the radio program *Philosophy Talk* (see www.philosophytalk.org). His Web site is www.csli.stanford.edu/~John/.

---

1. Consider, for example, Richard Dawkins, *The God Delusion* (New York: Bantam Books, 2006), or Dan Dennett, *Breaking the Spell: Religion as a Natural Phenomenon* (Penguin Group, 2006).

For those who don't believe in God, whether out of long habit or because of recent onslaughts, Earth-worship can provide a natural target for this potentially good human tendency. However, Earth-worship can also supplement the worshipful activities of believers. According to Catholics, worship can be a matter of adoration or veneration; only God should be the object of adoration, while saints can be venerated. Catholics can venerate the Earth along with the saints, while Protestants can venerate the Earth instead of them. If, like Plato and Bertrand Russell (in one of his moods), you worship the eternal forms—numbers and other unchanging things—it would not seem disloyal to add Earth to the list; although it is a contingent thing, it is one we'd like to last forever, without much change.

Some people already worship Gaia, the goddess of the Earth and the daughter of Chaos; that's fine, if sort of New-Agey. But I am not recommending that we think of the Earth as God, or as a god or a goddess, or as anything else than the planet we are lucky enough to live on. Considered just as it is, it seems to me a more suitable object of worship than any of the items, from God to the numbers, that I have just mentioned.

A proper object of worship, I think, should fit two criteria. First, it should be responsible for the existence and quality of life of the worshippers. The attitude of worship most basically combines giving thanks for existence and all the good things it brings with it, and expressing hope to be spared from the worst possibilities that it offers. If there were a God that created the heavens and Earth, He would fulfill this criterion. So would the universe as a whole, and, in some extended sense, even the laws of nature, and perhaps even the forms that Plato and Bertrand Russell worshipped.[2]

However, none of these things meets the second criterion. The practice of worship should make a difference to the thing worshipped, in a way that in turn provides some benefit to the worshippers. The object of worship should be something that the worshippers not only are affected by, but also have an effect on. And there should be a

---

2. Bertrand Russell, "A Free Man's Worship," in *Mysticism and Logic* (London: Longmans, Green & Co., 1918), 46–57.

feedback loop: the effects the worshippers have on the object of worship should enhance the good effects he, she, or it has on them.

One can see these criteria at work in the ordinary practice of prayer. We give thanks to God for creating us, and we ask for His help in various ways. The problem is that if we are praying to the God that we claim to be praying to, the all-powerful God of the monotheistic theologians and creeds, He doesn't fit the second criterion. We have no way of knowing what God's plan for the universe is, and how we fit into it, as individuals or as a species, and no evidence that prayers and the like have any effect on Him whatsoever.

Of course, that wasn't what God was supposed to be like. Historically, the God of Christianity and Islam developed from the God of the Old Testament, a powerful being who created the heavens and Earth. As powerful as He would have to have been, He does not strike the reader as all-perfect; He had the capacity to think, to err, to regret, to change His mind, to listen to prayers, to get angry, and the like. But the logic of monotheism leads relentlessly to the concept of an all-powerful God; and the logic of omnipotence, together with the facts of the world, leads relentlessly from there to the conclusion that we have no idea what this God has in mind and no idea how to affect him in ways He might appreciate, even if the idea of doing so makes sense. If, in addition, this God doesn't even exist, the whole practice of worshipping Him is clearly futile.

Our ancestors had a causal role in our existence, and if we are more or less happy with our existence, it seems reasonable to thank them, or at least be thankful they existed. But most of our ancestors are dead; we can at most affect the memories of them, and they are not in a position to do anything to improve our lot, or spare us from catastrophe. The same goes for the Founding Fathers, or any other dead mortals we might put in our pantheon.

Numbers, other abstract objects, the laws of nature, and the universe as a whole also don't meet the criteria. None of these things cares about us, or even has the capacity to think and plan on our behalf, or in any way be affected by our prayers, much less be affected in a way that will in turn lead to benefits for those who worship them.

There is one object that fits both criteria better than any other, and that is the Earth. All that we are we owe to the odd combination of properties that the Earth has, which produced not only human life, but also all the plants and animals, lakes and streams, mountains and valleys that make human life reasonably enjoyable for a part of most people's lives. Even if we give numbers, the laws of nature, the Big Bang, the universe as a whole, or an omnipotent God credit for our existence, these are all relatively remote causes compared to the Earth.

Unlike these other candidates, the Earth also passes the second criterion, because we have effects on it that feed back into the effects it has on us, as has become obvious in the past century or so. Nuclear warfare has the capacity to render the Earth basically uninhabitable rather suddenly. Degradation of the environment has the capacity to do so more gradually. Either way, human practices have the capacity to cut down by thousands or even millions of years the length of time that the Earth will provide us with an enjoyable place to live.

We can't think of the Earth, except metaphorically, as listening to and answering our prayers. It won't continue to provide us with a reasonable climate out of gratitude for Al Gore's efforts, or with clean water in appreciation of the work of the Sierra Club. So perhaps the precise mechanism of feedback isn't what some religions have had in mind. But the metaphorical attribution of perception, intelligence, reason, and other such attributes can be quite useful; we don't need to hesitate in thinking of the Earth as Mother Earth if we find it helpful, so long as we think of Her as a mother who respects only actions, not mere words; a mother who is less interested in being prayed to than in being left intact.

The idea of a taboo is a powerful one. There are some things that should be considered so out-of-bounds that they do not even enter into cost-benefit analyses. They ought not even to be options that sane people consider. One might have thought that destruction of millions of human lives through nuclear warfare was a good candidate, especially by adherents of the Ten Commandments, but apparently not, since from the end of World War II to the present time such things have been part of the contingency plans of governments led by Christians, Muslims, and Jews, as well as Hindus and "godless Communists." However, I'm pretty sure that none of the American presidents who used and contemplated using nuclear weapons, or using torture, for that matter, would have considered having sexual intercourse with their mothers or sisters, or eating the remains of dead people. The idea of a taboo still has some hold on us.

The destruction of the Earth, either suddenly or gradually, either in whole or in part, ought to be beyond the pale. It ought to be unthinkable. All sorts of practices ought to be taboo. The destruction of mountains to get coal; the pollution of water to carry off waste; the destruction of watersheds and aquifers to grow corn for cattle—such things should not merely be criticized. They ought to leave us aghast.

Philosophy can play some limited role in instilling such an attitude among the general public. A profession that can worry for more than a century about the uses of saying "$a = b$" can surely devote some efforts to developing convincing arguments for one important thing that we ought to believe on instinct: that the existence of the Earth in something like its present state is a contingent and changeable fact of enormous consequence. More certain than that God is the most necessary of necessities is the fact that the Earth is the most necessary of contingencies, the one big contingency that is responsible for all consciousness, all value, all happiness and all misery, all art and music and all philosophy. Developing a philosophical theology that gives the Earth its rightful place among human values will not make the Earth an object of worship among the general population. But it might help. Perhaps some introductory philosophy student of the future, if properly educated and inspired, could lead us into an era of Earth-focused zealotry that could make a real difference in the amount of time this greatest of all contingencies will provide reasonable human habitation. You never know.

Something Braver Than Trying to Save the World

Bill McKibben

Here are a series of facts about the world drawn from scientific observation in the last few years:

- Arctic sea ice has melted to record low levels, at a pace far beyond even the most dire predictions of a few years ago.
- 33 million acres of Canadian forest have died, thanks to a beetle now spreading south across the American border with similar effect. It no longer gets cold enough in the winter to kill the beetle, hence its eruption.
- High-altitude glaciers—which water much of Asia, Latin America, and the American West—are disappearing at a rapid rate, such that, for instance, the U.S. Park Service now says that Glacier National Park will be glacier-free by 2020. The effect will be much worse downstream from the Andes and Himalayas, where billions depend on the water stored in the snows above.
- The incidence of dengue fever—also known as breakbone fever—has increased in many countries by 200 percent or

---

BILL MCKIBBEN is an environmentalist and writer who frequently contributes to publications such as the *New York Times*, the *Atlantic Monthly*, *Harper's*, *Orion*, *Mother Jones*, *Rolling Stone*, and *Outside*. His Step It Up 2007 campaign has been described as the largest day of protest about climate change in the nation's history, and his 350.org campaign is the first attempt at a global grassroots organizing campaign on climate change. His most recent book is *Deep Economy: The Wealth of Communities and the Durable Future* (2007).

more in the last decade, as rising temperatures have allowed its vector, the *Aedes aegypti* mosquito, to expand its range.

And so on.

All of this has happened because human beings have—by burning coal and gas and oil—managed to raise the temperature 1 degree F. We have already put enough carbon in the atmosphere to guarantee more than another degree of increase, and the consensus estimate among scientists is that without enormous action soon to end the use of fossil fuel, we will see temperatures 5 or 6 degrees higher by century's end.

All of which means: the chance that we will in fact "leave to the future a world at least as rich in possibilities as the world that was left to us" is nil. As in, not going to happen. We have effectively ended the Holocene, the ten thousand years of climatic stability that allowed human society to establish itself, and then to flourish. The atmospheric concentration of carbon dioxide over that era wavered only slightly either side of 275 parts per million. Now it's at 390 ppm and rising. As NASA's James Hansen and other researchers have put it, that's too much. Any value over 350 ppm, they have said, is not compatible with the planet on which "human civilization developed and to which life on Earth is adapted." We're too high now. We ate the wrong stuff, our cholesterol soared, and we're having the strokes. That's why the Arctic is melting.

None of which lessens, even for a moment, our moral obligation, either to the people of the future or to the impoverished people of the present who will bear the heaviest burden despite having caused none of the trouble. In fact, we must do something braver than try to save the world we have known. We must accept the fact that the world we have known is going to change in hideous and damaging ways—and we must nonetheless work as hard as we can to limit that damage, to keep it this side of complete catastrophe, to save as many options for our descendants as are still possible. This, as I say, is hard—it's easier to defend a pristine rainforest than to save a woodlot that's already half cut over. Easier to rally support, easier to keep fired up. Once something's spoiled, it's easier to throw up your hands and walk away, which will be the great temptation for us. Still, we need to try. With a squad of young colleagues, I've lately been waging a campaign called 350.org, which tries to rally our leaders to make the agreements that would someday return us to an atmosphere that would at least give us a shot at survival. We're under no illusions. We know we can't make climate change go away.

But we know we can't go away either, not while there's time to keep it from sliding out of control. And in the process—with millions of others working on thousands of related causes around the world—we just may be able to tackle this question of future possibilities from another direction. The physical Earth will be degraded—there is no doubt of that. But perhaps the human world can still be made sweeter, deeper, more open.

We've simplified ourselves and our society as least as much as we've simplified the Earth—in the last dozen decades we've convinced ourselves that we're the center of everything, as a species but especially as individuals. In America, we've created the most hyper-individualized people the world has ever seen—we literally have no need of our neighbors for anything practical. As a result, we barely know them: three-quarters of Americans have no relationship with the person who lives next door. *We have half as many friends as the average person of fifty years ago.* We've turned ourselves into a very odd kind of ape, the first asocial one on record. And that's the world we're leaving the kids, the one where houses are mammoth, and the distances between them even larger. Where you have to fend for yourself. Where people grow, by every measure we have, steadily less happy, steadily more insecure.

That we can still change. As we finally, belatedly, take up the battle for a livable planet, we become part of a broader stream fighting for a different human future. A new connectedness—based in no small part on the synapses created by the advent of the Web—lets us sense the slightly different world that we might be able to midwife. In the United States, the fastest-growing part of the food economy is farmers' markets. Which have lots of good ecological consequences, which will be necessary as we raise the temperature and hence make it hard to keep growing industrial food. But farmers' markets are truly important for another reason: the average shopper has ten times as many conversations per visit there than at the supermarket.

Ten times as many conversations means ten times as many possibilities—for love, for action, for friendship, for ideas, for meaning, for connection.

And here's the thing. The fight against complete out-of-control climate chaos is the fight against cheap fossil fuel. Which is, automatically, the fight for farmers' markets, and for community wind towers, and for neighbors whom you actually need. For all the things that would make human flourishing a little easier. Even on a very much tougher planet.

Peace and Sustainability Depend on the Spiritual and the Feminine

Massoumeh Ebtekar

In the Name of Allah,
the Compassionate, the Merciful

Looking forward into the future with the experiences of the past, we face insurmountable obstacles in our endeavors to make the world a better place. Many observers believe that a one-dimensional materialist and patriarchal approach has been the hallmark of human civilization for the past century. Humankind has commenced the new millennium with many advances and achievements but also with experiences heavily laden with mistakes, miscalculations, and misjudgments. This is not to deny the great advances in science and technology or in other areas, but rather to emphasize that today unsustainable trends, double standards, violence, and war have overshadowed, and even threaten, those accomplishments.

The world today is not what we had hoped it would be. It does not embody what science and technology could have brought for the betterment of all. Although a few millions are better off, the majority of human populations still suffer from poverty, disease, and under-

---

MASSOUMEH EBTEKAR is a politician, environmental activist, and associate professor of immunology at Tarbiat Modares University in Tehran. As Iran's first woman vice president since the Islamic Revolution, she led efforts to tackle air pollution in Tehran and protect marine life in the Persian Gulf. She also served as the head of the Environment Protection Organization of Iran and was named Champion of the Earth by the United Nations Environment Programme in 2006. She is a Tehran City Councilor and president of the Center for Peace and Environment, a nongovernmental organization devoted to the promotion of just and sustainable peace and the protection of the environment.

development. Environmental degradation is on the rise. Prospects for the future are overshadowed by dark realities. The warming of the Earth's atmosphere will impede growth and human development, while the global recession in human values will take its toll on all. Beyond this, war, terrorism, and armed conflict have led to an increased sense of insecurity.

Corporate leaders, political leaders, and media leaders are all ultimately shaping the decisions that have taken our world to where we stand today. But we are heading in the wrong direction. The world is suffering from an acute case of mismanagement. The problem, it seems, is rooted mainly in both the quality of leadership and in the worldview of global decision makers. The major driving forces in today's world are material values of wealth, power, lust, and fame. Humanitarian incentives, including compassion and altruism and spiritual values, are in the dim and distant shadows. In addition to the low number of women in decision-making positions, there is general understanding that the lack of affection, love, and feminine attitudes in the governing of world affairs has led to an aggressive masculine grip over world affairs. Can reviving or reintroducing the spiritual dimension and the feminine archetype or personality trait bring a fundamental change in our approach to leadership? Can this change, therefore, change the manner in which we are managing our world today?

Current leadership has led the world to the verge of environmental destruction and to constant war and conflict, evoking double standards in dealing with nations, while poverty and disease take their toll and lead millions to despair. The unresolved and deteriorating conditions endured by the Palestinian people are a standing symbol of the unjust world order, as is the ongoing insecurity in many countries in Africa and the Middle East. This is probably why "change," the major campaign watchword of the current American president, gained so much popularity in the United States and throughout the world.

Many scholars believe that a balanced personality and genuine belief in spiritual and ethical values can serve as human attributes that define leadership qualities. There are different views on what constitutes a balanced personality, how it can be achieved, and how this balance can influence leadership. Religion and modern psychology point to a similar etiology for the personality and leadership crisis. The renowned Swiss psychoanalyst Carl Gustav Jung and the Mulla Sadra, one of the most prominent Islamic philosophers from Iran, present theories that are relevant to us today.

Jung defines the psychological balance required as a balance between two archetypal images: the anima and the animus. The anima is the feminine aspect of a male psyche—for example, gentleness, tenderness, patience, closeness to nature, and readiness to forgive. The animus is the male side of a female psyche: assertiveness, the will to control and take charge, and the fighting spirit. Unfortunately, for centuries, particularly in the Western world, it has been considered a virtue for men to suppress their femininity or anima. Even today, most women who have reached the helm of politics in their society have done so by suppressing their feminine qualities. The result has been the suppression of feminine attributes in national and international decision-making circles, leading to an escalation of confrontation and war.

War is possible when affections and compassion are forgotten and selfishness and arrogance are on the rise. When the internal balance is lacking, there is no inner peace, and when there is no inner peace, ethics fail and war prevails.

This war is both a war against other human beings and a war, a violence, against nature, against all forms of life and creation. The suppression of the anima has led to an untamed desire to conquer nature and impose human law on the ecosystem. Sustainable development policy is in strong need of an anima that could reverse the current mentality and promote an atmosphere of respect for nature. This might be the ultimate remedy to the climate change crisis that is threatening our future on Earth.

The question that still remains is how that inner balance can be achieved. Throughout history, thinkers and men of religion have pointed to the need for ethics and inner peace. The prominent Islamic philosopher Sadr al-Din al-Shirazi (Mulla Sadra) wrote a renowned treatise on this matter in his work *Asfare Arbae,* or *The Four Journeys.* Arguably the most important philosopher in the Muslim world in the last four centuries, and the author of over forty works,

Sadra was the culminating figure of the major revival of philosophy in Iran in the sixteenth and seventeenth centuries.

In the *Asfare Arbae,* Sadra speaks of the need to make four voyages. Those voyages may be likened to four stages of cognition required to achieve the inner peace and ethical equilibrium that all people, particularly those in leadership roles, need to attain.

The first journey is from one's self, or from creation, or the society, to the Truth, to gain a higher level of cognition and surpass the self and ego. It is the first level of cognition, a necessity in dealing with nature or society.

The second journey is in the Truth with the Truth. It is a journey to enlighten and provide spiritual uplift, highlighting the spiritual strength needed to lead people throughout life. The trek ensures the direction and orientation we so clearly lack in today's world; it can bring us back to the source of life, to nature and a new direction that would change our decisions and actions.

The third journey is from the Truth to the creation, or society; it is an attempt to penetrate the spiritual realm and to relinquish the selfish shortsightedness that has led to the excesses we observe today. The object is to establish the ethical standards that prevent aggression against nature and against others. Ethical standards can also prevent overconsumption, carelessness, and indifference. This is the revitalizing journey bringing the message of inner peace to the society, conveying a sense of direction for the young generation. It also enables leaders to regain the confidence they have lost in the hearts and minds of the youth.

The fourth and final journey is within creation and society along with the Truth. In this journey the elevated self merges with creation and with society, always carrying the banner of truthfulness. It is an indication of the intrinsic ties between the spirit and the material world—an essential message for the leaders and politicians who claim to be searching for solutions to the quagmires of global warming and the deterioration of conditions in the ecosystem.

If we are to fulfill our obligations to the future, people in positions of leadership—in business, media, education, and politics—need an internal peace to deal with their selfish and egoistic drives before they take the affairs of others in their hands. The inner journey and the revival of the anima can bring back the peace we need inside and the peace we need in the world.

The world is suffering from the lack of peace and justice, partly because of politicians who govern world affairs and lack inner equilibrium and peace. How can those who lack inner peace and spiritual prudence guide the ship of humanity through these troubled waters, to find and follow the eternal light that knows no ebb?

Should we right now and for all time affirm that it is our obligation to make all efforts to secure such peace, or should we keep faith in the notion that one day a sublime Savior who enjoys the inner equilibrium will come to lead humanity to peace with one another and to peace with nature?

A Life Worth Living

Dale Jamieson

Every action we take ripples into the future. I buy a tomato at the farmers' market, signaling to the grower that she should grow more tomatoes. I spend an hour sweating at the gym, increasing the likelihood that I will be healthy in old age. I fly across the country for a professional meeting, contributing slightly to climate change.

All of these actions affect the future, so they will affect people who will live in the future. Almost everyone claims to believe, at least officially, that when we are deciding what to do we should take the interests of future people into account. But on reflection, troubles appear. How can we now have obligations to people who do not exist? When it comes to my neighbor, what I do affects his welfare. But when it comes to future people, what I do affects their very existence. They are not "out there" in the future in the way in which a laborer in an Asian sweatshop is "over there." This matters when it comes to climate change because its most extreme effects will be felt by those who will live in the next century rather than by anyone who is now alive. Never before have the members of any species been forced to confront such long-term consequences of their actions.

Together we are changing the climate, but because no single nation,

---

DALE JAMIESON is Director of Environmental Studies at New York University, where he is also a professor of environmental studies and philosophy and affiliated professor of law. He is associate editor of *Science* and *Technology and Human Values*. He has published more than eighty articles and book chapters, is editor or coeditor of seven books, and has also authored a number of books, including *Morality's Progress: Essays on Humans, Other Animals, and the Rest of Nature* (2003) and *Ethics and the Environment: An Introduction* (2008).

person, industry, or action brings it about alone, no one feels wholly responsible. When the climate-changing impact of our daily activities is brought to our attention, we often sink into denial or depression. Most of us do not see ourselves as living luxuriously. We feel we are only doing what we must in order to get by. It is difficult for us to imagine how we would live without driving, heating our homes in the conventional way, and eating our normal diets. Not everyone has a subway that they can take to work, not everyone can afford to pay more for "green energy" from their local power company, and not everyone can afford to shop at Whole Foods. People are too busy just trying to make ends meet to garden, read labels, or engage in debates about sustainable living. One of the greatest paradoxes of our time is that the citizens of the richest, most consumptive nation in the history of the world see themselves as trapped by their own lifestyles. Their wealth and ability to command resources has only made them feel more dependent on forces outside themselves, rather than giving them a sense of freedom or a view of what might lie beyond the way in which we now live.

The politics of climate change are made more difficult by the fact that the victims of our behavior are not in a position to defend their own interests. During the civil rights movement, black Americans made their voices heard. During the Vietnam war, the Vietnamese made their presence known on the streets of America in the dead and wounded soldiers who were shipped back from that misadventure. However, most of the victims of climate change are not yet born. They will be the descendants of those who even now are almost invisible, living on the edge of survival, on the margins of the world. Some are already suffering from the effects of climate change, but it is difficult to apportion responsibility for misery in the lives of those who suffer so much and depend on the fickle forces of nature for their means of survival. Others, such as the inhabitants of small islands that will be swept away by rising sea levels, are already beginning to prepare for the end of their societies as they know them. However, these victims are not invited to present their case on the "fair and balanced" television networks that most Americans depend on for their news. Nonhuman nature will suffer most of all, but it is entirely excluded from our decision-making bodies and figures in our calculations only through a few interest groups that claim to speak on its behalf.

I doubt that knowledge of these injustices alone will motivate people to change their behavior. What will be required is forging a link between these injustices and the vision of a meaningful life that involves seeing ourselves as part of an intergenerational community.

What makes a life worth living? Many Americans would say that what makes a life worth living is doing what is right, and doing what is right consists in obeying God's commandments. This idea was current in the Greek world as well (though the Greeks spoke of "the gods" rather than "God"), and Plato systematically discussed this view in his dialogue *Euthyphro*. His conclusion was that anyone who held such a view was impaled on the horns of the following dilemma. If what makes something right is that it is commanded by God, then anything, no matter how horrific (e.g., murder, rape, torture) would count as right so long as God commanded it. But this is the view of cultists and terrorists who commit despicable acts of destruction in God's name. On the other hand, if what God commands is right independent of His commanding it, then we could do what is right and give meaning to our lives whether or not God exists.

Other ideas that have currency in contemporary America also had their analogues in the Greek world. The idea that success, fame, or celebrity is what is most important in life is reminiscent of the Greek idea that it is honor that gives life meaning. While honor is not the same as celebrity, fame, or success, it is similar to them in one very important respect: no one has it within themselves to be honored, famous, or successful. Whether one succeeds in achieving any of these goals depends on luck and the attitudes of other people. Thus, to suppose that life's meaning consists in such things is to take it out of our hands and make it hostage to luck or fate. Socrates, Jesus, and the Buddha unanimously rejected the idea that the meaning of life should be held hostage entirely to fortune. Whether my life is worth living is up to me. It doesn't depend on the attitudes of others, or the vicissitudes of fate.

The idea that it is success, fame, or celebrity that makes life worth living is an instance of a more general view that is ubiquitous in American society. On this view, life is an instrument whose value consists in its contribution to achieving some goal. This is the attitude that underlies the slogan, attributed to a widely admired football coach, that winning isn't everything, it's the only thing. For people who see themselves as devoted to progressive projects that began before they are born, projects that will persist long after they die, projects whose outcome is very much in doubt, these are not the right metrics for evaluating lives. Because we live in a society that is dominated by such values, it is easy to lose heart.

So what does make a life worth living? The views that I have been discussing all see the value of life as contingent on the attitudes or approval of others, or on fate or fortune. The contrary view is that what makes a life worth living is somehow internal to each person. Of course we don't want to exaggerate the independence of life's meaning from the vicissitudes of fate. Under conditions of extreme material deprivation, when each day is dominated by the struggle for bare survival, questions about meaning are not in the forefront. But most of you who are reading this essay live, like me, in an affluent Western society. For us, despite how it may sometimes seem, life presents itself as a field for choice, decision, and action rather than as a set of imperatives required for survival. It is against this background that each of us must decide what matters most.

I subscribe to the view that constructing meaning in our lives occurs in the context of our relationships to nature and to the world generally, and our responsiveness to new experiences, some of which can be revelatory. For me, the appropriate response involves finding a balance between such goods as self-expression, responsibility to others, joyfulness, commitment, attunement to reality, and openness to new possibilities. In the conduct of daily life what this resolves to is the importance of process over product, the journey over the destination, and the doing over what is done. If I were to summarize this view in a phrase, I would echo Aristotle's account that a life worth living is one that is devoted to valuable activities. Many of these activities are goal-directed, so to be sure, insofar as they achieve their ends, then so much the better. But the value of life turns on engaging in these activities, not on whether we succeed in fulfilling our purposes. What I am responsible for, on my view, is trying to make the world better. Whether or not I succeed is not entirely up to me.

Since this view is relational, it involves respecting our physical, social, and temporal environments. Consider first the idea of respect for nature. One reason for respecting nature is that it is in our interests to do so. The geoscientist Wallace Broecker compares our climate-changing behavior to poking a dragon with a sharp stick. If we anger the dragon of climate, we are in serious trouble. A second reason for respecting nature is that for many people and cultures nature provides important background conditions for our lives having meaning. It is easy to think of examples from history, literature, or contemporary culture. Blake's idea of England as a "green and pleasant land" is important both in literature and in English history. The cherry orchard in Chekhov's play of the same name defines the life of everyone in the community. Think of the role landscape plays in the lives of indigenous peoples. For that matter, think of how the Flatiron mountain backdrop defines Boulder, Colorado. A third reason for respecting nature flows from a concern for psychological integrity and wholeness. As Kant (and later Freud) observed, respecting the other is central to knowing who we are and to respecting ourselves. Indeed, failure to respect the other can be seen as a form of narcissism. Many of these same reasons for respecting nature apply to respecting those who have gone before and those who will come after. Seeing ourselves as related to others in this way is important to respecting ourselves and knowing who we are. It is also central to giving meaning to our lives, and such respect is also likely to help keep us from destroying ourselves.

Many questions remain about what it means to respect those who will follow. What exactly does this imply about the condition in which we must leave the world? Must we leave the world as rich in possibilities as the world that was left to us? And what exactly does this mean?

I will not try to answer these questions here. My point is simple, if radical. I have suggested that climate change poses a fundamental challenge to how we think of our lives having meaning, and that respecting those who will follow is part of how we might constructively respond to this challenge. Such a response does not depend on sweeping views about rights, duties, and responsibilities but only on modest views about what gives our lives meaning. What makes our lives worth living is the activities we engage in that are in accordance with our values, whatever happens in the world. If we live in this way, even if our cause isn't successful, we will have lived a life that is worthwhile because it will be a life that is authentically our own. That is not to say that, from time to time, motivation will not flag. Of course it will. But when these episodes are seen as part of a life that is engaged in valuable activities, they will not threaten the sense of meaningfulness that sustains us. Some may think that this is not enough to make a life worth living, but I cannot imagine what more we could ask than to live a life that is driven by our own values, that is directed toward a project of world-historical importance, to heal both ourselves and the Earth.

Who We Really Are

Thomas L. Friedman

In July 2007, I took part in a green technology conference in Colorado that brought together some of the world's top energy innovators and scientists. At the close of the conference, our hosts showed an old news clip.

Up on the screen came a slightly grainy video from the 1992 Earth Summit in Rio de Janeiro, Brazil. A twelve-year-old girl from Canada named Severn Suzuki was addressing the plenary session of the Rio summit. The camera would occasionally pan to the audience of environment ministers from all over the world, who could be seen listening to her every word with rapt attention—as we did. Suzuki's speech is one of the most eloquent statements I have ever heard about both the strategic and the moral purpose of a really green revolution at the dawn of the Energy-Climate era—from anyone of any age. It reads as well as it was delivered. Here is an excerpt:

> Hello, I'm Severn Suzuki, speaking for ECO—the Environmental Children's Organization. We are a group of twelve- and thirteen-year olds trying to make a difference: Vanessa Suttie, Morgan Geisler, Michelle Quigg, and me. We raised all the money to come here five thousand miles to tell you adults you must change your

---

THOMAS L. FRIEDMAN is a world-renowned author and journalist and a three-time Pulitzer Prize winner. His foreign affairs column appears twice a week in the *New York Times* and is syndicated to one hundred other newspapers worldwide. His most recent book, *Hot, Flat, and Crowded: Why We Need a Green Revolution—and How It Can Renew America* (2008), was a number-one best seller. He has served as a visiting professor at Harvard University and has been awarded honorary degrees from several U.S. universities.

ways. Coming up here today, I have no hidden agenda. I am fighting for my future. Losing my future is not like losing an election or a few points on the stock market. I am here to speak for all generations to come. I am here to speak on behalf of the starving children around the world whose cries go unheard. I am here to speak for the countless animals dying across this planet because they have nowhere left to go. I am afraid to go out in the sun now because of the holes in the ozone. I am afraid to breathe the air because I don't know what chemicals are in it. I used to go fishing in Vancouver, my home, with my dad until just a few years ago we found the fish full of cancers. And now we hear of animals and plants going extinct every day—vanishing forever. In my life, I have dreamt of seeing the great herds of wild animals, jungles and rain forests full of birds and butterflies, but now I wonder if they will even exist for my children to see. Did you have to worry about these things when you were my age? All this is happening before our eyes and yet we act as if we have all the time we want and all the solutions. I'm only a child and I don't have all the solutions, but I want you to realize, neither do you. . . . You don't know how to bring the salmon back up a dead stream. You don't know how to bring back an animal now extinct. And you can't bring back the forests that once grew where there is now desert. If you don't know how to fix it, please stop breaking it! . . .

At school, even in kindergarten, you teach us how to behave in the world. You teach us: not to fight with others, to work things out, to respect others, to clean up our mess, not to hurt other creatures, to share—not be greedy. Then why do you go out and do the things you tell us not to do? Do not forget why you're attending these conferences, who you're doing this for—we are your own children. You are deciding what kind of world we are growing up in. Parents should be able to comfort their children by saying "Everything's going to be all right," "It's not the end of the world," and "We're doing the best we can." But I don't think you can say that to us anymore. Are we even on your list of priorities?

My dad always says, "You are what you do, not what you say." Well, what you do makes me cry at night. You grown-ups say you love us, but I challenge you. Please make your actions reflect your words. Thank you.

Every time I listen to that speech, I get a little chill—especially from the line, "You are what you do, not what you say." For me, the beauty, power, and virtue of Suzuki's words is in their raw reminder of what a real green evolution is all about. It is not about Earth Day concerts. It is not about special green issues of magazines. It is not about 205 easy ways to go green. It is not just the latest dot-com gold rush or marketing fad, either. It is a survival strategy. It is about what we do in response to the truly massive challenge that we face to preserve the natural world that has been bequeathed to us. [. . .]

The environmental law expert John Dernbach once remarked to me that in the final analysis, "the decisions Americans make about sustainable development are not technical decisions about peripheral matters, and they are not simply decisions about the environment. They are decisions about who we are, what we value, what kind of world we want to live in, and how we want to be remembered."

We are the first generation of Americans in the Energy-Climate era. This is not about whales anymore. It's about us. And what we do about the challenges of energy and climate, conservation and preservation, will tell our kids who we really are.

ETHICAL ACTION

How can we fully express our human virtue in the decisions we make?

Ask not, What shall I do? Ask, What kind of person do I want to be? Then act as that person would act. You are what you do. To be the sort of person who helps move the world toward shared, sustainable flourishing will require a strong set of virtues:

- A sense of wonder, to perceive and value the extraordinary beauty and mystery of the thriving world.
- Compassion, to feel the suffering of both human and nonhuman animals caused by climate change and ecological collapse.
- Imagination, to envision new and sustainable ways to provide for human needs without plundering the planet.
- Independence of mind, to distinguish true from false, to distinguish real needs from created markets, to understand how to make good moral decisions under conditions of uncertainty.
- Integrity, to do what one thinks is right, even if it means making decisions that are radically different from the decisions one's friends and neighbors make, decisions contrary to what is well-advertised or easy.
- Justice, to honor the needs of other people and other species as highly as one's own, and to respect in others the rights one claims for oneself.
- Courage, to do what needs to be done even if the lonely odds are against you.

So now. Choose one virtue. Make a decision (what to purchase, how to travel, where to donate time) that embodies that virtue. Now choose another virtue. Make a decision that embodies both of them. Continue. Virtues are habits of the mind and heart. Habits are developed by practice, over time.

7

Do we have a moral obligation to take action to protect the future of a planet in peril?

Yes, because all flourishing is mutual.

Maybe the Western intellectual tradition has made a metaphysical mistake. Maybe we have horribly misunderstood the nature of the world and our place in it. All people hold a worldview, a set of answers to questions such as, What are humans? What is nature? What is the relation of humans to nature? Our answers to these questions matter. For better or for worse, we act and we judge our actions in the context of our understanding of the world.

So, what if we who are descendants of the Western intellectual tradition have been wrong all these years? The Western tradition has asked us to accept three entwined assumptions. First (for a variety of reasons), that we are fundamentally separate from nature. Second (and consequently), that we are superior to nature. Third (and further consequently), that our superiority makes nature unimportant.

But ecological science and most of the non-monotheistic religions of the world tell us that Westerners are dangerously wrong on all three counts. Like everything else on Earth, they say, humans are born into interdependent, life-sustaining relationships with other animals, with plants, with sun and moon and rock. Each has its place, its role, in the thriving of everything else—not a hierarchy, but a dance. Human flourishing dances with the flourishing of nature, like a little girl dances on the feet of her father. Accordingly, human life is utterly dependent on other earthly elements. If so, then people who profess to be concerned about people but not about nature are profoundly misinformed.

Say we are egoists. Say we deeply care about only ourselves. Or say we are anthropocentrists, concerned only with the interests of people. Then we should be deeply concerned about the thriving of all parts of the systems on which our lives depend, from the smallest ecosystems to the grandest workings of time and wind, from the past to the future. Work to ensure the flourishing of nature *is* work to likewise ensure the flourishing of humans, because all flourishing is mutual.

An American Indian Cultural Universe

George Tinker

wita wadonthekide, My Relatives:

An American Indian theological perspective on our current environmental crisis moves far beyond the typical liberal or even radical eurowestern argument, that the world around us is dear to human beings, important for human life on the Earth, and thus should be protected and cared for by human beings. The anthropocentrism already implicit in this eurowestern concern is not a part of our American Indian worldview.[1] Rather, an Indian environmental concern begins with a deeply embedded sensitivity to our relationships with all life-forms, meaning all persons—if we can, as Indians do, interpret the english word "person" much more broadly, to include other-than-human persons.

I began this essay by calling you, all you readers whom I have never met, "My Relatives." We humans are all related, regardless

---

GEORGE "TINK" TINKER (wazhazhe udsethe/Osage Nation) is the Clifford Baldridge Professor of American Indian Cultures and Religious Traditions at Iliff School of Theology in Denver, Colorado. As a member of the Osage Nation, he has worked in the Indian community for two decades as a director of Four Winds American Indian Survival Project in Denver. He is past president of the Native American Theological Association and a member of the Ecumenical Association of Third World Theologians. His publications include *American Indian Liberation: A Theology of Sovereignty* (2008).

---

1. One local neocon talk-show pundit offered this comment in a newspaper op-ed piece: "Call me human-centric if you like, but in the final analysis, the only reason to preserve the balance of nature is to sustain human life. In the absence of humans, what would it matter if the earth existed?" Mike Rosen, "Warming Watermelons," *Rocky Mountain News,* July 7, 2006.

of our nationality or skin color. I do not mean, however, to signal some New Age romanticized notion of relatedness. The language of relatives indicates something much more dynamic and embedded in the American Indian world. It is one thing to claim that we are all related, but quite another to realize the full extent of those relationships—and the responsibilities that come with maintaining them. The consequent moral responsibilities are the reason why the Indian idea of interrelatedness is the one thing that all modern armies and modern warfare must disallow. They must dehumanize the enemy in order to generate the ability to kill another human being and to secure the military advantage and eventually victory. The Indian notion of relatedness, however, treats even enemies with the keen respect due to all life-forms. Even enemies are relatives.[2]

To more fully understand the Indian worldview when we talk about relatives this way, we need to expand our discussion to include persons other-than-human. Thus, "my relatives" include many more than all you readers or all two-legged folk of the world. Indeed, it necessarily includes all of life on our planet: the four-legged persons, the flying persons (from birds to butterflies, and even flies), and all those people called the living-moving ones (that is, the mountains and rivers, the trees and the rocks, the corn that we plant to sustain our lives, and the fish in the lakes). Now we can begin to appreciate the moral ethic involved in praying for all our relatives—including especially those other-than-human relatives.

Ultimately our understanding of our relationship to all that lives in the world around us is an understanding of a shared Earth. When Indian people take from the Earth, we always feel a need to return something of value back to the Earth. So, for instance, we might need cedar leaves to use ceremonially as a medicine; we would use the smoke of the cedar to purify or might use a cedar tea for other medicinal purposes. Yet before we can take the cedar leaves for our use, we would always offer something, perhaps tobacco, back to the cedar tree persons as a way of thanking the cedar trees and doing

2. We should hasten to remind ourselves, however, that calling all two-legged folk "relatives" does not automatically mean that Indians have a responsibility to admit every colonizing New Age seeker into the intimate privacy of an Indian ceremony. That is not what the phrase would signify, even as New Agers rush in headlong to lecture us on the meanings of our own language—always in their terms and to their advantage, of course.

our part to maintain harmony and balance. And yes, before picking the tobacco some offering would be made back to the tobacco plant persons to thank them for their gift.

The reality in our world is that human beings cannot live without taking from our relatives. We might even go so far as to argue that human living requires some perpetration of violence in the world. For humans to eat means that we necessarily kill close relatives, whether we kill buffalos or deer, or take the life of sisters corn or squash. These acts of violence disrupt the harmony of the world around us; they create imbalance that must somehow be repaired. Thus, it is important to Indian people to remember how to perform those ceremonies needed to re-create balance in the world, to maintain balance in our relationships with those other-than-human people around us.

For an old Osage village to feed and take care of itself, for instance, would usually require the killing of fifty or sixty buffalos three times a year, in our spring, summer, and fall hunts. We are told that there was a ceremony to be performed before each of these communal hunts and that the ceremony was in all respects nearly identical to the ceremony (called the "war" ceremony by white interpreters) performed before a military contingent could leave the village to defend the people from an enemy invader. While those military forays might be completed without even killing an enemy, we knew ahead of time that the hunt would indeed kill some of our sisters and brothers of the buffalo nation. In either case, the ceremony was a twelve- or thirteen-day public ceremony to make sure that any violence we committed was done with utmost respect for those relatives who might be killed. Another useful social device was to set aside one whole clan of Osages whose task it was to nurture and protect our relationship with the buffalo nation. Because of their responsibility for maintaining balance in the world with our buffalo relatives, those who are in the *thoka udsethe* (Buffalo Bull clan) are proscribed from eating buffalo meat—except in the context of a religious ceremony. Because the people of this clan are "buffalo people," for them to eat buffalo meat would be an act of cannibalism. The rest of the Osage Nation counted on this clan to spiritually maintain our vital relationship to this important source of nourishment and protein and to help the Nation maintain harmony and balance even when we necessarily engaged in the violence of hunting.

In most Indian national communities there was an annual ceremony that functioned more generally to restore balance. These ceremonies, like the Plains Indian sun dance or the southeastern Green Corn Dance, were concerned for the balance of the whole of the world and are sometimes referred to as world renewal ceremonies. In most Indian national community contexts, the killing of any one (human or other-than-human) was not allowed during the ceremony because of the nature of the ceremony itself. Three times in four days at one Lakota sun dance I attended, a stray rattlesnake crept out of the canyon and entered the arbor. At the first instance, white visitors who ran to get something to kill the snake had to be restrained and told that they were acting inappropriately. Each time, two fire keepers carefully carried the snake out of the arbor and down to the bottom of the canyon. They left it there with offerings of tobacco and gentle words asking the snake to stay away until the ceremony was over.[3]

In our balancing of the world around us there is much more at stake than just our own village or (Osage) national well-being. If we act recklessly and thoughtlessly we could easily put the whole of the world out of balance—for others as well as for ourselves. This is, of course, precisely what we are experiencing globally in this eco-crisis of global warming. Having denied or ignored the relatedness of all being and the responsibilities that come from that relatedness, we violently take from our relations. We have not maintained harmony or balance in the world, or a vital spiritual relationship to what nourishes us.

Imbalance, not balance, is the order of the day in our globalizing political-economic context of climate change and global warming, just as it is in military and political coercions. Had George Bush engaged in a twelve- or thirteen-day ceremony (the typical length of the Osage prebattle ceremony) before attacking Iraq or Afghanistan—especially one that might have recognized his enemies as relatives—I might have had a modicum of respect for his wars. If Georgia-Pacific or any other paper or lumber corporation performed religious ceremonies prior to clear-cutting a forest, if they had spoken to the trees as relatives explaining why their death was necessary, and if they had

---

3. At another sun dance the cooks had to be asked to remove the flypaper they had posted to catch flies and keep the flies out of their food preparation. Killing flies was not an option; rather, they had to be tolerated and allowed to take their share of the food.

returned something of value back to the forest, then perhaps I might have less of a guilty conscience in using their products. Instead, we have a mining industry that returns hazardous waste to the environment in the form of methyl mercury or waste from cyanide used to process gold ore. Indeed, it seems that the only notion of balance is the accounting concern for the bottom line in the profit column.

American Indians, of course, are not temporally oriented in our worldview, nor do we inherently have a concern for something that amer-european folk would call "the future." Indeed, our constant and continual concern is the here and now. It is a concern for place, our particular place in the world, our land; and it is a concern for managing ourselves within the spatial reality of place. Even the celebrated concern for generations yet unborn—to the seventh generation, as some national communities would proclaim (particularly those of the Iroquois Confederacy)—has to do with how we live in this place in the present moment. Balance and harmony in the future begin with harmony and balance in the now—our concern and the concern of every new generation of Indian peoples.

If we begin with our place in the world and our concern for all our relatives in the world, human and other-than-human, then notions of justice and peace flow naturally from that spiritual center.[4] If I know

4. Some two decades ago I was deeply involved in trying to convince the World Council of Churches (WCC) that they would be better served to look very closely at indigenous cultures (and, from my perspective, particularly at American Indian cultures) for new and powerful insights for living out their concerns for global environmental well-being. Their project at the time was a worldwide dialogue called "Justice, Peace, and the Integrity of Creation." The WCC wanted to build on its important 1980s and 1990s work toward justice and peace (with the Global South having insisted on the priority of "justice" in the word order) so that they might now include a concern for the environment (a largely Northern Hemisphere concern, at the time). Hence, "the integrity of creation" language was tacked onto the end of the phrase so as not to decenter the Global South's concern for justice.

My argument was that it would ultimately prove more satisfying to reframe the process as "creation, justice, and peace." With this word order I was arguing for a new euro-christian theological notion of creation that would understand concern for creation as generative of justice and then peace. It would see justice and peace emerging naturally out of a heightened order of concern for the whole of our planet and all our relatives. I suppose this was extremely difficult to conceive within the context of the WCC, since eurowestern Christians have committed so much to the notion of anthropocentrism. Moreover, their sacred writings, with the prominent notion of dominion in Genesis 1, could never quite allow the American Indian notion of egalitarian interrelatedness with all living things that I have in mind.

empirically that I am related to all these persons in the world, then it becomes more difficult for me to hurt any one of these persons, human or other-than-human. It becomes more difficult to engage in any war to destroy one's enemy when we understand the enemy as our relatives. In the same manner, the commitment to global balance and harmony that is implicit in a spiritual focus on the well-being of all my relatives makes it more difficult to engage in environmentally destructive behavior in the trivial interests of (personal) profit. We understand that a corporation's responsibilities go far beyond the so-called fiduciary responsibility to stockholders, a dodge used by every corporate board of directors to validate their company's eco-destructive behavior. In the cultural universe of Indian peoples, we look at the world and all its constituent parts at any moment in time and recognize our relatives in each constituent. Our prayers each day are a prayer that our relatives in this world might live, from trees to dart-fish.

wita wadonthekide cane (For All My Relatives)

No Separation Between Present and Future

Fred W. Allendorf

> To study the Way is to study the self.
> To study the self is to forget the self.
> To forget the self is to be enlightened by all things.
> To be enlightened by all things is to remove the barriers between one's self and others.
>
> DOGEN

> A human being is a part of the whole, called by us "Universe," a part limited in time and space. He experiences himself, his thoughts and feelings as something separated from the rest—a kind of optical delusion of his consciousness. This delusion is a kind of prison for his consciousness, restricting us to our personal desires and to affection for a few persons nearest to us. Our task must be to free ourselves from this prison by widening our circle of compassion to embrace all living creatures and the whole of nature in its beauty.
>
> ALBERT EINSTEIN

FRED W. ALLENDORF is Regents Professor of Biology at the University of Montana and Professorial Research Fellow at Victoria University of Wellington, New Zealand. He has published over two hundred articles on evolution, population genetics, and conservation. He coauthored his most recent book, *Conservation and the Genetics of Populations* (2007), with Gordon Luikart. Allendorf has served as president of the American Genetic Association and director of the Population Biology Program of the National Science Foundation.

The author wishes to thank Arnie Kotler for his comments and Michel Colville for her comments and encouragement.

The Buddha was an evolutionary ecologist. He taught that everything is connected (emptiness) and that everything is constantly changing (impermanence). He came to these realizations not through science, but rather through deep contemplation. Biology has verified these conclusions through our modern understanding of ecology and evolution.

The Jewel Net of Indra is a metaphor developed by the Mahayana Buddhist school to illustrate the Buddha's concept of interconnectedness or emptiness (all things are empty of a separate self):

> Far away in the heavenly abode of the great god Indra, there is a wonderful net which has been hung by some cunning artificer in such a manner that it stretches out infinitely in all directions. In accordance with the extravagant tastes of deities, the artificer has hung a single glittering jewel in each "eye" of the net, and since the net itself is infinite in dimension, the jewels are infinite in number. There hang the jewels, glittering like stars in the first magnitude, a wonderful sight to behold. If we now arbitrarily select one of these jewels for inspection and look closely at it, we will discover that in its polished surface there are reflected all the other jewels in the net, infinite in number. Not only that, but each of the jewels reflected in this one jewel is also reflecting all the other jewels, so that there is an infinite reflecting process occurring.[1]

The ecologist David Barash[2] discussed the parallels between Zen Buddhism and ecology. He felt that the concept of the interdependence and unity of all things was fundamental to both the practice of Zen and the science of ecology. In addition, both share a common nondualistic view of the fundamental identity of subject and surroundings.[3] A bison cannot be understood in isolation from the prairie; understanding requires study of the bison-prairie unit. Barash concluded that "the very study of ecology is the elaboration of Zen's nondualistic thinking."

---

1. F. H. Cook, "The Jewel Net of Indra," in *Hua-yen Buddhism* (University Park: Pennsylvania State University Press, 1977), 1–19.

2. D. P. Barash, "The Ecologist as Zen Master," *American Midland Naturalist* 89 (1973), 214–217.

3. F. W. Allendorf, "The Conservation Biologist as Zen Student," *Conservation Biology* 11 (1997), 1045–1046.

The Buddha taught that nothing is fixed, that all is impermanent. Everything is constantly changing. According to these teachings, the river is a metaphor for life. It is a successive series of different moments, joining together to give the impression of one continuous flow. It flows from cause to cause, effect to effect, one point to another, one state of existence to another, giving an outward impression that it is one continuous movement, whereas in reality it is not. The river of yesterday is not the river of today. The river of this moment is not going to be the river of the next moment. So goes life. It changes continuously, from moment to moment.

Everything is changing. We can see this in the life of an individual: birth, childhood, maturity, and death. This same constant change applies to the continuity of life on Earth. We are different from our parents, and our children are different from us. Over time, this change from one generation to the next results in different species, the process of evolution.

The recognition of the interdependence of all life in space and time is the end product of the scientific process to gain knowledge of the world. However, knowing intellectually that we are interdependent with other species is not itself enough to change our behavior. Zen is a process in which this intellectual knowledge is a starting point to lead to deep understanding and transformation of our behavior.

Through meditation and the cultivation of mindfulness, Zen works toward the realization that self and world are not separate. Vietnamese Zen master Thich Nhat Hanh offers the following guidance: "If we want to continue to enjoy our rivers—to swim in them, walk beside them, even drink their water—we have to adopt the nondual perspective. We have to meditate on being the rivers so that we can experience within ourselves the fears and hopes of the rivers. If we cannot feel the rivers, the mountains, the air, the animals, and other people from within their own perspective, the rivers will die and we will lose our chance for peace."[4]

The deep feeling that we are of the future is more difficult to cultivate than the feeling that we are of the river. Our daily life depends on drinking the river, both metaphorically and literally. It is easy to

---

4. Thich Nhat Hanh, *Peace Is Every Step* (New York: Bantam Books, 1991), 105.

see that if we don't take care of the river, we won't be taking care of ourselves. If we allow the river to become polluted, we will suffer from that pollution.

How can we cultivate a similar deep awareness that we are of the future? If we can free ourselves from the delusion of separation, we will act naturally to prevent harm in the future. For help, we can turn here to a traditional Zen koan, a paradoxical statement used as an aid to gaining awareness:

When did your life begin?

We usually begin counting our "age" on the day we were born, but we existed as an embryo and a fetus up to nine months before our birth. Perhaps our life began at the moment the sperm from our father and the egg from our mother united. However, at this time, our mother's egg had not yet completed the meiotic division that eventually determined the chromosomes our mother contributed to our genome. Moreover, the egg and the sperm that united to form our zygote were already alive. Our life did not begin at fertilization; it was passed on to us from our parents. This transmission of life from generation to generation has been going on for eons, since the beginning of life on Earth nearly 4 billion years ago. Our life began some 4 billion years ago. This is not a metaphor; this is reality.

My own experience with this koan began in morning meditation next to a mountain lake on a backpack trip in Montana. I began in that moment and traced my life back in time. I initially contemplated my parents and those few ancestors of which I was aware. As an evolutionary geneticist, my mind traced the DNA that had been passed down from generation to generation since the beginning of human time. I seamlessly traced the DNA back to the ancestors of our species on the savannah of Africa. And beyond. My journey did not end until I reached the beginning of life on Earth.

Through continued deep contemplation of this koan we can experience the concept of "non-self" in time by tracing our ancestry in evolutionary time. Looking deeply into this seemingly simple question, we can see that our life had no beginning, other than the origin of life on Earth. How can something that had no beginning have an end? If our life did not have a beginning, other than the origin of life itself, then it cannot have an end, other than the end of life itself. Our life will end only when life on Earth itself ends.

Gathas (short verses used to bring the energy of mindfulness to each act of daily life) are one traditional form of Zen practice used to increase our awareness. The following gatha, written by Thich Nhat Hanh, can be used before every meal:

> In this food
> I see clearly the presence
> of the entire cosmos,
> supporting my existence.[5]

We *can* see the entire universe in our breakfast cereal if we take just a moment to reflect. The ocean is there: the rain that watered the grain was carried from the ocean by clouds. The sun is there: the grain could not grow without energy from the sun. The Jurassic ecosystem in which the dinosaurs dwelled is there: plants that fed the dinosaurs 200 million years ago were transformed into the fossil fuel that was used to harvest the grain and to carry it to the table. Gregor Mendel is there, along with the plant breeders who developed the strains of grain. Such moments of reflection strengthen our appreciation of our interdependence to countless beings, near and far.

Regularly using the following gatha, perhaps every time we hear a clock strike on the hour, can help us become of the future:

> In this moment,
> I see every moment;
> my life had no beginning;
> it will not end.

Allowing this gatha to deeply penetrate our awareness can help us to become free from the delusion of separation. If we succeed, we will naturally act in a way that will prevent harm in the future.

---

5. Thich Nhat Hanh, "Look Deep and Smile: The Thoughts and Experiences of a Vietnamese Monk," in *Present Moment Wonderful Moment: Mindfulness Verses for Daily Living* (Berkeley, Parallax Press, 1990, 2006). Reprinted with permission of Parallax Press.

A Transformational Ecology

Jonathan F. P. Rose

> The dogmas of the quiet past are inadequate to the stormy present. The occasion is piled high with difficulty, and we must rise with the occasion. As our case is new, so we must think anew and act anew.
>
> ABRAHAM LINCOLN,
> 1 December, 1862

We live in an extraordinary human-made world of communities made of buildings, the systems that connect our buildings, and the systems that make the stuff with which we build and fill our buildings. Every aspect of the world that we have created is dependent on the Earth's climate, soil, fresh water, oceans, and the sun's radiance. But our contemporary culture has a blind spot—it fails to recognize the deep interdependence between our human-made systems and natural systems. This has given rise to a disturbing imbalance.

The ecological issues facing us are human-caused issues. In Judaism, the practice of Tikkun Olam calls on us to repair the fabric of the world. Yet the fabric of the world was quite whole before humans

---

JONATHAN F. P. ROSE chairs the Metropolitan Transportation Authority's Blue Ribbon Commission on Climate Change and leads Jonathan Rose Companies LLC, which repairs the fabric of cities, towns, and villages, while preserving the land around them. He is a leader in the Smart Growth, national infrastructure, green building, and affordable housing movements. He serves as trustee of the Urban Land Institute, the Natural Resources Defense Council, and Enterprise Community Partners.

came. And so the fabric that we are repairing actually is the fabric we ourselves have destroyed.

Our disharmony with nature is not nature's fault, except that nature evolved humans who could choose not to see the whole, and thus could destroy it. Our disharmony comes from flaws in the way that we think. In essence, the ecological issues before us are ontological issues, issues having to do with our beliefs about the nature of the world.

Einstein noted, "We can't solve problems by using the same kind of thinking we used when we created them." And yet we are trying to solve our ecological problems with the same kind of thinking that created them. To begin to heal our ecology, we need to transform the way that we apprehend our ecology.

The human species is blessed with great intelligence, with access to both intuition and rationality. The intuitive parts of our brain see the whole; our rational brain sees parts. We have developed a culture that has made extraordinary progress with the rational brain, by maximizing the potential of the parts, but with little connection to the whole. These powers have dramatically expanded the human impact on its environment. However, the human ability to directly perceive the consequences of its actions is not as finely attuned, thus humans have often acted without fully understanding their impact on the environment.

The root cause of the ecological issues we face arises from our definition of self. The ecological self is the ecology of the Earth. The perceptual boundary between us and the larger ecology is a mental construct. We need to transform this mental construct and function from a view of the whole, rather than of parts. That would give rise to a transformational ecology, based on a transformation of our view from one of parts to one of the whole.

The disconnect between the parts and the whole is particularly prevalent in our economics, which places value in the component, the transaction, the independent item rather than in the whole system. And so there is a disconnection at all scales between the things that every culture claims to deeply value, and the values we express in our economic structures and daily actions. The consequence of this disconnect is not conceptual; it is fatal. We need to connect what we ecologically know with what we value.

The core driver of climate change and biodiversity loss is consumption. All species consume; it is the nature of life. The issue at hand is that many humans consume amounts that are out of balance with the Earth's carrying capacity.

For many of us, information on the state of the Earth's ecology is essentially secondhand. To many who live in the global South, the droughts and floods, heat and sickness that come from climate change are real, are felt daily. But for the decision makers and consumers of the global North, the effects of climate change come as data, as news, rather than as directly experienced suffering.

Yet humans (and many other species) do have a remarkable ability to perceive the joy and suffering of others. We call this compassion. Wikipedia defines compassion as "a profound human emotion prompted by the pain of others. More vigorous than empathy, the feeling commonly gives rise to an active desire to alleviate another's suffering." Compassion, cooperation, and altruism are social behaviors that emerge from evolution. Altruistic societies are more "fit"; for human societies, compassionate behavior is essential for evolutionary success.

Compassion is most palpable when we deeply love another being. And this love moves us to deeply desire the well-being of the beloved. We also have the capacity to love more than just one other. Many of the world's societies are organized around the individual's deep interconnections to extended families or tribes. And this love can be extended even more widely. Developing ecological compassion is one of the gateways to a transformational ecology.

The ethics of many religions brings us to see that we are part of a whole rather than individuals. In Buddhism, to see oneself as an individual and not part of the whole is considered an illusion. One's goal is to recognize the impermanence of things and the total interconnection of things. Out of this one sees that all life on Earth coevolves, that no thing acts independently, that everything is part of a whole. Evolution arises out of the interdependent interactions of species and the ecologies that they are part of.

Religions attempt to give individuals an operating manual for how to function as part of a larger whole. Both religious and secular ethics call for us to see this larger whole. For example, in Judaism, the kosher rules are designed to inspire thoughtful consumption, to see that every act of consumption can be sanctified. This calls on us to pay attention not only to the product of life but also to the process of life.

And in Christianity, the principle of interconnectedness and ethical responsibility is captured in the phrase "As you sow, so shall you reap." In essence, every action has a reaction: that which we do in the world continues to ripple through multiple generations.

We live in an interesting time in history, when the world's economy is rapidly in decline, when many are reducing consumption to what they need rather than what they want, finding that the reduction is giving rise to a new sense of value and values. This is the time of transformational potential. Can we give rise to a human world that is more compassionate, balanced, and whole, a world that interacts in a healthy way with the Earth's systems? If not now, when?

Why Should I Inconvenience Myself?

Mary Catherine Bateson

Where does the individual fit in as we mobilize to preserve the viability of our home planet for the indefinite future? Why should you or I as individuals inconvenience ourselves or limit our consumption when others are not doing so and our separate impacts are so small a part of the whole?

These questions arise only because of two ideas: first, that there is such a thing as an autonomous individual, and second, that the interests and rights of the individual are the primary basis of ethics. These ideas have been pivotal in Western culture, and yet they support behaviors that have led us to environmental emergencies that threaten much of life on Earth. It is possible that the time has come to reconsider the entire concept of the individual and the concept's ethical consequences. If the focus on the individual is to be retained in some form, as it surely will be, how can it be modified to support responsibility and limitations on individual self-interest?

I have come to believe that the idea of an individual, the idea that there is someone separate from relationships, is simply an error. We create each other, bring each other into being, by being part of the

---

MARY CATHERINE BATESON is a writer and cultural anthropologist who divides her time between New Hampshire and Massachusetts. The author or coauthor of many books and articles, she lectures across the country and abroad and serves as president of the Institute for Intercultural Studies in New York City. Bateson is professor emerita of anthropology and English at George Mason University. Her most recent book is *Willing to Learn: Passages of Personal Discovery* (2004).

matrix in which the other exists.[1] Our experience of separate bodies offers a metaphor for understanding systems, but that metaphor must be extended to understand ourselves as part of wider relationships that unfold beyond the immediate context into the life of this planet, not only as the mother that bears us, but also as the future life to which we give birth and which we must care for in ever-closer patterns of interrelationship, as Earth and air, plants and animals, our kind and others play their complementary sustaining roles in the whole.[2] If we approach it from this point of view, we can use our own bodies to think inward about systemic organization, in which cells are partially independent systems coupled together as parts of organs, each of which may be self-regulated and interdependent within the complete organism. We can then think outward of individual organisms (or persons) within ecosystems (or social systems), interacting and interdependent, never entirely autonomous, always parts nested within larger wholes. Over two centuries ago the United States achieved independence from England, yet interdependence has continued and expanded and is now global. The economic downturn of 2009 was a telling example of interdependence and of the way in which one subsystem within an economy (the mortgage and housing sector) can trigger a progressive destabilization affecting an entire nation, and of how one component of a global system (the United States) can destabilize the whole.

All the same, independence remains a strong value. In the United States, books on child rearing devote considerable attention to what is called "independence training," a theme ranging from toilet training to allowances, from shoelaces to careers. Child rearing in the United States is oriented toward independence and, ultimately, separation from the family of origin. Through most of human history, children have slept alongside parents or siblings, but to many Americans it seems good to give each child a separate room and to put infants to sleep alone in cribs with an electric monitor. The goal of development

1. Mary Catherine Bateson, *With a Daughter's Eye: A Memoir of Margaret Mead and Gregory Bateson* (New York: William Morrow, 1984), 117.
2. Mary Catherine Bateson, *Our Own Metaphor: A Personal Account of a Conference on Conscious Purpose and Human Adaptation* (New York: Alfred A. Knopf, 1972), 285.

seems to be to leave home and stand on one's own two feet, self-reliant and private.

The literature on aging mirrors the same value, with its emphasis on independent living and avoiding dependence on others—especially, perhaps, one's own children. Given this emphasis, the increasing fragility of marriage is no surprise. In traditional societies, intact nuclear families have multiple stakeholders, for they, more than individuals, are the subsystems from which society is built. But in America today, who will argue against the claim to individual self-realization?

Ironically, no amount of ideology can change the profound dependence of each human person not only on other human beings, but on air and water and other species as well—the invisible "others," the great "cloud of witnesses" we depend on. But whereas "independence" is a word enshrined in history, "interdependence" is an uncelebrated academic neologism. Efforts to train children to cooperate and share are outweighed by the emphasis on independence, and quickly shift into training them to compete in environments where sharing is often defined as cheating. We have gone further than other cultures in defining kinds of ownership: whereas some societies have regarded land as common to all, we have invented the ownership of airspace above buildings, of genes and even species, of ideas, of tunes (not yet of single notes), of affection, and even, in the shameful past, of human beings.

But the real issue is neither independence nor dependence, nor even interdependence. It is learning to define ourselves through relationships, as part of something larger, and recognizing that the goals of individuals must sometimes give way to, or at least harmonize with, larger entities. Human beings are components of larger systems, of ecosystems and ultimately the biosphere—components with the capacity to disrupt but also to analyze the properties of these systems and learn that they too are valuable and deserve respect. Single lives are part of a winding evolutionary skein stretching through millennia and encompassing the entire surface of the planet.

Thus, it is a fallacy to ask why you or I as individuals should inconvenience ourselves or limit our consumption when others are not doing so and our impact is so relatively small. What is critical is that each is more accurately a part of the whole than a separate being.

If the concept of the individual is to give way to a recognition of relatedness, how must the concept of individual rights correspondingly change? Various efforts have been made to argue for animal rights. These arguments tend to focus on mammals and birds, species with which humans perhaps like to compare themselves and for which they feel empathy—whales, for example, or sea turtles. We don't hear much in the West about insect rights, and although declining populations of bees are beginning to attract attention, the concern is largely instrumental. We don't generally speak about the rights of plants. What is more serious, perhaps, is that we do not hear about the rights of oceans or marshes or jungles, which are treated as containers (habitats) for the species that capture the imagination. Yet arguably these too are living systems of which the vertebrates that inhabit them are parts. We may make an effort to protect the whales; but if the plankton in the oceans are destroyed by changes in acidity, the food chain will collapse, not only for the whales but for many other species as well. On this account, rights may belong more appropriately to systems than to individual species.

It's important to note that the rhetoric of rights is generally a symmetrical one, based on some concept of comparable value, by analogy with the idea of human equality. This may explain why we find no paradox in assigning rights to the so-called "charismatic megafauna," but not to insects or plants. Sentience proposes similarity, similarity proposes empathy, empathy proposes compassion. Although the tonality is entirely different, there is a surprising relationship between the Western fascination with rights, argued in political terms, and the emphasis on spiritual development toward compassion in Buddhism. Each is based on recognition.

An alternative to the rhetoric of rights—or perhaps its necessary complement—is an asymmetrical rhetoric of responsibility or stewardship. The concept of stewardship has been important in the effort to persuade human beings who have grown up in the Abrahamic traditions of Judaism, Christianity, and Islam that the nonhuman natural world was not created solely for human use and convenience, that indeed we should take active responsibility for protecting it (if only for our descendants). This approach affirms that human beings do have power over other species and other aspects of the natural world (often for harm), and may indeed have a certain kind of superiority. It's a bit reminiscent of noblesse oblige, which works best when convenient. But it has one important advantage in that it does not put the entire emphasis on one level of systemic organization.

The recognition of *similarity* challenges the human individual to feel compassion for individual dolphins or bald eagles. In contrast, the rhetoric of *responsibility* is based on difference; without the protection of a shepherd, a protection that can be provided only if the shepherd differs from the sheep, the sheep may scatter and be devoured by wolves. Furthermore, the successful shepherd must pay attention to the grass, the sheep, the wolves, his own health, and the market for wool or mutton, which requires more than mirror neurons. In effect, responsibility may extend more easily to entire species or habitats or ecosystems than equality does. We may need to claim a certain superiority in order to embrace responsibility as an alternative to irresponsible exploitation. An enlightened anthropocentrism is potentially practical.

However, there are dangers in a rhetoric of responsibility that rests on difference. Again and again, difference has been used to justify domination, whether by one sex or one race or one religion. Given two alternatives or two objects that are dissimilar, we move all too quickly to regard one as superior to the other—potentially seeing the other as there only to be used and perhaps discarded. The responsibility that an adult may take for a child translates, between adults, into paternalism on the one hand and infantilization on the other. It is easy to forget that the adult depends on and learns from the child as well as vice versa, and that humans are deeply dependent on and learn from the entire complexity of a biosphere that we only partially understand. But for all the risks we can recognize in human society, ecosystems and social systems thrive on difference.

The dangers of emphasizing difference can be addressed only by looking at relationships as moving in two directions for mutual benefit, and this is not achieved by a series of negatives. We cannot express a sense of responsibility for the planet solely by what we do *not* do; in fact, commandments in a "Thou shalt not" form cry out to be broken. Instead, a positive value that can take various forms has to be proposed. For the suburbanite about to fire up his SUV to drive to the store for a loaf of bread, shopping can be transformed into a more mindful task, whether by focusing on local or organic produce or by the challenge (are we smart enough for this?) to plan out the week ahead, so that the process of shopping becomes a way of habituating the self to more careful consumption. Carpooling can help us become accustomed to the pleasures and inconveniences of interdependence. In either case, the trip to the market is responsibly and positively reconceptualized in time and space.

There is a third possibility as well: if you run out of bread today, instead of driving to the store, borrow from a neighbor (even if it's embarrassing) and be ready to reciprocate tomorrow (even if it's inconvenient), working over time toward a mutual interdependence. It takes time to dismantle the structures of self-sufficiency that have been built up and to create each other as neighbors instead of as separate individuals. But new habits lead to new ways of seeing, and it may be that we can discover the reality of our dependence on the living world around us only by practicing mutual dependence with others of our own species.

Extra! Extra! New Consciousness Needed

Angayuqaq Oscar Kawagley

Scientists, environmentalists, politicians, and governments around the world have been attempting to find a way to resolve the problems associated with global warming, but so far to no avail. To an indigenous way of thinking, the problems have been approached in the usual fragmented, reductionist way of the modern world, when what is needed is a spiral system that brings our efforts to a deep and cumulative simplicity of healing the wounds to Nature and stabilizing global warming.

We already know that there are people from all societies that recognize the connectedness of all life. But for many, the process stops there, without the encouragement that comes from drawing on prior knowledge, stories, and sense makers of Nature. We are not very adept at reading the signals that can help us in living in harmony with

---

ANGAYUQAQ OSCAR KAWAGLEY was born at Mamterilleq (now known as Bethel, Alaska). His parents died when he was two years old, so he was raised by a grandmother. Yupiaq was his first language, as his grandmother could not speak English. The grandson caught the tail-end of the Upper Stone Age from which his grandmother and Yupiaq people emerged into the modern age. He is now researching and trying to find ways in which his Yupiaq people's language and culture are used in the classroom to meld the modern to the Yupiaq thought-world. Dr. Kawagley, who holds a PhD from the University of British Columbia, serves as associate professor of education at the University of Alaska Fairbanks College of Liberal Arts. He has been executive director of several nonprofit corporations, including ESCA Corporation, an earth science and remote sensing consulting company, and Calista Corporation, a native regional corporation; and project director of the Indian Education Act Program, Anchorage Borough School District.

Nature, which is a self-regulating and self-organizing entity. A few people do use the signs and values from Nature to work toward an ecological balance, but the majority are cursorily unaware and lack direction and guidance to this magnificent notion.

We, the Yupiat, believe that the Ellam Yua (God) is in Nature. Therefore Mother Earth has a culture. This is why we as Native people emulate Nature. We see God in Nature and know that everything Nature has made is a vehicle for teaching us how to make a life and a living for ourselves. Our subsistence way of living is a process of actualizing a lifeway that encourages altruism. Altruism requires that we give utmost respect and honor to everything of Nature, as each element does its job as required by the Ellam Yua. So, the amoeba has a job to do and does it well, and this calls for our respect. The wind, *anuka,* gives its name to us not as a subject but as "being the wind." This gives it the force of life—it is living and therefore requires our utmost respect.

The above are merely reminders to the Indigenous people that we are part of a living ecological system. This requires that we learn the language of the surrounding environment to really know that place. When we know our place, we have our identity. The Yupiaq hunter is one who is well versed in the languages of the elements of Mother Earth, including the plants and animals. He/she is keenly observant and understands the language, whether it is silent, vocal, or visual. That Indigenous person is keenly in touch with the environment as a matter of survival.

An Indigenous person living in and with Nature is constantly reminded of being a part of a living world. This is the primary reason Indigenous people must be directly involved in the processes dealing with climate change. The modern world seems to have a militarized orientation toward making changes to mitigate the effects of climate change. Perhaps if the Indigenous people are directly involved they will be able to enlist their ancestors' knowledge and experience in dealing with such change. We must collectively find ways to work with Nature to change our ways of living.

This requires a change in consciousness. As Albert Einstein said, the consciousness that was used to construct a system cannot be used to make changes to it. We must change from a materialistic, techno-mechanistic worldview to one in which we embrace a living universe and Mother Earth. Our attitudes and intentions of dealing with a living body will change to one of giving respect, honor, and dignity to all resources of the living Earth. This is the ultimate gift of the Indigenous people to the global world.

Just a Few More Yards
Edwin P. Pister

"Just a few more yards and we'll be there," I gasped as we neared the California Department of Fish and Game pickup truck and its aeration equipment, which would provide temporary safety for my precious cargo. I stumbled on in the semi-darkness across the desert marshland with its abundance of grassy hummocks, rodent burrows, and downed barbed-wire fences. "We" in this case was myself and about eight hundred oxygen-deprived fish that constituted the entire remaining world population of an entire vertebrate species, *Cyprinodon radiosus*, the Owens pupfish, which I was carrying in two five-gallon buckets as their remaining habitat dried up downstream behind us late in the evening of 18 August, 1969. I had scooped them from their last remaining habitat as their pond dried to playa.

I am a retired aquatic biologist who spent a career, now approaching sixty years, working out of the town of Bishop in California's eastern Sierra Nevada. My job has been to save from destruction arid land ecosystems and their component organisms. I started this work following World War II as a University of California (Berkeley) graduate student while employed by the U.S. Fish and Wildlife Service and, later, by the California Department of Fish and Game. At Berkeley I

EDWIN P. "PHIL" PISTER is the executive secretary of the Desert Fishes Council and has been referred to as "the father of native fish restoration in America." He is a retired fisheries biologist from the California Department of Fish and Game and the winner of many conservation awards, including the President's Fishery Conservation Award from the American Fisheries Society and the Edward T. LaRoe III Memorial Award from the Society for Conservation Biology.

was fortunate to study under A. Starker Leopold, eldest son of Aldo Leopold, father of wildlife ecology at the University of Wisconsin. Since Starker and his dad were very close in their thinking, I was exposed early on to ethical concerns as they relate to the environment. I was fortunate to have read drafts of *A Sand County Almanac* prior to its publication in 1949.

Shortly following retirement, and in a moment of nostalgia, I began to review my daily journals. These included essentially everything I had done for more than half a century, beginning in 1950. Anyone acquainted with bureaucracy will be aware of the diversity of items such journals might contain—everything from routine vehicle maintenance to attending international meetings and testifying in legal hearings that ultimately led to the U.S. Supreme Court.

Of more than ten thousand such journal entries, the one for 18 August, 1969, the day I carried these fish to safety, stands out as the most dramatic. Such traumatic events tend to imprint themselves indelibly in one's brain. I recall being fearful and keenly aware that if I tripped in the dark night of that 200-yard journey, the Owens pupfish species would almost instantly become extinct: an entire species gone forever. Fortunately for the pupfish (and me), fate smiled on us, and the species remains extant, though still endangered.

In 1950 our nation was emerging from World War II, a war that for four long years had required major sacrifice and effort by America's citizens. Americans wanted to relax a bit, and it was my job as a California Department of Fish and Game aquatic biologist to provide good angling in the eastern Sierra recreation area for millions of people in southern California, primarily through the stocking of large numbers of hatchery-reared rainbow trout—a species foreign to eastern Sierra drainages.

Encouraged and supported by environmental concerns expressed by Rachel Carson in her landmark *Silent Spring,* and by the ecological and ethical principles learned from the Leopolds, I soon realized that this approach to fisheries management was in direct conflict with what I had been taught and was actually detracting from the integrity of the biological resource. Aldo Leopold, in a corollary to his famous Land Ethic presented in *A Sand County Almanac,* reminded us that "a thing is right when it tends to preserve the integrity, stability, and beauty of the biotic community. It is wrong when it tends otherwise."

Though I recognized the irrefutable truth of Leopold's counsel, implementing the required management changes under the eye of my superiors within the Department of Fish and Game promised to be both politically and philosophically difficult. However, for my entire thirty-seven-year career I was lucky enough to live and work at least three hundred miles from my nearest supervisor. This allowed me much freedom of action. With the strong encouragement of academic colleagues Robert Rush Miller of the University of Michigan, Carl L. Hubbs of Scripps Institution of Oceanography at the University of California, San Diego, and W. L. Minckley of Arizona State University, I began to evaluate the native fish populations under my jurisdiction—those that had evolved here in the Death Valley hydrographic system of the extreme western Great Basin.

The results of my surveys were alarming. Of our (only) four native fishes (drainage systems in the desert support a depauperate fauna), I found that three were either endangered or seriously threatened. One, the Owens pupfish (*Cyprinodon radiosus*), was thought to be extinct when described by Miller in 1948, but a relict population was discovered by Hubbs, Miller, and me in July 1964, thereby beginning a rather frightening recovery process. It was during this time that I made a major change in my career direction and began devoting virtually all of my effort toward the recovery and safety of these species, as well as the protection of native fishes of the entire American Southwest and northern Mexico, taking comfort in Aldo Leopold's observation that nonconformity is the highest evolutionary attainment of social animals.

Water will always be in short supply in desert regions, exacerbated by climate change and the inexorable growth of major desert cities. We are fighting a delaying action in our native fish recovery efforts, hoping to preserve this precious resource until society begins to share concerns that, to many of us, are terribly obvious.

But I see bright spots on the horizon. New perspectives are being brought into both state and federal fish and wildlife management agencies by new employees who are better educated, not simply trained. I am encouraged (and amused) by an observation of German scientist/philosopher Max Planck: "A new scientific truth does not triumph by convincing its opponents and making them see the light, but rather because its opponents eventually die, and a new generation grows up that is familiar with it."

Aldo Leopold, in discussing the evolution of the nation's ecological conscience, stated: "I have no illusions about the speed or accuracy with which an ecological conscience can become functional. It has required 19 centuries to define decent man-to-man conduct and the process is only half done; it may take as long to evolve a code of decency for man-to-land conduct. In such matters we should not worry too much about anything except the direction in which we travel. The direction is clear." Recognizing that this was written by Leopold more than sixty years ago, and having observed the American conservation scene in relation to the nation's growth since then, it seems reasonable to question whether we are moving fast enough as a society to temper our consumptive appetites and population numbers sufficiently to preserve such things as pupfish and, ultimately, ourselves. In a very real sense all our planet's life-forms are in two buckets as poorly understood forces carry us into a future of either safety or oblivion. The warnings have been sounded, but will we heed them? Leopold warned us that one of the penalties of an ecological education is that we live alone in a world of wounds. It appears that our ultimate salvation lies in making these wounds apparent to everyone, and to somehow underscore their urgency.

Several years ago I summarized some of my previous thoughts in a *Natural History* essay titled "Species in a Bucket," which clarified the fragility of many of our species, ending with the following: "That August day twenty-three years ago had been a very humbling experience for me. The principles of biogeography and evolution I had learned many years before at Berkeley had taught me why the pupfish was here; it took the events of those few hours in the desert to teach me why *I* was. Such are the reflections of a biologist who, for a few frightening moments long ago, held an entire species in two buckets, one in either hand, with only himself standing between life and extinction."[1]

---

1. E. P. Pister, "Species in a Bucket," *Natural History* 102, no. 1 (1993), 14–19.

Given the slow and difficult growth of a national ecological conscience since then, time and patience seem to be running out. The world may not have time to wait for another generation to evolve an ecological conscience. It seems reasonable to question whether we are moving fast enough as a society to "preserve the integrity, stability, and beauty" of the planet's life-forms, tempering our consumptive appetites and population numbers sufficiently to preserve such things as pupfish and, ultimately, ourselves.

In a very real sense, we and all our planet's life-forms are slopping in two buckets while wild and poorly understood forces carry us into the future. The warnings have been sounded, but will we respond soon enough, and with urgency sufficient to the work? There are no Fish and Game trucks and aeration systems waiting at the end of the trail to save us. We must do that work ourselves.

Why Sacrifice for Future Generations?

Kimberly A. Wade-Benzoni

Some of the most important issues that we face in society today involve long time horizons and thus have implications for future generations of people. In the case of global-scale environmental change, decisions made today will constrain the options of future generations for decades to come. Part of the problem is that the interests of present and future generations are not always aligned. Relaxing controls on greenhouse emissions, for example, creates cheaper energy in the short term, but higher costs in the future. In cases like this, benefits accrue immediately to the present generation, while associated burdens are passed along to people in the future. Intergenerational justice may require a reversal of that pattern; it may be necessary for the present generation to make sacrifices in order to protect the interests of future generations.

What factors make it difficult for people in the present generation to sacrifice their own self-interests for the benefit of future others? And, more important, what factors lead people to act with generosity and goodwill toward future generations?

KIMBERLY A. WADE-BENZONI is an associate professor of management and Center of Leadership and Ethics Scholar at the Fuqua School of Business at Duke University. Her research on intergenerational and environmental issues has received competitive awards and funding from the U.S. Environmental Protection Agency, the National Science Foundation, and Kellogg Research Centers, among other organizations. She has published numerous academic articles, and she coedited the book *Environment, Ethics, and Behavior: The Psychology of Environmental Valuation and Degradation* (1997) as well as a special issue of the journal *American Behavioral Scientist*.

In the areas of philosophy and law, scholars question the extent to which present actors are morally obligated to protect the interests of future others.[1] And economists are trying to determine that balance between the interests of present decision makers and future others that produces optimal levels of efficiency.[2] In contrast to these normative approaches, psychological research has taken a descriptive approach,[3] identifying the factors that affect actual decision-making behavior: What are the psychological factors that influence how much members of present generations are willing to sacrifice their own self-interest for the benefit of future others, in the absence of economic or material incentives?

A central barrier to intergenerational beneficence is the inherent psychological distance between decision makers and the future consequences of their decisions—a distance that is both temporal and personal. First, the consequences of intergenerational decisions are removed from the decision maker through the time delay that exists between the decision and the future consequences of that decision. Research indicates that the greater the time delay between an intergenerational decision and the consequences of that decision to future generations, the less people act on the behalf of future generations.[4]

---

1. See, for example, B. Barry, *Theories of Justice* (Berkeley and Los Angeles: University of California Press, 1989); D. A. J. Richards, "Contractarian Theory, Intergenerational Justice, and Energy Policy," in *Energy and the Future*, ed. D. MacLean and P. G. Brown (Totowa, NJ: Rowman and Littlefield, 1981), 131–150; E. B. Weiss, *In Fairness to Future Generations: International Law, Common Patrimony, and Intergenerational Equity* (New York: Transnational Publishers, 1989).

2. For example, L. J. Kotlikoff, *Generational Accounting* (New York: Free Press, 1992); P. R. Portney and J. P. Weyant, *Discounting and Intergenerational Equity* (Washington, DC: Resources for the Future, 1999).

3. For reviews, see K. A. Wade-Benzoni, "Giving Future Generations a Voice," in *The Negotiator's Fieldbook*, ed. A. Schneider and C. Honeyman (Washington, DC: American Bar Association, 2006), 215–223; L. P. Tost, M. Hernandez, and K. A. Wade-Benzoni, "Pushing the Boundaries: A Review and Extension of the Psychological Dynamics of Intergenerational Conflict in Organizational Contexts," in J. J. Martocchio, ed., *Research in Personnel and Human Resources Management*, vol. 27 (Greenwich, CT: JAI Press, 2008), 93–147; K. A. Wade-Benzoni, and L. P. Tost, "The Egoism and Altruism of Intergenerational Behavior," forthcoming at *Personality and Social Psychology Review*, 2009.

4. K. A. Wade-Benzoni, "Maple Trees and Weeping Willows: The Role of Time, Uncertainty, and Affinity in Intergenerational Decisions," *Negotiation and Conflict Management Research* 1 (2008), 220–245.

People discount the value of commodities that will be consumed in the future, reflecting an inborn impatience and a preference for immediate over postponed consumption. As time delay increases, people have greater difficulty fully understanding the consequences of decisions. Beyond these cognitive limitations, however, motivational effects—such as the immediate pain of deferral—also make it difficult for people to delay benefits for the future.

The psychological distance that is already present in the time dimension is compounded by the fact that it is *others,* rather than oneself, who will be affected in the future by one's decisions. When making trade-offs between the well-being of the self and that of others, there is a tension between self-interest and the desire to benefit others. Although people may care about the outcomes of others, trade-offs between one's own and others' well-being are typically skewed to the point where very little weight is put on the consequences for others.[5] The impact of individuals' decisions on themselves is generally far more immediate than their impact on other parties. So in terms of intergenerational sacrifices, the costs to oneself of changing behavior for the benefit of future generations are more immediate than the future benefits to others not only in the temporal sense, but also because they affect the decision maker directly.

In light of the combination of temporal and interpersonal distance inherent in intergenerational decisions, we might expect the prospects for future generations to be quite grim. Research on intergenerational decisions, however, has uncovered a number of factors that can help promote intergenerational beneficence despite the inherent barriers.

First, the psychological distance between oneself and future generations can be reduced by increasing what has been called "affinity" to future generations. Affinity is a function of perceived oneness between oneself and others. When affinity with future generations is high, the distinction between the interests of present and future generations becomes blurred. This psychological connection can help people to vicariously experience the benefits and burdens left

---

5. G. Loewenstein, "Behavioral Decision Theory and Business Ethics: Skewed Tradeoffs Between Self and Other," in *Codes of Conduct: Behavioral Research into Business Ethics,* ed. D. M. Messick and A. E. Tenbrunsel (New York: Russell Sage Foundation, 1996), 214–227.

to future generations. Consequently, greater levels of affinity with future generations promote more intergenerational beneficence.[6]

Philosophers and theorists have cited the lack of immediacy of future consequences (i.e., psychological distance) as one reason why people often do not act on behalf of future generations. But they have also pointed strongly to the absence of traditional bonds of reciprocity as a factor thwarting sacrifice for future generations.[7] In Western culture, one generation typically does not benefit from the sacrifices it makes for future generations.

In conventional interpersonal contexts, reciprocity is the mutual reinforcement by two parties of each other's actions.[8] In intergenerational contexts, reciprocity takes a more generalized form in which people can "reciprocate" the good or evil left to them by previous generations by behaving similarly to the next generation.[9] In other words, people can pass on benefits (or burdens) to future generations as a matter of retrospective obligation (or retaliation) for the good (or bad) received from past generations. The beneficence of prior generations can create a sense of indebtedness, such that people want to "pay it forward" by acting generously to the next generation. Thus, intergenerational reciprocity can come into play as either a facilitator or a barrier to intergenerational beneficence, depending on the behavior of prior generations. The phenomenon of intergenerational reciprocity highlights how our intergenerational behavior has more far-reaching implications than we may realize. Not only does it affect the next generation directly; it also influences how the next generation will treat subsequent generations.

Perhaps most critically, research has revealed the central role of legacies in the psychology of intergenerational decision making. The

---

6. Wade-Benzoni, "Maple Trees and Weeping Willows."

7. N. S. Care, "Future Generations, Public Policy, and the Motivation Problem," *Environmental Ethics* 4 (1982), 195–213.

8. T. Parson, *The Social System* (New York: Free Press, 1951); A. W. Gouldner, "Reciprocity and Autonomy in Functional Theory," in *Symposium on Sociological Theory*, ed. L. Goss (New York: Harper & Row, 1959), 241–270; A. W. Gouldner, "The Norm of Reciprocity," *American Sociological Review* 25 (1960), 161–167; P. M. Blau, *Exchange and Power in Social Life* (New York: John Wiley & Sons, 1964).

9. K. A. Wade-Benzoni, "A Golden Rule over Time: Reciprocity in Intergenerational Allocation Decisions," *Academy of Management Journal* 45 (2002), 1011–1028.

notion of a legacy is meaningful only in a context, such as an intergenerational one, where a person's behavior has implications for other people in the future. Intergenerational beneficence presents an opportunity to create an enduring sense of personal life-meaning by establishing a legacy. A legacy is a means by which people can extend themselves into the future to create a positive, ethical, and lasting impact.

Legacies provide a symbolic form of immortality because they enable people to create something that will outlive themselves.[10] Indeed, older individuals, typically around and after the time of retirement, start to engage in work that will solidify their long-term legacies.[11] Intergenerational beneficence can help people to feel like a part of something that will outlast their own individual existence. Acting on the behalf of future generations thus paradoxically represents a dramatic form of self-interest even as it helps people to strive for immortality. The desire to extend oneself beyond mortal life is a deep and strong impetus for generative action that helps to align the interests of the self and future others. Believing that we have made a difference by leaving the world a better place than it was before we became a part of it helps us to gain a sense of purpose in our lives and buffers the threat of meaninglessness posed by death.

People need to feel that life is meaningful and purposeful. They need to believe that they have made a useful contribution to the world. Intergenerational beneficence can help to fill that need by enabling people to make a connection with something that will continue to exist after they are gone. Being able to live on, even if only in impact and memory, is an important motivator of intergenerational beneficence.

10. J. Greenberg, S. Solomon, & J. Pyszczynski, "Terror Management Theory of Self-Esteem and Social Behavior: Empirical Assessments and Conceptual Refinements," in *Advances in Experimental Social Psychology*, vol. 29, ed. M. P. Zanna (New York: Academic Press, 1997), 61–139; K. A. Wade-Benzoni, "Legacies, Immortality, and the Future: The Psychology of Intergenerational Altruism," in *Research on Managing Groups and Teams*, vol. 11, ed. M. Neale, E. Mannix, and A. Tenbrunsel (Greenwich, CT: Elsevier Science Press, 2006), 247–270.

11. D. P. McAdams, A. Diamond, E. de St. Aubin, and E. Mansfield, "Stories of Commitment: The Psychosocial Construction of Generative Lives," *Journal of Personality and Social Psychology* 72 (1997), 678–694.

Hope and the New Energy Economy

Jesse M. Fink

I have been thinking about my grandchildren a lot lately, and I don't even have any. But when my kids are older, I probably will, and I keep wondering what life is going to be like for them and for their children. For most of my life, I have been an environmentalist, thinking that nature itself was at the center of my concerns. But I am coming to realize how much of my concerns are really about future generations of people—not just my own family, and not just a generation or two, but generations rippling outward, over a longer period of time. I'm generally an optimistic person, so I have to envision a changing world, an adapting world, not an awful world.

I have spent my professional career creating new businesses in emerging technology sectors, trying to catalyze relationships that align economic growth with environmental stewardship. I see all around me signs that business as usual is changing. The signs fill me with hope. The way we produce and consume energy is the root cause

---

JESSE M. FINK is an entrepreneur committed to solving environmental challenges, particularly climate change. He works in partnership with financial markets, academia, policymakers, and NGO communities. Searching for investment solutions, he cofounded MissionPoint Capital Partners, a private investment firm focused on financing the transition to a low-carbon economy. His professional career involved developing innovative business models, including cofounding Priceline.com and serving as its founding Chief Operating Officer. Jesse received a BS from SUNY-ESF, where he met his wife Betsy, and an MBA from Syracuse University. Together they manage the Betsy and Jesse Fink Foundation, directing catalytic environmental grant making and mission-related investing. Betsy created Millstone Farm, an incubator for adaptive solutions for community-based energy and agriculture.

of global climate change. How energy is produced and consumed is largely determined by our economy. It follows that we can address global climate change by making changes in the nature of our economy. What is needed is a new energy ethic, driving new policies and regulations that will align the interests of commercial and financial markets with the requirements of a thriving Earth. We have a historic opportunity to reinvent our energy economy, and it does not depend on destroying the world in which our grandchildren will live.

A New Energy Ethic

There is good scientific evidence that incremental changes to the Business as Usual Curve will not reduce our CO_2 emissions sufficiently. But there are other signs that a new energy ethic is emerging that will result in the dramatically new economy my grandchildren will experience, where communities of change are local but globally networked, and the interests of the worlds of commerce, financial markets, government, education, and philanthropy are aligned with the requirements of a thriving ecosystem.

Two factors have made it difficult in the past for business as usual to address the problems of climate change. First, thinking long-term is a challenge in the world of business and finance. Commercial markets have a much shorter rhythm than natural cycles, and "long-term" investment is foreign to an increasingly accelerated cable news culture, where investors' horizons are shortened by the immediate monitoring of condensed business cycles and never-ending patterns of consumption. This has led to short-term thinking, condensing business decisions from annual or quarterly projections to overnight horizons. The commercial world will always work within short spans of time, but longer-term thinking is central to capital formation and structure, for investment in growth and sustainability over time. The longer-term perspectives that are good business practices also reflect wider horizons of "value" in the broadest sense, including what matters in the long run, and what means are worth pursuing for what ends.

The second factor is the mistaken though widespread belief that current economic viability requires sacrificing longer-term environmental sustainability. I am convinced that it does not. A false dichotomy between economic growth and environmental sustainability may have separated ecosystems and their resources from our economic lives and misguided us into narrow conceptions of prosperity as limitless accumulation of material goods. But all this is changing. Businesses are realizing that their own sustainability and their own prosperity will be accomplished by aligning their interests with longer-term consideration of environmental sustainability, and of mitigating climate change in particular.

Despite these bars to change, we are living through a major transition right now. While transitions can wreak havoc, we are realigning the interests and institutions whose historical mismatch has led to our environmental predicament. It is possible for me to hope because I can envision a world in which economic activity involving individuals and institutions will dramatically change in a short period of time. We can create a very different economy, based on bold changes in how we think and live as individuals, how communities are shaped, and how innovative business and capital markets will take root, creating new jobs, activity, and prosperity.

What does this change require? First, consumers who are empowered to make good choices based on information about their energy use. Second, new ways to finance companies that assist in the transition to a low-carbon economy. And finally, mission-related investments wherein investors can grow their capital while funding energy-efficient projects. These can be achieved by the cooperative efforts of the educational, business, philanthropic, and governmental sectors of society. The relation of these goals and institutions might be understood this way:

Education and the New Economy

Energy awareness at all levels of public education can bring us together to act in our communities, whether they be local, national, virtual, or global. Many of us are too unaware of the science and economics of energy to make informed decisions as citizens and consumers. The connections between energy use and global climate change are rarely taught to us in school, and as a result we leave our decisions

```
                    Changing
                    Consumer
                    Behavior
```

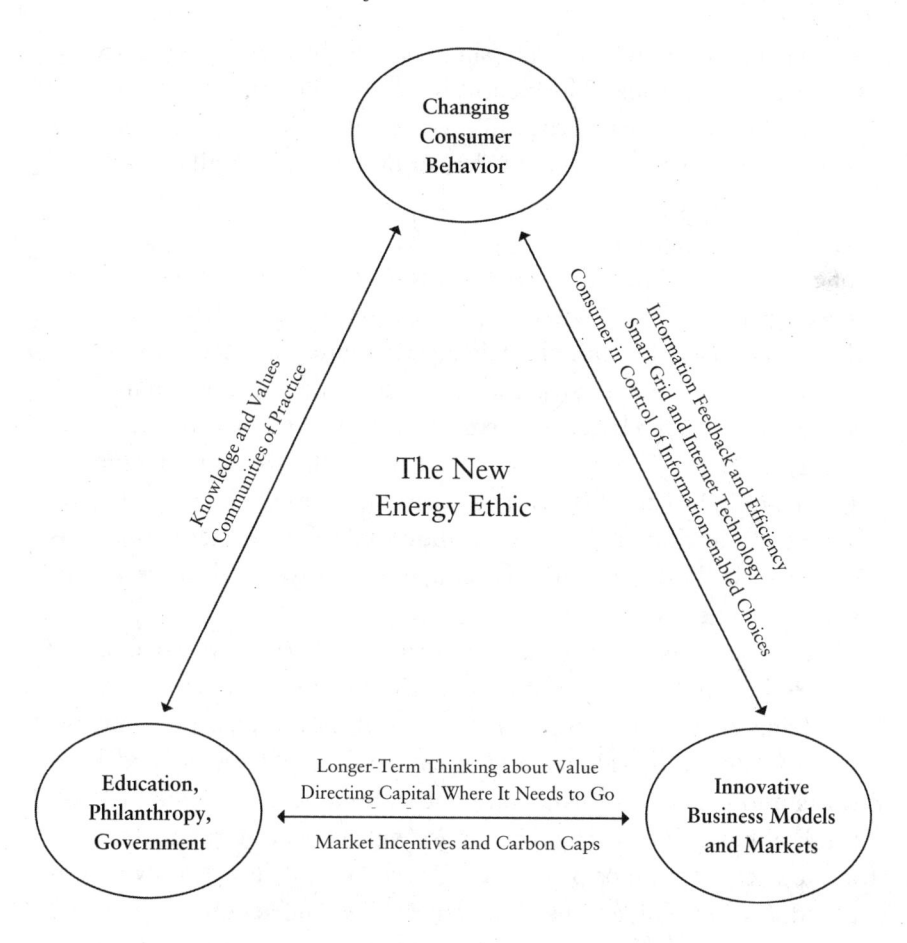

Diagram: Triangle with "The New Energy Ethic" at center.
- Top vertex: Changing Consumer Behavior
- Bottom-left vertex: Education, Philanthropy, Government
- Bottom-right vertex: Innovative Business Models and Markets
- Left edge label: Knowledge and Values / Communities of Practice
- Right edge label: Information Feedback and Efficiency / Smart Grid and Internet Technology / Consumer in Control of Information-enabled Choices
- Bottom edge labels: Longer-Term Thinking about Value / Directing Capital Where It Needs to Go / Market Incentives and Carbon Caps

about public policy to others. Values are formed and decisions are made without the knowledge that even a middle school student could digest, even though knowledge about the effects of our energy use can be life-changing. We have begun to address questions in school about where our food comes from, but we have barely begun to address our national inability to understand where our energy comes from, or the consequences of our individual habits of consumption.

This, too, is beginning to change. I see it every day, as I direct a philanthropic foundation that supports environmental education initiatives. Because we are empowered to make choices when we have information at our disposal, and because new choices are becoming available to us, energy education must become a national priority if we are to accelerate the promising changes in the energy economy that are taking place. Policy choices will be adopted with a new urgency when an informed citizenry feels a call to action based on what they know. Our thinking and behavior as consumers will change and will be supported by new economic arrangements. Educational institutions will play a central role in easing an abrupt transition to longer-term thinking about nature, resources, energy, and environmental degradation. It will change our behavior as individuals and in the marketplace. But not all the education will take place in schools.

I happen to be on an airplane as I write this paragraph, on my way home from a conference on "smart grid" technology and the new energy economy. The overarching theme of this showcase is that we are beginning to act differently as consumers; as we have new information at our fingertips, we are empowered to make energy use decisions that will usher in a new paradigm of how we produce, transmit, and use energy.

Behavior changes start with information. When we have smart meters in our homes, we can measure and monitor every activity involving energy consumption. Starting with the measurement, we begin to create the ability and incentive for reducing wasted kilowatts and cost. It is in our power to dramatically change how we think about using energy, bringing a much longer-term perspective into our conception of prosperity. Understanding how energy is produced, transmitted, and used can have a huge impact on our personal and global demands for energy. We will never again look at the light switch or thermostat the same way.

Local and global distinctions will change as everyone becomes both a producer and a user, and new networks will emerge in an attempt to keep energy use close to the source of production, preventing losses in efficiency through transmission over long distances. As distributed generation and efficiency become the hallmarks of how we produce and use energy, a new model of networks and grids will emerge, one that recoups a sense of community among people who trade, share, compete, and cooperate. We will have at our fingertips instantaneous information about the energy we consume, and the power to use more or less at any moment.

Innovative business models and markets will develop as the new energy economy creates a real opportunity for all different types of capital to be deployed and invested. Traditional capital providers such as venture capital, private equity, and financial institutions have the opportunity to invest in growth-oriented companies providing solutions. New and unique financing gaps also create an opportunity for mission-related investing, which enables families, foundations, endowments, and pension funds to align their longer-term capital investments with specific mission results and outcomes. These sectors of investors should provide an important bridge to spur investment and innovation.

There are many, many examples of businesses that provide glimpses of what a realignment of interests might look like, where private capital is already generating market-rate returns in new investments in a cleaner energy economy. The investment firm I helped to build has as its mission the financing of companies that will assist in the transition to a low-carbon economy. An entire financial industry is emerging that will shift capital in this direction in a bigger way and is bringing such endeavors to a scale that is helping to bend the Business as Usual Curve into the new energy economy.

It's true that individual investors, both conservative and risk-taking, often see mission-related investment as a philanthropic frill. But responsible investing need not be less profitable; and it could become even more profitable if there were, for example, bond markets directed specifically at energy initiatives. Even as things now stand, we are on the brink of significantly widening such opportunities.

There is a natural role for leadership here by those investors whose perspective is longer-term and focused on responsibility, yet who are also savvy about protecting their resources over longer periods of time, even for future generations. Philanthropy, educational institutions and other endowments, and pension funds can take leadership by directing capital to the right places with mission-related investments that give more than an economic return. They can be growing their own capital while funding energy efficiency projects in low-income communities, or renewable energy projects in rural communities, or even microloan finance programs for household energy efficiency. Mission-related capital can be the bridge to accelerate the transition to a low-carbon economy.

What has been missing is the opportunity to more robustly deploy more capital in investments that will further a mission. When the new financial industry provides mission-related investors with the investment vehicles to use their capital to promote their mission without sacrificing returns, the influx of capital will unleash a spirit of entrepreneurial initiatives, from green-collar jobs to new capital markets.

Government and Philanthropy: Redirecting Capital for Innovation

Government and philanthropy can help bring about bigger changes in the business of how energy is provided, how it is used by communities, and how we think about markets and resources—indeed, how we think about our wants and needs and the very concept of prosperity.

Currently, proposed federal and state legislation is already advancing polices that will reflect the costs of carbon in its price, motivating innovation. A tighter cap on emissions will matter more in stimulating market-based solutions than whether the legislative mechanism turns out to be a system of cap and trade, tax incentives and dividends, or a combination. This will play an important role in our coming to appreciate the deeper connections between global climate change and the economic arrangements of our daily lives and institutions.

Foundations and nonprofit endowments are now attending carefully to whether their philanthropic donations, their "investments" in the work of others, are getting the mission return they expect. Over time, they have become more demanding, asking for accountability from the charities and endeavors they support. On the other side of the house, in protecting their assets over the long run, foundations and endowments have become very sophisticated investors, deploying their capital to maximize returns.

Foundations and nonprofits will, for example, lend money to energy efficiency projects, or invest in start-ups and existing companies that are promising alternatives to past practices, and get a financial return that is equal to other investments with the same risk. We should be able to invest in the new energy economy as we have in low-income housing, or even highways and water treatment plants.

This exists now in the equity markets but has yet to be developed in the debt markets. If we really want to accelerate the transition to a low-carbon economy and a new energy economy, we need to do more on this front, and do it quickly. But every day at the office, working with government officials and NGOs, I see this growing, and it provides yet another reason for hope. I believe families, foundations, endowments, and pension funds will be the leaders in this area. They exist for the future, and that is something the rest of us need to incorporate more into our daily lives, our values, and the way we conduct business.

Reasons for Hope

I predict that when we recognize the environmental damage and the long-term depletion of our resources that are caused by our inefficient and dangerous carbon emissions, those costs will become reflected in the price we pay for energy. With new policies governing our markets, consumers will be empowered to make better choices. The new economy will redirect its capital to job- and wealth-creating entrepreneurs, commercial enterprises, and investment opportunities. Having worked in just these commercial and capital markets for many years, I believe components of the economy are not merely compatible with a paradigm shift, but are indeed necessary for it, and that such a shift is happening right now.

Business and commerce will continue, and their focus will remain largely on short-term horizons. As the rules by which they play are changed by policies that align their interests with energy efficiency and environmental stewardship, new capital formation and structures are redirecting investments to a clean energy economy. As businesses and consumers are empowered with the knowledge to make different choices, new opportunities are becoming affordable to both.

The world in which I envision my children and grandchildren living will not be easy. Technology will not be able to remove the greenhouse gases we have left in the atmosphere, and they will feel its effects in a dramatic way. We are beginning a new era that will not come to full fruition while we are alive. But it is beginning, and we have reason to hope that the world my children leave to their children will be on a different path than the one we inherited. We have no choice but to reconsider what prosperity means to us and to the way we do business, and to embark on a new future immediately. Hope should not come too easily if it is to be meaningful, and we will have to work together as we never have before. But there is good reason for hope, and without it we have nothing.

I would like to thank the many people who have influenced my thinking and helped to organize these thoughts. In particular, Joan Briggs, who brilliantly manages our Foundation and has enough soul for the world; my childhood friend Scott Brophy, now a grown-up discussion partner and professor of philosophy at Hobart and William Smith Colleges; Mark Cirilli, my partner at MissionPoint Capital Partners, a financial leader in the low-carbon economy who has blazed the path with innovative solutions; and my wife, Betsy Fink, who is willing to take on big environmental issues by successfully catalyzing local education and awareness solutions.

ETHICAL ACTION

How can we protect the flourishing of all life?

If it's true that we can't destroy our habitats without destroying our lives, as Rachel Carson said, and if it's true that we are in the process of laying waste the planet, then our ways of living will come to an end—some way or another, sooner or later, gradually or catastrophically—and some new way of life will begin. But how can we even begin to protect all of the Earth's flourishing life? The job is unimaginably big. In the face of a catastrophe on the scale of global warming, what is the use of trying?

The answer may come from scientific studies of other devastations. When Mount St. Helens erupted in 1980, the ash-laden blast leveled forests for dozens of miles around. Scientists thought it would take dusty centuries for life to spread in from the edges of the rubble plain. But now, only thirty years later, the mountain is carpeted in moss and purple lupine and laced with the tracks of deer and fox. What the scientists learned is that when the mountain blasted ash across the landscape, the devastation never touched some small places hidden in the lee of rocks and trees. "Refugia," scientists called them: places of safety where life endures. From the refugia, small animals emerged blinking onto the blasted plain. And from a thousand, ten thousand, maybe countless small places of enduring life, meadows returned to the mountain.

This suggests that if destructive forces are building under our lives, then our work is to create refugia of the imagination. *Refugia:* places where new ideas are sheltered and encouraged to grow.

We can create small pockets of flourishing, and we can make ourselves into overhanging rock ledges to protect their life, so that the full measure of possibility can spread and reseed the world. Doesn't matter what it is; if it's generous to life, imagine it into existence. Create a bicycle cooperative, a seed-sharing community, a wildlife sanctuary. Write poems for children. Sing duets to the dying. Tear out the irrigation system and plant native grass. Imagine water pumps. Dig a community garden in the Kmart parking lot. Learn to cook with the full power of the sun at noon.

We don't have to start from scratch. We can restore pockets of flourishing lifeways that have been damaged over time. Breach a dam. Plant a riverbank. Vote for schools. Introduce the neighbors to each others' children. Celebrate the solstice. Write a story in an old language. Slow a rivercourse with a fallen log.

Maybe most effective of all, we can protect refugia that already exist: they are all around us. Protect the marshy ditch behind the mall. Ban poisons from the edges of the road. Save the hedges in your neighborhood. Boycott what you don't believe in. Refuse to participate in what is wrong. There is power in this—an attention that notices and celebrates thriving where it occurs, a conscience that refuses to destroy it. These acts will be the wellspring of the new world. From sheltered pockets of moral imagining, and from protected pockets of flourishing, new ways of living will spread across the land.

Here is how we will start anew: not from the edges over centuries of invasion, but from small pockets of good work, shaped by an understanding that all life is interdependent and driven by the uniquely human gift—practical imagination, the ability to imagine that things can be different from what they are now. "Your calling," philosopher Frederick Buechner said, "is at the intersection of your great joy and the world's great need." Go to that place. Do that work.

8

> *Do we have a moral obligation to take action to protect the future of a planet in peril?*
>
> **Yes, for the stewardship of God's creation.**

In the Judeo-Christian tradition, God created the world in six days. At the end of each day, God stepped back, surveyed the day's work, and pronounced it "good." Good, period. Good without qualification. Not good to the extent that it serves human needs, not good because God himself will find it useful. Just good. Period. No footnotes. Similarly for Muslims. Allah gives a single command, "Be," and it is. In indigenous religions around the world, deities created the seas, the deserts, the mountains, the flowing water and life-giving springs. All of this Earth—and not just the planet, but the firmament, the stars in their wanderings—are shaped by divine hands. We walk to work on God's creation. We pour it into our gas tanks. We feed it to our children, and wash our cars with the work of divine hands. We breathe God's air.

Imagine. How should one live in a world created by God omniscient, omnipotent?

Reverently. Gratefully. Respectfully. Surely all these. But also this: we are called to care for the Earth, to act as stewards of divine creation. To keep it safe and nourish its well-being. To make honorable and grateful use of the divine gifts, so that they may flourish even as we use them.

We can pretend that God would want us to vandalize the planet, trample it and mine it out, on the grounds that we are created in God's image and so have special standing on Earth. We can pretend that we know better than God how to arrange the workings of the world, and alter and simplify them without consequence. But if God is perfect, and God created the world, it might be reasonable to conclude that His Creation has value beyond what humans can create. To maintain that value, to protect and steward it—that is the work and the privilege of humankind.

A Manifesto to North American Middle-Class Christians

Sallie McFague

Preamble

It is time for an Ecological Reformation. The Protestant Reformation and Vatican II brought the importance of the human individual to the attention of Christians. It was a powerful revolution with many impressive religious and political results. But our current version of this model—the individualistic market model, in which each of us has the right to all we can get—is devastating the planet and making other people poor. This model is bankrupt and dangerous. We now need a new model of who we are in the scheme of things and, therefore, how we should act in the world.

The Individualistic Model

The model of human being as individual, which comes from the Protestant Reformation and the Enlightenment, is deeply engrained in American culture. It is the assumption of our Declaration of Independence, symbolized by the phrase "life, liberty, and the pursuit of happiness." These goals are oriented to *individuals*—to their

---

SALLIE McFAGUE is the E. Rhodes and Leona B. Carpenter Professor of Theology Emerita at Vanderbilt University and Distinguished Theologian in Residence at Vancouver School of Theology. As an eco-theologian, she is a prolific author whose most recent publication is *A New Climate for Theology: God, the World, and Global Warming* (2008).

rights and desires. Likewise, American Christianity has been focused on individual well-being, either as salvation of believers or comfort to the distressed. This model of human life supports the growth of deep-seated assumptions about who we are and what we can do: we are a collection of individuals who have the right to improve our own lives in whatever ways we can. We see ourselves as basically separate from other people, while acknowledging the right of others also to improve themselves to the best of their abilities. This picture of ourselves is so deeply embedded in our culture that for most people it is simply "the way things are." It is seen as a description of the way human beings should respond to other people and to nature. But it is not a description; it is a model, a *way* that increasingly is proving to be harmful to most of the world's people and to nature.

The individualistic model of human life, which is now joined by twenty-first century market forces, has resulted in rampant consumerism by North Americans. We feel we have the *right* to whatever level of material goods and comfort we can amass for ourselves. The top 20 percent of the world's people now account for 80 percent of private consumption. We middle-class North Americans are a large part of that 20 percent. What began centuries ago as a reformation to free the human individual religiously and politically has resulted in our present model of human life, which is on course to ruin the planet.

Destruction and Denial

However, this ruinous course is not common knowledge because the market forces benefiting from the individualistic model do not want people to slow down consumption or to question what overconsumption is doing to our planet and to poor people. There is very little public discussion of the key consequences of this model: its contribution to global warming, to the increasing gap between the rich and the poor, to the extinction of other species, and to the rapid decline of natural resources. We are being kept in denial about the seriousness of these major global issues by powerful business lobbies and timid politicians, *but also* by our own reluctance to disrupt the most comfortable lifestyle that any people on Earth have ever enjoyed. Even the recent economic meltdown has not revised our view: the solution to the financial crisis appears to be more growth, more consumer spending, more destruction of natural resources.

Global Warming

Yet once the curtain is drawn aside and we take a peek at the deeply disturbing consequences of one critical issue driven by our voracious consumerism, we may begin to question our model of human life. Climate change is now up close and personal. The reports coming from the United Nations Intergovernmental Panel on Climate Change are grim indeed. We now know that we may be approaching the tipping point when out-of-control global temperature increases will change life on planet Earth beyond any of our imaginings. Floods, droughts, hurricanes, tornadoes, and other weather events are becoming more frequent and extreme, as well as large-scale Arctic melting. Climate is the largest, most complex, and most important of all Earth systems. Even minor changes have implications for food production, disease, water scarcity, air quality, desertification, biodiversity—as well as justice between peoples. Since global warming is the result of energy use, especially the burning of fossil fuels, the 20 percent of the world's people who use five times the energy of the rest of the world's population should pay special attention to this issue. We are creating a worldwide problem that heavily impacts others, specifically poor people and nature. Our cars and airplanes especially, but also the multitude of other ways energy makes our lives comfortable and privileged, must come under scrutiny. *Should* we live this way if it is achieved at the expense of nature's well-being and that of other people? *Can* we continue to live this way?

A cold, hard look at the individualistic market model of human life says, "No—we should not and cannot." We *should* not as Christians because this model is contrary to our clearest and deepest traditions: that the Redeemer is also the Creator, that God is on the side of the oppressed, that God loves the world and dwells in it. We *cannot* continue to live this way because our planet is no longer able to support such a model of human life. If all people on Earth were to live as North Americans do, we would need four more Earths to produce sufficient energy.

The Ecological Model: Who We Are

The individualistic market model has failed us: it has limited religious viability and is proving to be dangerous to our planet, both environmentally and economically. We need another model of human life: we need an Ecological Reformation. An Ecological Reformation would base its model of human life on how reality is understood in *our* time. The individualistic model arose several hundred years ago and is no longer supported by the science of our day. The picture of reality emerging from cosmology, evolutionary biology, and ecology focuses on relations and community, not on individuals and objects. According to this picture, everything in the universe has emerged from a Big Bang 15 billion years ago when a tiny bit of matter exploded and over billions of years became the galaxies and stars and—about 4 billion years ago on our planet—life itself. All life grew from one cell into millions of species, into the rich, diverse, and infinitely interesting forms we know—from mushrooms and mice to wheat and giant cedars, from fungi and frogs to chimpanzees and human beings. *We are all related;* we all came from the same beginning. There never has been a grander, more awe-inspiring creation story, and it is available to Christianity and to other religious traditions as a way to reimagine God the Creator in twenty-first-century terms.

This story also provides us with a new model of human life, one that is based on the best science of our day—in other words, on reality as presently understood. In this story, human beings are not individuals with the power to use nature in whatever ways we wish. Rather, we are *dependent* on nature and *responsible* for it. In a sharp reversal, we do not control nature, but rely utterly on it. In this picture, human beings are products of nature and depend on it for our every breath and bit of food. We cannot live for more than a few minutes without air, a few days without water, a few weeks without food. The rest of nature does not, however, depend on us; in fact, if human beings were to disappear from the Earth tomorrow, all plants and animals would be better off. We are among the neediest, the most vulnerable of all Earth's creatures, dependent on nature's gifts every moment of our lives.

New House Rules: What We Should Do

Our radical dependence on nature means that we are also responsible for it. As the species currently laying waste the planet—and aware that we are doing so—we must accept responsibility for our actions. The ecological model of human life tells us not only who we are but also what we must do; it gives us guidelines on how we should act. In other words, it is a *functional* creation story, one that has practical implications for how we live at personal and public levels.

We could call these implications our new "house rules." The common creation story tells us that the Earth is our home—it is where we evolved and where we belong. It also tells us what we must do for all of us to live decently and happily here. House rules are what one pins on the refrigerator as guidelines for sharing the space, the food, the resources of the home. The basic rules are (1) Take only your share; (2) Clean up after yourself; and (3) Keep the house in good repair for future occupants. The ecological model comes with some definite house rules, clearly seen in the fact that "ecology" and "economics" come from the same word root having to do with laws for living in a household. The basic rule is that if everyone is to have a place at the table, the limits of planetary energy must be acknowledged. Energy that is consumed is not re-cycled; it goes into the atmosphere as carbon dioxide (one of the main causes of global warming). The house rules of our home set limits to growth—of both our consumer desires and our greenhouse emissions. We need, then, to become "ecologically literate," to learn what we can and cannot do if our home is to continue to exist in a sustainable way. We must fit our little economy into the Big Economy, Earth's economy, if our economy is to survive. Thus, the twin crises of the environment and the financial markets point in the same direction—to a model of human living that respects the laws and limits of the planet.

Christianity and the Ecological Model

As Christians we need to do all this and more. The Protestant Reformation and Vatican II supported the model focused on the well-being of the human individual. Protestants and Catholics should also support the ecological model. It *is* the picture of reality in our time, and Christianity has always been most effective when it has reconstructed its doctrines in light of reality as currently understood. The new model is also religiously rich and suggestive; it has enormous worship and liturgical potential. But of greatest importance, it is in profound agreement with the two deepest traditions in Christianity—the sacramental and the prophetic. This new model, which could be summarized by a version of Irenaeus's watchword—the glory of God is every creature fully alive—provides Christians with new ways to say that God is *with* us on the Earth and that God is *for* us, especially the oppressed. The new model suggests to Christians that the way to picture God's presence with us is the eschatological banquet to which all are invited, all people and all other creatures.

The Abundant Life

The ecological model, then, suggests a new vision of the "abundant," the good life. It is not and cannot be the consumer model of individual gain; it must be a shared life where the rich must live more simply, so that the poor may simply live. We must envision models of the abundant life based not on material goods, but on those things that really make people happy: the basic necessities of food, clothing, and shelter for themselves and their children; medical care and educational opportunities; loving relationships; meaningful work; an enriching imaginative and spiritual life; and time spent with friends and in the natural world. In order to move toward this good life, we will need to make changes at every level, personal, professional, and public—how we live in our houses, how we conduct our work lives, and how we structure our economic and political institutions. It is a life that for us North Americans may well involve limitation and significant change at our level of comfort. Christians might see it as a form of discipleship, a cruciform life of sacrifice and sharing burdens.

A Call to Action

The Ecological Reformation is the great work before us. The urgency of this task is difficult to overstate. We do not have centuries to turn ourselves around and begin to treat our planet and our poorer brothers and sisters differently. We may not even have the next century. But the scales are falling from our eyes, and we see what we must do. We must change how we think about ourselves, and we must act on that new knowledge. We must see ourselves as both radically dependent on nature and as supremely responsible for it. And most of all, we North American privileged people who are consuming many times our share at the table must find ways to restructure our society, our nation, and the world toward greater equitability. Christians should be at the forefront of this great work—and it is a *great* work. Never before have people had to think about the well-being of the entire planet. We did not ask for this task, but it is the one being demanded of us. We Christians must participate in the agenda the planet has set before us, and do so in public and prophetic ways, as our God "who so loved the world" would have us do.

God's Passion in the Bible: The World

Marcus J. Borg

Write about what you know, they say. And so as a Christian whose academic specialty is biblical scholarship, I write about resources within the Bible for an environmental ethic.

The record of Christianity with regard to the nonhuman world, the world of nature, is, at best, mixed. Some forms of Christianity emphasize the afterlife so much that the Earth doesn't matter very much. Some forms see "the world" primarily as a source of temptation. Forms that expect the second coming of Jesus soon commonly see the world as something that will be destroyed and replaced by a new world. These forms do not generate an environmental ethic—in them, the nonhuman world doesn't matter very much.

Moreover, the wedding of Western Christianity to Western culture has intensified the notion that we humans are entitled to exercise dominion over nature. For many Christians since the Enlightenment, selected biblical passages have been used to legitimate the notion that nature is for us and our use. It does not have intrinsic value, only instrumental value.

The most commonly quoted passage is Genesis 1:26–28. There, humans are given "dominion over the fish of the sea, and over the

---

MARCUS J. BORG is Hundere Distinguished Professor of Religion and Culture Emeritus at Oregon State University. Known as one of the leading historical Jesus scholars of this generation, he is the author of seventeen books, four of which have become bestsellers. His most recent publication is *Jesus: Uncovering the Life, Teachings, and Relevance of a Religious Revolutionary* (2008). His works have been translated into nine languages, and he lectures widely in North America and overseas.

birds of the air, and over the cattle, and over all the wild animals of the Earth, and over every creeping thing that creeps upon the Earth." The passage concludes with God's words addressed to the first humans: "Be fruitful and multiply and fill the Earth and subdue it."

In the same way, Psalm 8 has been understood to give humans dominion. Verse 4 asks, "What are human beings that you are mindful of them?" Verses 5–8 answer: we are but "a little lower than God," and God has given us "dominion over the works of God's hands; God has put all things under our feet, all sheep and oxen," and everything else.

To this day, "dominion" Christians emphasize these verses. They see the modern Western dominion of nature as the will of God. This voice from the Bible is emphasized because it accommodates Western Christianity to the central dynamic of Western culture for the past few centuries.

And so it is important to realize that there is another voice in the Bible, one that has been muted in Western Christianity and Western culture ever since we began to "master" nature. This voice emphasizes that the world—all of it, including the nonhuman world—matters to God. Indeed, it matters passionately to God.

As we turn to this other voice, we need to remember that there was no perceived environmental crisis in the ancient world as there is in our world. The domination of humans over nature was minimal, and their frail mastery was seen as a boon.

We also need to remember that the authors of the Bible addressed the issues of their day. For them, the central issue was the domination of humans by other humans—injustice and unnecessary human misery embedded in social systems. Nevertheless, there are major biblical themes that can be the foundation of a robust environmental ethic for our time.

To begin with the beginning: the story of creation in the first chapter of Genesis emphasizes again and again that the whole of the created world is good. The refrain resounds throughout the story: after each day of creation we are told, "And God saw that it was good." At the end of the story, on the sixth day, we are told, God saw everything that had been made and pronounced it "very good."

This is the same chapter in which humans are given "dominion" over nature. Yet it is important to know that the word used is the same as the one used to characterize the relationship of shepherds to their sheep. That relationship includes caretaking and protection. The good shepherd is not one who devours all the sheep.

The biblical understanding of creation is not just about origins but about ownership. To whom does the Earth belong? Psalm 24 emphatically answers, "The Earth is the Lord's and all that is in it, the world, and those who live in it."

To say the obvious, we have most often treated the world as if it belonged to us—to us as a species, though to some of us more than to others. But the Bible's affirmation is quite different: the Earth belongs to God, not to us. The Earth is the Lord's, not ours—not ours to use as we wish.

Moreover, in the Bible nature is sometimes seen as the revelation of God, the sacred, terms that I use interchangeably. The nonhuman world is seen as filled with the glory of God—with the radiant presence of God. Isaiah 6:3 affirms, "The whole Earth is full of God's glory."

So also in the concluding chapters of the book of Job, which contain perhaps the most magnificent nature poetry in the Bible. In Job 38–42, the main character is given a spectacular display of the grandeur of nature. Then he addresses God and exclaims, "I had heard of you by the hearing of the ear—but now my eyes see you" (42:5). He has seen God—and what has he seen? The magnificence of nature.

In the New Testament, Jesus sees in nature the lavish generosity of God. God feeds the birds of the air and clothes the lilies of the field with a glory exceeding that of Solomon (Matthew 6:25–33, Luke 12:22–32). Nature not only is created by God and pronounced good by God, but also reveals God.

John 3:16, perhaps the most famous verse in the New Testament, proclaims, "For God so loved the world . . ." What does God love? Not just me. Not just you and me. Not just Christians. Not just human beings. Rather, God loves *the world*.

I conclude by speaking of God's passion. The word "passion" among Christians has both a narrow meaning and a broader meaning. The narrow meaning refers to suffering, as in the story of Jesus' suffering and crucifixion. The broader meaning is what we mean when we ask of somebody, "What is your passion?" What do you care most intensely about, what is your dedicated commitment, what is your abiding concern?

The answer of this alternative voice within the Bible is clear: God's passion—what God is most passionate about—is "the world." We as humans are spoken of as the crown of creation. But that does not mean that we are apart from creation, like lords reigning over creation. Rather, the Bible calls us to participate in God's passion—and God's passion is the whole of creation. The world—not just human beings—matters to the God of the Bible.

God's passion for the world includes its future. To use a nonbiblical but theological word, eschatology in the Bible is not about the end of the world, not about the destruction of the world, but about its renewal and restoration. The world is good—God will never destroy it. We could—and if we follow that path, God will not intervene to rescue us. Instead, the future of the world will be the result of our collaboration (or lack of collaboration) with God.

Archbishop Desmond Tutu, paraphrasing Augustine, another African bishop from sixteen centuries ago, speaks of divine-human collaboration in a memorable aphorism: God without us will not; we without God cannot. For the Bible, the world and its future matters to God, and we are called to participate in God's dream for the Earth.

Our Obligation to Tomorrow

Seyyed Hossein Nasr

As we contemplate our responsibility toward the future, we must first of all recall that we can think and act in the present only, because we have been endowed with a consciousness and a memory that enables us to be aware of the past and to some extent anticipate and foresee the future on the basis of our knowledge of the past and our thoughts and actions in the present. We cannot, however, determine and foretell the future completely, as the failure of strictly controlled social engineering and futuristic predictions by human beings in recent times have amply demonstrated. As for eschatological predictions of the future, they are based not on human conjecture but on revelation. Moreover, they purposely contain an ambiguity from the human point of view and always leave a space open for free human action, as we can see in the Puranas in Hinduism, the Revelation of John in the Christian Bible, and the Qur'an and Hadith in Islam. At the same time, these and other religious texts emphasize that we weave the fabric of our future through our actions here and now, although there is always a vertical dimension that allows us to transcend the horizontal web of causes and effects, or what the Hindus call "karma."

SEYYED HOSSEIN NASR is president of the Foundation for Traditional Studies and a professor of Islamic studies at George Washington University. He is one of the world's leading experts on Islamic science and spirituality and has authored over five hundred articles and fifty books, including *Knowledge and the Sacred* (1989), *Religion and the Order of Nature* (1996), and *Man and Nature: The Spiritual Crisis of Modern Man* (1998). The Seyyed Hossein Nasr Foundation propagates spirituality through perennial philosophy and traditional teachings contained in the Qur'an.

Furthermore, traditional religions emphasize that the future is determined not by our actions alone but also by the Transcendent Cause, which can penetrate in a manner that is beyond our understanding and control into the spatiotemporal reality that we call the world.

This Divine intervention does not absolve us, however, of our responsibilities in the present or toward the future, nor from the need to preserve our spiritual heritage. Rather, we are held responsible for our own individual future as well as the world's future, to the extent that we can affect it. In the Qur'an human beings are called God's vicegerents on Earth, responsible for the preservation of the trust He has left in our hands, a trust associated with the world of nature as well as the human world in which He has placed us, not to speak of the religion that has been revealed to us. We have responsibility in God's eyes to the extent of our possibilities for the care of ourselves and of our neighbors, whether they be human beings, animals, plants, or minerals, the deep seas or the rivers or mountains. We are also responsible for the preservation of what we have inherited spiritually, intellectually, artistically, and socially from the past as well as for the preservation of the natural world in all its wonders, the natural environment with the remarkable harmony of its ecological systems that we have inherited from the past.

For most of us, what we have inherited from the past determines "our world," the world in which we think, plan, and act. The ordinary person is often passive vis-à-vis the "world" even if there is aggression on the human level and also against nature, as we see today all around us. In fact, this aggression is itself the fruit of the worldview inherited from the earlier modern world, which still dominates the present-day world and molds the manner of most people's thoughts and actions. It is only the spiritual hero who, instead of being dominated and determined by the world around him, dominates and determines that world, but in a positive manner. And it is such a person who must be the vanguard and guide of those who will be able to understand the significance of the traditional worlds whose heritage is so crucial for us today and in the future, at the very moment when so many try to deny their importance while clinging to the errors of the modern worldview. Moreover, it is they who can see clearly the spiritual significance of the natural world, beyond all immediate material benefits, and the absolute necessity to protect nature against the greed and shortsightedness of so many present day human beings if there is to be any future at all for human life on this Earth, which was a mirror of paradise when we human beings came upon the scene.

Human beings are conscious that they will die, that there is a time when they will no longer be alive here on Earth. This basic knowledge is shared by those who believe in the reality of the spiritual world and the afterlife as well as those who do not. The majority of people in both categories have always felt a sense of responsibility toward the future as far as concerns their own lives, their family, or in certain cases their country or the fruits of their creativity and achievement. Today, when everyone speaks much of freedom and little of responsibility, even this natural sense of responsibility toward the future of one's own world has weakened. And yet it seems that the future is crying ever louder to make us aware of the unprecedented responsibilities we have toward it.

But why should we respond to this call? Why be responsible for the future? For those who follow a religious tradition, the answers to these questions come from the teachings of their religion, and they do not have to provide philosophical responses, although many religious philosophies and theologies have dealt extensively with this matter. But even for most of those who deny the Sacred and accept only a completely secularized world, there are reasons for feeling responsible for the future that derive from the innate nature of being human. Even an atheist wants to guarantee a good future for himself or herself in later life and desires that his or her children and grandchildren live in a sane and healthy world. Moreover, even those who deny the category of the sacred often refer to the sacredness of life and the responsibility to preserve it, although strictly speaking that sentence has no scientific meaning if we accept the scientistic worldview. In any case, the believer and nonbeliever alike feel responsible to the extent possible for the survival into the future of human life, of the great works of art and thought belonging to the past and present, of the ecological health of the planet and of their own heritage to be bequeathed to their family and community. As for those who believe in some form of messianism and the immanent approach of eschatological events, they also in a sense believe themselves responsible at least for their own future, even if they understand the future in a different way. Perhaps they should heed the advice of the Prophet of Islam, who said that in the eyes of God it is a blessed act to plant a tree, even if it be a day before the end of the world.

We cannot be moral human beings without having a sense of responsibility for the future within the orbit of our possibilities. Only the most selfish of human beings could repeat the famous French saying "Après moi, le déluge" (After me [let there come] the deluge). How we feel about this responsibility depends on who we are and how we think, and this in turn is colored to a large extent by what we inherit, learn, and often choose to adopt from the past. The heritage of the past includes the perennial wisdom of traditional civilizations and their mode of living in harmony with both Heaven and Earth, as well as five centuries of modernism that began in Europe and has now spread to the whole globe—modernism that has been marked by the rise of a secularist humanism, the desacralization of nature leading to a science based on the domination of nature, the Industrial Revolution, and the rape of the natural world all joining hands in landing us at the edge of a precipice, the fall into which could mark the end of human life on Earth. In such a situation we cannot but exercise discernment about what we have inherited from the past, what we are responsible for preserving, imitating, and cherishing, and what we must discontinue if we are to survive in the future.

Those in previous generations who had a sense of responsibility usually thought in terms of their own lives and that of their extended families and peoples or tribes, the survival of their religions and traditions, and the continuation of their intellectual traditions, along with their arts and sciences. They, unlike us, did not have to worry about global warming, the loss of species, the destruction of whole ecological systems, the pollution of the air they breathed and the water they drank, and the poisoning of their food. They did not feel responsible for the preservation of the natural world for future generations. In light of the horrendous errors made due to the hubris of modern man, combined with mastery of a science based on complete human domination over nature considered in only its material and not its spiritual aspect, the dimensions of our responsibility toward the future must now change completely if we are to have a future at all. There must be a prioritization of our responsibilities. Above our concerns for the material welfare of ourselves, our families, our communities, and our nations must stand our concern for the welfare of the natural environment and the need for our immediate response to the agonizing cry of the planet, suffering in the hands of a humanity in rebellion against Heaven and busy with the corruption of the Earth in the name of such lethal and dangerous concepts as continuous material progress, endless economic growth, and ever-greater so-called development.

We do not know exactly what the future will bring us. Even for those who do not believe in God or the possibility of His affecting His creation, there are too many factors involved for us to be able to predict the future precisely. But to anyone with enough intelligence to know that if you light a piece of cotton it will burn, it is obvious that barring a Divine intervention (which in any case is beyond us and does not absolve us of our responsibilities), we are marching toward a catastrophe that threatens all human futures. This itself is a new experience to which we must be completely attentive. Indeed, we are responsible for how we interpret and use our heritage from the past that also affects to a large extent our patterns of thought and action in the present, for of course our mode of being and consciousness now determines our actions in the present and therefore affects the future.

At the present moment of history, however, at this perilous time when planetary survival is threatened, the normal time line seems to have been reversed. There seems to have come into being a process in reverse, moving from our projections of the future to our present-day actions. Since our intelligence has the power of anticipation, we can in a sense hear anew the cry of the future, beckoning us to heed its call and fulfill our responsibilities toward it. And it is this call to which we should be attentive; this call more than anything else should determine the responsibilities we bear in our thoughts and acts at the present moment and toward our heritage from the past that we must interpret in light of our responsibilities to the future. Without hearing this cry of the future and fulfilling our responsibilities to it in the present on the basis of the judicious use of the traditional wisdom that we have inherited from the past and that, along with the rejection of the errors of yesterday, continues to be of the greatest pertinence today, there will simply be no future for the human species here on Earth. But as traditional Muslims say, "And God knows best," especially when it comes to the question of the future, which is ultimately in His hands but toward which we also bear a very heavy responsibility as His vicegerents here on Earth.

The Biblical Mandate for Creation Care

Tri Robinson

In the 1960s I was a student on a university campus along with thousands of other baby boomers. Everything was in cultural flux, and our generation was confused, scared, angry, opinionated, and passionate. We not only wanted change, we demanded it. We were wrapped up in a ruthless war that no one seemed to understand, our president was being exposed for dishonesty, and our environment was showing signs of becoming nonsustainable. We were looking for meaning and truth—but most of all we wanted authenticity. During that time, many of us found our answer not in religion, but instead in an authentic faith in Christ. History later referred to us as the "Jesus movement," a movement of young, Bible-believing, nondenominational evangelicals. We put our faith and hope in God, and we were no longer angry or scared—although we were still very opinionated and passionate, just about different things. Unfortunately, our passion for the environment was lost in a new focus on the second coming of Christ. Today, the things we once feared as young seekers are now happening: the world population has doubled, and the once seemingly plentiful natural resources of fresh water, soil, air, and energy sources have begun to decline. The extinction rate of endan-

---

TRI ROBINSON is the founding pastor of the Vineyard Boise Church in Boise, Idaho, a growing fellowship of over three thousand. He has served on the national board for the Association of Vineyard Churches USA and as a regional overseer for more than one hundred churches. His most recent book is *Saving God's Green Earth: Rediscovering the Church's Responsibility to Environmental Stewardship* (2006).

gered species has escalated, and even the global climate has started to change, threatening the future of our planet.

I am a Christian, and not just a Sunday Christian, but a passionate evangelical Christian pastor of a thriving church. I believe in the Bible with all of my heart and have diligently tried to mold my life around its truths. I believe that Jesus is coming again but sincerely can't claim to know when. I believe that his Kingdom (that is, the Kingdom of God) has come to Earth as it is in heaven. I believe that it came because of his ultimate sacrifice on the cross for humankind. I believe that he has called his people to stewardship in his Kingdom, to care for the people he loves, specifically the poor and broken, and for the Earth he lovingly created. I believe that this responsibility is not just a suggestion but a commission and a mandate without option.

The Bible is an amazing book, and it has a miraculous characteristic: you can read it over and over again and never get tired of it, always extracting something different with each reading. God has a way of illuminating his never-changing truths with new understanding. As a committed student and Bible teacher I have read and taught the story of the flood numerous times. I always read and saw it the same way until one morning in 2005. As I read Genesis 9 that morning, the Lord began to reveal something I had never seen before. This portion of scripture tells of God's covenant with Noah to never again destroy the whole Earth with a flood. God said, "I establish my covenant with you: Never again will all life be cut off by the waters of a flood; never again will there be a flood to destroy the Earth" (Genesis 9:11). As I carefully examined the passage I realized for the first time that God's covenant was made not between God and Noah or even God and humanity, but between God and his entire creation. God said, "I now establish my covenant with you and with your descendants after you and with every living creature that was with you—the birds, the livestock and all the wild animals, all those that came out of the ark with you—every living creature on Earth" (Genesis 9:9–10). Not only did he say it, but he repeated it six times in a row. He really meant it, and to me that was profound.

As God laid out the conditions of this covenant agreement he spoke a number of mandates about which everyone should know. First of all, this was to be a covenant of blessing to humanity. If his mandate was observed, the world's environment would provide for all humanity and would allow humans to inhabit and populate the Earth. He said, "Everything that lives and moves will be food for you. Just as I gave you the green plants, I now give you everything" (Genesis 9:2). Secondly, his covenant demanded responsibility and an accounting that we would use the Earth, but not abuse it. He called us to the sanctity of all life. He said, "I will demand an accounting from every animal. And from each man, too, I will demand an accounting for the life of his fellow man" (Genesis 9:4). Also, his covenant was not to end with Noah's generation or even with the Old Testament law, but was to be everlasting—for all generations to come. He said, "This is the sign of the covenant I am making between me and you and every living creature with you, a covenant for all generations to come: I have set my rainbow in the clouds, and it will be the sign of the covenant between me and the Earth" (Genesis 9:12–13).

This changed everything for me. No longer could I separate my passion for the Kingdom of God from my commitment to care for the environment. I had to tell everyone that Christians not only should care about creation, but had been mandated by God to be leaders in a worldwide environmental movement. I shared this mandate with my church, I wrote a book on caring for creation (*Saving God's Green Earth*), and I became an advocate of Genesis 9 on radio and television. Together with other like-hearted people I started a ministry called Let's Tend the Garden. I wrote tirelessly for all kinds of publications and spoke nationally to anyone who would listen. I told everyone that caring for creation is not an option but a commission, especially for those who value and believe God's word in the Bible. Many have listened, and others who shared this conviction have joined in an effort to change the minds and practices of Christians worldwide. Over one-third of the world's people say they profess Christ as savior and believe that the Bible is true. That is nearly 2.5 billion people who, if they recaptured this God-given mandate, could unite to make a lasting difference due to obedience to an ancient but culturally relevant truth.

Will Religions Guide Us on Our Dangerous Journey?

Martin S. Kaplan

We have been living in an era of selfishness dominated by short-term thinking—two- and four-year electoral cycles, quarterly and annual financial results, and values that reflect material consumption and immediate gratification. With little regard for the future, current generations are using up more than our share of the Earth's resources, as prosperous people throughout the world express their values by purchasing as many things as possible, whether needed or not.

The long term—a future beyond our own lives and those of our family—is almost beyond our comprehension. Analysts of population growth generally conclude their estimates around the year 2100, as if we either cannot project further or have no responsibility beyond then. But 2100 is within the lives of many of our children and grandchildren. What will their lives be like at the end of this century? And what will their expectations be in 2100 for their own children and grandchildren?

Public discourse at the beginning of the twentieth century was marked by an unshakable belief in progress toward the rational solution of all problems—now simply a disconcerting and charming

---

MARTIN S. KAPLAN, a retired partner of the international law firm Wilmer Cutler Pickering Hale and Dorr, received the 2009 Thomas Berry Award in recognition of his role in developing the field of religion and ecology. He has been an active leader in interreligious affairs with the American Jewish Committee, chair of the Massachusetts Board of Education, and a leader in many nonprofit organizations and foundations relating to the environment, education, the arts, and human rights.

notion. No one at that time could have had even a glimmer of awareness of what the world would be like now, a hundred years later.

And here we are, at the start of the new millennium, still thinking about the future as if that means ten or twenty years forward. But as we are aware of the remarkable history of humanity, are we capable of looking further down the road? Can we ensure that human life will survive another millennium? Can we contemplate the state of the world five hundred years in the future? Two hundred years? Are we not morally required to do so?

We live in a world we did not create and must not destroy. Environmental issues respect no borders. Human life on our planet is interdependent with other living organisms. Global warming can destroy the world as we know it, accelerating the ongoing crash of biodiversity that can lead to a crash of all life. Prime Minister Anders Fogh Rasmussen of Denmark, speaking at the 2008 United Nations General Assembly, called this the "grinding catastrophe of global warming." John Holdren, the science adviser to the president, has made the frightful statement that "the human response to climate change will be threefold: mitigation, adaptation, and suffering, and we will have a great deal of all three."

Scientists, policymakers, and economists all agree that the dangers can be abated. But the missing ingredient has been the willpower of governments and leaders to recognize the gravity of the risks, to understand the applicable science, and to address the problems—all of which are necessary to implement possible solutions.

However, world leaders are finally waking up to the dangers of climate change and to the ongoing destruction of the intricately balanced biosphere that we share with other species. With a deep awareness of the requirements of human life and indeed all life on Earth, thoughtful institutions have begun to address issues that relate to climate and the environment. The hope that humanity can, and indeed will, sustain life on this planet rests with those individuals and institutions that will marshal the intellect and resources to study and address the central environmental challenges that will determine the human future.

Throughout history, people have turned to their ancient religious faiths for guidance and support when confronted with crises. But the responses of organized religions to great cultural and social movements and shifts have not always been positive. Witness the Christian churches' divided approach to the civil rights movement in America in the 1950s and 1960s, and the tensions within all religions as societies try to respond to demands for women's rights and gay rights, movements that face stone walls of resistance from many religious leaders and people of faith. People hear confusing, contradictory, or hostile responses from their religious leaders to the challenges of change.

The response to global warming surely will bring immense change to life as we know it. How will religious faith communities react? All faiths should share a desire to preserve the lives of all living organisms, humans included, and the Earth as a biosphere. As the crisis grows, will people find sources of moral guidance in their religious institutions?

Aside from those who believe that the ultimate catastrophe of climate change may be the promised apocalypse, will any religious traditions find reason to argue *against* saving human life, all species, and the biosphere? Will religious institutions be concerned only when it affects their own faithful? Will people look to religion solely for their own individual salvation, support, or comfort? This is the individualistic or selfish aspect of connection to a religion: the human habit of using religion only for themselves. Will people seek guidance and support only for their own communities, whether defined by faith, ethnicity, or geography? Or for all peoples? And all species? Will this vary according to the worldview of each religion?

To put it starkly, if and when 120 million people are endangered, even more than at present, by the rising waters in Bangladesh, and refugees seek welcome elsewhere, will religious leaders worldwide rise to the challenge and provide moral leadership for supportive action? Will the ministering classes of all faiths respond with a recognition of the importance of their role, just as scientists have done in warning us all of the impending and, indeed, present threats? Will religious leaders respond to such disasters better than they have in Bosnia? Rwanda? Darfur? New Orleans? Or will they disregard the plight of others as not their concern, or as not caused by human activity? Will the reaction by religious leaders in America to rising waters in Bangladesh be different from their reaction to rising waters in Florida?

These are the challenges facing the human species. We hope that our religious faiths will provide the needed leadership so that societies and nations will respond with willpower and vision to reduce the suffering that Holdren warned of, empower efforts to mitigate the effects of climate change, and help people to adapt to potentially disastrous change.

Religions are the most ancient formulators of culture and values in the world. They are the primary source of ethics for humans around the planet. In spite of worries (or hopes) that religions would someday disappear with increased secularization, their power and influence in all societies throughout the world has in fact grown over the last half century. Faith communities have immense potential to provide the value structures to change consciousness and behavior as we face existential threats to the very survival of the biosphere and human life.

Just as no one in 1500 could foresee the shape of human destiny five hundred years hence, we at least now know that five hundred years is a measurable time frame in the human existence. If life is to continue as we know it until the year 2500, the necessary changes will be monumental. Will religions guide us on our dangerous journey?

ETHICAL ACTION

How can we protect God's creation?

Maybe it's time to think again about "sanctuary." It's not just the nave of a church; it's a place of safety. Are the sweeping lawns of your church a safe place for divine creation? Can they become a true sanctuary, bursting with birdsong and native plants, protected forever from poisons and bulldozers? And after your church becomes a true sanctuary for divine creation, what about the lands owned by parishioners? Imagine a sanctuary movement of a different sort.

And what about this notion of "preaching to the choir"? Every week we come together to sing God's praises. But what can we do to protect the songs of creation? Can the choir become the protectors of birdsong, the protectors of marshland rich with the calls of frogs? There are many kinds of music in the world. Is it enough to sing about them as they vanish? If those with beautiful human voices don't act to protect the voices of the birds and the frogs, who exactly will?

Can an omnipotent being weep? Can we even begin to imagine the ferocity of divine grief when the last varied thrush vanishes or when the last flecks of colored fish fade from a coral reef? Can there be a dishonoring of God greater than this—to disregard the love of God for the small lives He has shaped with His hands? People of the church know how to spread the word of God. Spread this word—that the Earth is the Lord's Creation and our work in the world is to keep it safe.

9

> *Do we have a moral obligation to take action to protect the future of a planet in peril?*
>
> **Yes, because compassion requires it.**

Of all the virtues that a human being can possess, the greatest may be compassion. Compassion: to "feel with," to imagine ourselves in another person's place. Understanding the joys or sufferings of others, the compassionate person is joyous or suffers too. Thus the truly compassionate person acts in the world, striving to create conditions that bring forth joy and to prevent or diminish conditions that create pain.

Among the calamities of environmental degradation and climate change is an increase in human suffering and the suffering of other sensate beings. Climate change disrupts food supplies, destroys habitat, reduces or contaminates drinking water, spreads disease, increases the terror of storms, floods entire nations, and slides villages into the sea. The price of the reckless use of fossil fuels will be paid in human and animal suffering. Even worse, the more blameless people are for creating the conditions that lead to suffering, the more likely they are to suffer.

If virtuous people are compassionate, if compassionate people act to reduce suffering, and if climate change will cause terrible suffering around the world, then we who call ourselves virtuous have an inescapable obligation to the future to avert the effects of the coming ecological storm.

Winter Wheat

Libby Roderick

1.

When I was young I thought that failure was impossible
All wrongs would be righted in my time.
Now I am old I see that failure is impossible
I pass the torch to you. Will you hold it high?

For we are sowing winter wheat
That other hands will harvest
That they might have enough to eat
After we are gone.

We will plant shade trees that we will not sit under
We will light candles that others can see their way
We'll struggle for justice though we'll never see it flower
Our children's children will live in peace one day.

2.

When I was young I thought there are no human enemies
People must be taught to hate and fear.

LIBBY RODERICK is an internationally acclaimed singer and songwriter, poet, activist, teacher, and lifelong Alaskan. She is also a faculty member at the Institute for Deep Ecology. The global impact of her song "How Could Anyone" has been featured on CNN, in *Reader's Digest,* and in the Associated Press. In 2003, NASA played her song "Dig Down Deep" on the planet Mars as encouragement to the robot "Spirit." Her six studio albums include *A Meditation for Healing* (1998) and *How Could Anyone* (2005).

Now I am old I know there are no human enemies
But only souls in pain and we can reach them if we dare.

So we are sowing winter wheat
That other hands will harvest
That they might have enough to eat
After we are gone.

Each generation passes like the leaves
On old oak trees whose roots are strong
Each new generation bursts out like the spring
And they will be the ones to carry on.

3.

When I was young I dreamed the Earth was healed and whole
   again
Creatures, trees, and rivers free and wild.
Now I am old I dream the planet healed and whole again
That dream's reborn forever in the heart of each new child.

So we are sowing winter wheat
That other hands will harvest
That they might have enough to eat
After we are gone.

We will plant shade trees that we will not sit under
We will light candles that others can see their way
We'll struggle for justice though we'll never see it flower
Our children's children will live in peace one day.

We will plant shade trees that we will not sit under
We will light candles that others can see their way
We'll struggle for justice though we'll never see it flower
Our children's children will live in peace one day.

Our children's children will live in peace one day.

We Are Called to Help the Earth to Heal

Wangari Maathai

I reflect on my childhood experience when I would visit a stream next to our home to fetch water for my mother. I would drink water straight from the stream. Playing among the arrowroot leaves, I tried in vain to pick up the strands of frogs' eggs, believing they were beads. But every time I put my little fingers under them they would break. Later, I saw thousands of tadpoles: black, energetic, and wriggling through the clear water against the background of the brown Earth. This is the world I inherited from my parents.

Today, over fifty years later, the stream has dried up, women walk long distances for water that is not always clean, and children will never know what they have lost. The challenge is to restore the home of the tadpoles and give back to our children a world of beauty and wonder.[1]

In 1977, when we started the Green Belt Movement, I was partly responding to needs identified by rural women—namely, lack of firewood, clean drinking water, balanced diets, shelter, and income. Throughout Africa, women are the primary caretakers, holding significant responsibility for tilling the land and feeding their families.

WANGARI MAATHAI was born in Nyeri, Kenya, and was the first woman in East and Central Africa to earn a doctorate. She is the founder of the Green Belt Movement, an environmentalist, a civil society and women's rights activist, and a parliamentarian. In 2004 she became the first African woman to receive the Nobel Peace Prize for her contribution to sustainable development, democracy, and peace.

---

1. Preceding material excerpted from Wangari Maathai, Nobel Lecture, Oslo, 10 December 2004, p. 6. Available online: http://greenbeltmovement.org/a.php?id=34.

As a result, they are often the first to become aware of environmental damage as resources become scarce and incapable of sustaining their families.

The women we worked with recounted that now, unlike in the past, they were unable to meet their basic needs. This was due to the degradation of their immediate environment as well as the introduction of commercial farming, which replaced the growing of household food crops. But international trade controlled the price of the exports from these small-scale farmers, and a reasonable and just income could not be guaranteed. I came to understand that when the environment is destroyed, plundered, or mismanaged, we undermine our quality of life and that of future generations. A degraded environment leads to a scramble for scarce resources and may culminate in poverty and even conflict.

Tree planting became a natural choice to address some of the initial basic needs identified by women. So, together, we have planted over 30 million trees. In the process, the participants discover that they must be part of the solutions. They realize their hidden potential and are empowered to overcome inertia and take action. They come to recognize that they are the primary custodians and beneficiaries of the environment that sustains them.[2]

It is thirty years since we started this work. Activities that devastate the environment and societies continue unabated. Today we are faced with a challenge that calls for a shift in our thinking, so that humanity stops threatening its life-support system. We are called to assist the Earth to heal her wounds and in the process heal our own—indeed, to embrace the whole creation in all its diversity, beauty, and wonder. This will happen if we see the need to revive our sense of belonging to a larger family of life, with which we have shared our evolutionary process.

In the course of history, there comes a time when humanity is called to shift to a new level of consciousness, to reach a higher moral ground. A time when we have to shed our fear and give hope to each other.

2. Preceding three paragraphs excerpted from Maathai, Nobel Lecture, Oslo, 10 December 2004, p. 2. Available online: http://greenbeltmovement.org/a.php?id=34.

That time is now.[3]

The extreme global inequities and prevailing consumption patterns continue at the expense of the environment and peaceful coexistence. The choice is ours.

I would like to call on young people to commit themselves to activities that contribute toward achieving their long-term dreams. They have the energy and creativity to shape a sustainable future. To the young people I say, you are a gift to your communities and indeed the world. You are our hope and our future.[4]

Africa is the continent that will be hit hardest by climate change. Unpredictable rains and floods, prolonged droughts, subsequent crop failures, and rapid desertification, among other signs of global warming, have in fact already begun to change the face of Africa. The continent's poor and vulnerable will be particularly hit by the effects of rising temperatures, and in some parts of the continent, temperatures have been rising twice as fast as in the rest of the world.

In wealthy countries, the looming climate crisis is a matter of concern, as it will affect both the well-being of economies and people's lives. In Africa, however, a region that has hardly contributed to climate change, it will be a matter of life and death.

Therefore, Africa must not remain silent in the face of the realities of climate change and its causes. African leaders and civil society must be involved in global decision making about how to address the climate crisis in ways that are both effective and equitable.[5]

But the environment degrades slowly, and the changes may not be noticed by the majority of people. If they are poor, selfish, or greedy they will be more concerned about survival or satisfying their immediate needs and wishes than worrying about the consequences of their actions. Unfortunately, the generation that destroys the environment may not be the one that pays the price. It is the future generations

---

3. Preceding three paragraphs from Maathai, Nobel Lecture, Oslo, 10 December 2004, p. 4. Available online: http://greenbeltmovement.org/a.php?id=34.

4. Preceding two paragraphs from Maathai, Nobel Lecture, Oslo, 10 December 2004, p. 5. Available online: http://greenbeltmovement.org/a.php?id=34.

5. Preceding three paragraphs from Wangari Maathai, "On Climate Change." Available online: http://greenbeltmovement.org/c.php?id=17.

that will confront the consequences of the destructive activities of the current generation.

Unless we change course, the coming generations will inherit an impoverished environment that will mean a hungrier, less fertile, and more unstable world. We have a responsibility to protect the rights of generations, of all species, that cannot speak for themselves today. The global challenge of climate change requires that we ask no less of our leaders, or ourselves.[6]

6. Wangari Maathai, "Africa Must Be Heard on Climate Change," 2007. Available online: http://greenbeltmovement.org/a.php?id=251.

An Invisible Killer

Ming Xu and Xin Wei

Climate change comes invisibly and slowly, so many people underestimate its damage to our life and ecosystems. It happens on a global scale that is far beyond our normal sight, thus leaving it unnoticed by the common people. Climate change is a hidden killer that destroys crops and animals, sucks water resources, causes more natural disasters such as floods and droughts, spreads infectious diseases and invasive species, melts glaciers and permafrost, submerges coastal cities and wetlands, and reduces biodiversity and other ecosystem functions and services. Rapid climate change has been with us for decades, and numerous events over the world have provided evidence of the guilt of its killers. The experience of a typical farmer from southern China offers an eloquent example:

MING XU is a professor and research group leader at the Institute of Geographical Sciences and Natural Resources Research, Chinese Academy of Sciences. His research focuses on climate change and ecosystem processes, especially carbon cycle and climate change impact, vulnerability, and adaptation. He is also interested in process-based ecosystem modeling and climate change policy. He has been a leading scientist on a number of major research projects in the United States and China, including international cooperation projects. He is the director of the Global Carbon Project (GCP) Beijing Office.

XIN WEI is a master's student at the Institute of World Literature of Peking University. Her research interests are eco-literature, biography and autobiography studies, cross-cultural studies, and African literature. She is one of the organizers of China's first World Ecocriticism Conference on Ecological Literature and Environment Education, held at Peking University in 2009. She is also the executive director, lyrics author, and translator of this conference's theme song, "Eco-World Green Home."

Fulin Wang, a forty-five-year-old farmer in southern China, is one of the victims of climate change. Fulin and his family live in a small village in Hunan province where their ancestors had farmed for hundreds of years. They traditionally grow rice in the lowland valleys and fruit trees on mountain slopes. With an increasingly warming and drying climate in the past decades, the rice production has declined substantially due to the enhanced plant respiration consumption and the damage from more severe pest and insect diseases and more frequent summer heat stresses. Meanwhile, the farming cost, such as fertilizers, fuels, pesticides, and herbicides, has more than doubled in the past ten years, making rice farming less and less profitable and even losing money in some years.

Given these difficulties, Fulin had hoped to make a living on his mountainside orchards, where he used to grow oranges, cherries, and tea. In 1997 he rented about ten acres of "barren lands" from the village and planted cherry trees on the hillsides. With the cherry trees growing bigger and bigger through his hard work year after year, Fulin hid the happiness in his mind. He even discussed with his wife on how to spend the money made on cherry sales: renovating his old houses, buying himself a new motorcycle, and leaving a major portion as the betrothal "gift" for his twenty-year-old son.

He never thought the invisible killer was around him and his cherry trees. Eight years after he planted the cherry trees, he hardly harvested any fruits, though the cherry trees grew very well and bloomed in some years. He thought the trees were not big enough to bear fruits. But he was very surprised one day, when he was watching TV, to learn that the farmers had a very good harvest of cherries in his neighboring provinces, while he harvested only about two hundred pounds of cherries out of thousands of trees. He started consulting horticulture experts and found that the cherry trees require a certain period of chilling during the winter to overcome bud dormancy. Otherwise, the trees may produce erratic and prolonged flowering, but bear few fruits. The annual mean air temperature in the area where Fulin lives has increased 1.6 degrees C, with winter temperature increased more than 2 degrees C in the past three decades, which cannot meet the chilling requirements for cherry trees to bear fruits.

Learning that this human-induced global warming might continue for the coming decades due to the persistence of the greenhouse gases in the atmosphere, Fulin cut all the cherry trees four years ago and replanted with eucalyptus trees, a pulp tree species introduced from Australia, to take the advantage of global warming. Driven by the fast-growing paper industry, eucalyptus plantations expanded rapidly in southern China and moved northward to marginal lands, including Hunan province, where the eucalyptus trees cannot survive cold winters. Having experienced the warmer and warmer winters in the past years, and learning of global warming from various sources, Fulin and the villagers believed that the freezing risk was gone. They converted almost all the woodlands including orchards in the village into eucalyptus plantations in a couple of years. They did not know that the killer was still with them all the time. And this time it used the other side of the sword: a record-cold ice- and snowstorm that hit southern China from mid-January to early February in 2008. All the eucalyptus trees in the village died as a result of the storm, and afterward Fulin cried hopelessly under the dead trees. He was confused by climate change and he did not know what he should grow. He was worrying more about his family, particularly the future of his son.

Climate change is beyond global warming. It comes with many weapons, such as heat waves, cold fronts, flooding, droughts, hurricanes, storms, and rising sea levels. Climate change is a challenge that tests the capacity of human knowledge, technology, and cooperative spirit. Climate change involves so many ecological and socioeconomic processes and their interactions at different scales. So the prediction of climate change and its impact will always come with uncertainties because of all kinds of limitations. However, these uncertainties should not be the excuse for taking no action. We cannot wait for fifty years to get an adequate understanding and meanwhile miss the opportunity to fix the problem. Uncertainties also mean opportunities for us to improve our actions and measures on climate change in the future. The most difficult challenge is nothing but the greedy and selfish nature of human beings.

Climate change is solvable only if we humans work together across nations, cultures, religions, and socioeconomic status. Solutions to climate change call for cooperation among governments and international organizations. Solving climate problems can also start with us as individuals, families, and communities by reducing our carbon footprints every day. Climate change is the most important threat we face, and there is a threshold beyond which the climate system will be irreversible, its damage becoming permanent and the cost of fixing it too expensive to afford. Therefore, we should start fixing the problem right now rather than leaving more threats to our future generations and putting their lives at a much higher risk. The gambling attitude is very dangerous, and we cannot afford any mistakes because we have only one Earth and there is no second chance if we fail. Therefore, acting *now* is our moral obligation to the future of our planet's unique life-sustaining climate system and to the future of human beings and other life-forms on Earth.

Climate Change Is a Moral Problem for You, Right Now

James Garvey

In the next few paragraphs, I hope to convince you of just one thing: climate change is a moral problem for you, right now. This conclusion is an easy one to miss. It's not hard to see that the governments of the world have a moral obligation to take meaningful action on climate change, but in the clamor to that conclusion you can shuffle right past the fact that you have a moral obligation to act as well. The conclusion matters a great deal, too. If everyone recognized their obligations clearly and acted on them rationally, our lives would be much different than they are now. Our world would be different as well. We'll start by getting the moral dimension of climate change into clear view.

We know that our world is changing: temperatures are rising, the sea level is rising too, sea ice is thinning, permafrost is melting, glaciers are in worldwide retreat, habitats are shifting, precipitation patterns are moving, and on and on. We know that human beings are suffering and that they will continue to suffer as a result of climate change. Things will become more difficult as the planet heats up. We know that our fellow creatures are already suffering too. There is a lot of unnecessary suffering under way and on the horizon. It's this suffering that makes climate change a moral problem.

Science can give us the facts, but we need something more if we

---

JAMES GARVEY is secretary of the Royal Institute of Philosophy, which is dedicated to the advancement of philosophy in all its branches through the organization and promotion of teaching, discussion, and research of all things philosophical. His most recent book, *Ethics of Climate Change: Right and Wrong in a Warming World* (2008), explores the moral demands of climate change.

want to act on the basis of those facts. The something more has at least a little to do with what we think is right, with justice, with responsibility, with what we value, with what matters to us. You cannot find that sort of thing in an ice core. You have to think your way into and through it. It helps to start small, with everyday thoughts about doing the right thing.

Think a little about the connection between an individual's capacities and morally demanded action. You would have some explaining to do if you saw a child drowning in a river and just walked past without doing anything to help. You would have a lot more explaining to do if you were a healthy lifeguard with a free afternoon. The difference has something to do with a connection in our thinking between being well placed to do what's right and doing what's right. If it's easy for me to stop a thief (because I'm a martial artist) or give up some time to help the homeless (because I have plenty of free time) or contribute some cash to the poor (because I have a very large disposable income), then my failure to do these good things stands out a little. You can apply these everyday thoughts to governmental action on climate change.

The industrialized, developed, rich world certainly has the capacity to take action on climate change. It has the economic might, the room for reduction, the technological know-how, the influence, the brains, the brawn (put it however you like) to do a great deal. It could move mountains. It could take effective steps to reduce its own emissions dramatically and enable the developing world to leapfrog into green energy production, all the while protecting our planet's carbon sinks. It could help pay for the costs of adaptation, too. The rich world has the capacity to do something about the changes ahead, changes that will result in human suffering. The fact that the rich world has failed to take meaningful action stands out a little, maybe more than a little.

If walking past a drowning child is wrong, particularly when one is well placed to help, then the West is doing something wrong by carrying on with business as usual or by taking only limited action on climate change. It amounts to walking past, to doing nothing, in the face of human suffering. It stands out even more given the West's capacity to do the right thing. It amounts to letting people suffer and die when one has the capacity to help. It is a kind of moral outrage.

It's not difficult to see the actions of the rich, developed world as morally lacking in this connection. What's much more difficult to look in the face is the possibility that we are doing something morally wrong too, something morally outrageous in our everyday lives. You can have thoughts about yourself and your own life that are analogous to thoughts about the developed world's failure to act. Don't take all of this personally if it stops you taking it seriously. For what it's worth, it's not just you but me and everyone else living a life enmeshed in a fossil-fuel-burning world—almost all of us in the West and many of us elsewhere too.

Line up your thoughts about your situation with the thoughts we've just explored about governments. You've got the capacity to do what's right, just as the West has the capacity to take strong action on climate change. Compared to most people on the planet—certainly if you are a citizen of the developed world—you have the economic might, the room for reduction, the technological know-how, the influence, the brains, the brawn (put it however you like) to reduce your carbon footprint dramatically. Compared to a lot of people on the planet, your emissions are probably enormous, and so are your wealth and power and freedom. If you come to the conclusion that the West is doing wrong by doing nothing much about climate change, apply the same thinking to your own life and see what you get. Think again about that lifeguard. Among other things, you might be led to the conclusion that climate change is not just a moral problem for governments. It's a problem for you, too.

Philosophers are exercised here and there by thoughts about so-called "applied moral problems," questions associated with cloning, abortion, genetically modified crops, euthanasia, just war, and so on. You can look at those problems from a safe distance and hope that you'll saunter through your life without encountering abortion or euthanasia. Probably no one is going to clone you. You won't have to decide whether or not to invade a neighboring country. But climate change is a moral problem for you, right now.

It's a moral problem because there's a certain and dramatic connection between your everyday choices and unnecessary human suffering. Most of the things you do in the course of an ordinary day involve the use of energy, the burning of fossil fuels, the thickening of the blanket around our planet, the changes to our climate, and, finally, the suffering of human beings and the destruction of countless plants and animals, even entire ecosystems. Your everyday choices play a part in an enormous moral wrong. What you do has a connection to suffering, both now and in the future. If that's true, then every one of your everyday choices presents you with a moral problem. It's why climate change is not a distant problem or someone else's problem. It confronts you countless times each day.

The fact that your part in the wrong might be very large compared to other people on our planet and the further fact that you are comparatively well placed to take action can make the wrong stand out a little, maybe more than a little—just as a similar wrong stood out when we thought about the developed world and its failure to act. It can lead you to the conclusion that your life might be a kind of moral outrage. It might have to change right now. Failing to take action is on a par with just walking past, with doing nothing in the face of human suffering.

This line of thinking can also lead to one further thought, maybe a surprising thought. It's not easy to live a green life in the fossil-fuel-burning world that we've got. If you are enmeshed in this world, and you have a moral obligation to escape that connection to suffering, maybe it follows that you have to change not just your life, but your society as well. You have to reduce your own emissions, certainly, but perhaps you also have an obligation to join with others to put pressure on governments and businesses and ensure that the world we are in changes too.

Maybe your life has to change a great deal. Maybe you ought to play a part in changing our world too. What is the right thing to do? The question is on the table. It's your question, and you have to decide how to answer it. Climate change is a moral problem for you, right now.

From Engagement to Emancipation

Sulak Sivaraksa

Even if those of us who are socially engaged Buddhists talk about the simple magic or the normal miracles of everyday life, of a flower, of each breath, or of life in harmony with nature and other sentient beings, we are not automatically against a real revolution, a great break, change, or rupture in the way things are—however "natural," "necessary," or "commonsensical" they seem to be.

We not only practice social engagement, which can be merely about action and not change, but we also work for social transformations to bring an end to oppressive and exploitative relations. We are interested in emancipation—in universal emancipation—in a time when hopes for revolutionary and progressive changes are in retreat, if not completely abandoned. Thus, to be more precise, shouldn't we see ourselves as promoting emancipatory Buddhism? Or is this really taking socially engaged Buddhism to a different level, beyond as opposed to behind the present stage? The greatest delusion is to be oblivious to or to affirm as bliss the unjust and oppressively inegalitarian present.

Even if we socially engaged Buddhists incessantly talk about non-

---

SULAK SIVARAKSA is the founding director of the Sathirakoses-Nagapradeepa Foundation, a Thai nongovernmental organization guided by a spiritual, environmental, and activist vision. He also cofounded the International Network of Engaged Buddhists. He was nominated for the Nobel Peace Prize in 1994 and has been honored with the Swedish Right Livelihood Award, the Unrepresented Nations and Peoples Organization Award, and the Millennium Gandhi Award. Sulak has written more than one hundred books and monographs in Thai and English, most recently *The Wisdom of Sustainability: Buddhist Economics for the 21st Century* (2009).

violence, it does not mean that we cannot or will not struggle militantly and courageously for our cause, for universal emancipation. Nonviolence does not mean that no one will be disturbed physically and mentally, or that everyone will be left at ease. Nonviolence does not mean that we will not foster dissent. It does mean that we will take the trouble to make troubles when and where they matter. Unexpected disruptions can also be democratic or democratizing. We militantly engage in nonviolence for emancipatory causes.

Even if we socially engaged Buddhists talk a lot about spirituality and inner peace, it does not mean that we are apolitical or politically disinterested and passive. Even if we socially engaged Buddhists stress the importance of compassion, generosity, and loving-kindness as a ground for nurturing ethical relations with others (that is, interbeing), we realize that in itself it is insufficient for ethics. The same goes with upholding and maintaining the rule of law. To fill the gap, politics is necessary. We can be passionate politically. Here by politics we refer broadly to collective struggle and power-sharing for emancipatory causes. Politics should not be reduced to mere dialogues and persuasion, to representation and voting. Politics without spirituality or ethics is cold and blind. Spirituality without politics is simply inconsequential—it regresses into a form of New Age self-help and escapism. So we spiritualize politics and politicize spirituality. Politics makes spirituality or ethics possible, so to speak. Buddhadasa Bhikkhu calls it "Dhammic Socialism."

What is pejoratively called Western Buddhism, however, seems to take the opposite view and practice. It is said to be all about cultivating personal salvation, forgetting that at the heart of Buddhism there is no notion of personal salvation. How can there be freedom in such an inegalitarian world? Or what does freedom really mean in such a world? At best, Western Buddhism is said to be all about making a virtue out of indifference. We socially engaged Buddhists see that the problem of Western Buddhism is far more insidious. The escapism of Western Buddhism is really an escape into the present, is really the affirmation of the prevailing injustices and oppressive structures. The Western Buddhist says, I know the present is exploitative, bad, and so forth, but at least I am maintaining my sanity through meditative practices. In other words, the Western Buddhist is really practicing meditation and cultivating inner peace in order to preserve the present, warts and all. In this respect, it is complicit in violence. Being a projectless spirituality, Western Buddhism does not advocate universal emancipation and is thus empty. This is where socially engaged Buddhists part company with Western Buddhists.

As socially engaged Buddhists we confront suffering. We cannot be indifferent to suffering. But the problem is not that simple. Whose suffering do we attend to? Do we attend to all sufferers equally? If not, why not? Do we feel superior and condescending when we deal with the suffering of others? Do we pity others simply in order to affirm our superiority? Here are we regressing into narcissism? Furthermore, we have to ask, suffering from what? Of course, we can broadly answer that the root causes of suffering are greed, hatred, and delusion. But still we have to probe deeper: Is this suffering merely an individual problem? Definitely not. It is also structural and institutional—hence the importance of politics. We may be participating in structural violence (e.g., of capitalism and capitalist relations) even if at the individual level we are peace-loving and compassionate. We must learn to develop ethical responsibility for structural violence. What is the politics of confronting suffering? How do we confront suffering? How do we reduce or minimize suffering? What are the meaningful acts that will bring an end to suffering? What kinds of subjects/agents are needed to bring about these transformations? What are the possible ways of or political projects aimed at reducing suffering? What if they conflict with one another? What framework do we use when we confront suffering? Do we end up with a better understanding of suffering? Or is it just an empty feeling devoid of understanding (as to the sources of suffering)?

These questions cannot be adequately answered here, and they are not for one individual to answer. Collective efforts are needed to refine and answer them. The idea is to propel socially engaged Buddhism beyond the engagement paradigm and to seriously embrace emancipation. And in our effort to confront these questions, let us certainly remember that these obligations, these efforts to thwart structural violence, apply to the nonhuman world as well. Moreover, let us remember that these efforts extend into the future as well—they are not temporally bounded; they extend for all time.

The Architecture of Language, Parts 9 and 10

Quincy Troupe

9.

great language is a shower of words inside a blizzard
of tongues full of rhythms & syllables, is a flow,
is snowstorms of meaning coming & going everywhere our ears
turn, hear rainstorms, tornadoes, lightning bolts unzipping
    clouds
towering around the calm, savage eye of hurricanes, is a flow
coming & going, bringing new systems of music,
new ways of listening connected to hearing,
structures carrying a host of evolving languages, roiling tongues
inside cross-fertilized speech of immigrants & new poetry,
is a flow also located in the evil eye of katrina, rita

QUINCY TROUPE is the author of seventeen books, including eight volumes of poetry, the latest of which is *The Architecture of Language,* recipient of the 2007 Paterson Award for Sustained Literary Achievement. He received the 2003 Milt Kessler Poetry Award for *Transcircularities: New and Selected Poems* (Coffee House Press, 2002), selected by *Publishers Weekly* as one of the ten best books of poetry published in 2002. He is professor emeritus of creative writing and American and Caribbean literature at the University of California, San Diego, was the first official poet laureate of the State of California, and is editor of *Black Renaissance Noire,* a journal published by the Institute of African American Affairs at New York University. Troupe's poetry and prose have been translated into many languages, and he has read his work throughout the United States, Europe, Africa, Canada, the Middle East, the Caribbean, and Latin America. Troupe lives in New York City and Goyave, Guadeloupe, with his wife, Margaret.

swirling in from the gulf of mexico carrying thunder & death,
foaming with cataclysmic omens, terrors beyond any
    understanding,
categories beyond any knowing what horror will bring

through cracks of daylight destruction unfolding, rancid bodies
bloated, floating in toxic water, is a language, is a flow
            we hear but do not know how to recognize

thirty-foot storm surges speaking in tongues more violent
than any language we think we hear or know,

is a flow disjunctive beyond any application of money,
is perhaps a cosmic spiritual payback
for bug-eyed children who mirror the language of hunger,
murder, no sympathy, or empathy for blues people festering
in a place full of heat, water, mosquitoes, poisonous snakes,
high prices for gasoline for cars thirsty for petro, is a flow
of anarchy spreading like a plague in this place, is a form of
    language
ignited by category-five winds & angry seawater foaming salt
& screaming in the voodoo language of the sea goddess, erzulie,
hougans blowing calamities ashore through their mouths
of long bamboo horns, is perhaps a payback
for all the terrors released in this flow

when coffins, mummified corpses, leering skeletal bones
unearthed by katrina's savage flooding tongue
are scattered like dead leaves & broken branches all over
louisiana's devastated countryside, it tracks the fall posthaste
of america's once promise of greatness,
lost here in this macabre jumble of unknown spirits
evicted from their graves with no names, identity,
no race-ticket or skin color has privilege in this space
of spirits displaced, scattered from former resting places
of chipped tombstones—also scattered—
their skulls reminding us of broken teeth of ex-fighters, junkies,
these corpses grinning teeth set in jaws of cracked bone,

is a powerful language screaming for redemption,
if we look deeply into this moment it reveals our true selves
in these spaces we live in corrupted by greed,
skin color, class, religion, power at all cost,
                                            is a definite language
suffocating in claustrophobia, ethnophobia, no connection
to the real world, to the flow (caca) flow of humane language,
poetry inside the most profound beauty of utterance
sings still inside the deepest grain of that flowering word,

sound by sound, word by word, speech evolves into beautiful
architecture, creates a scaffolding of cross-fertilized utterances
crossbeamed inside poetic sentences fused with music,
where metaphors spring from deepest sources of community,

these are the seeds that will link, bond us together

10.

after tongues of fierce winds howl full of calamitous journeys,
shipwrecked adventures swept astray by circular history,
where are we going feeble question marks of human embryos
bloating after the wind's anger has died down, turned soft &
   gentle
as a sweet tongue of breeze caressing the passion of naked lovers,
where are we going led by troglodytes in dark suits selling
   wolf-tickets,
tiny metal american flags festooning their lapels,
violent little chicken-hawk men in seats of power,

where are we going running all this bric-a-brac fear of hitler's
germany, murmuring a language full of evil secrets, murder,
mysterious as the ocean's salty sandpaper tongue of screeching
felines carrying sounds evoking warnings,
where are we going on this stormy sea full of huge treacherous
   rocks,
100-foot waves looming toss our premonitions like matchsticks,
thunderclaps of vowels flooding the peaceful conversations

we try to evoke within a spiritual connection,
where are we going carrying this toxic speech full of static
buzzing like hornets or flies through this pestilential
space where we live out our lives full of fear,
trepidation, & hateful loathing,
where are we going on this stormy sea,
heading into a space full of huge sharp rocks,
behind time instead of on time,

where are we going, going, going

ETHICAL ACTION

What is the work of compassion in a time of climate change?

Compassion, literally "shared feeling," depends on the moral imagination, the ability to imagine yourself in another's place and to feel that person's joy or despair and desperation. Moral imagination, in turn, is nurtured by all kinds of creative and natural storytelling that bring us into other peoples' hearts and minds—poems and interviews, photo-essays and films, novels and statistical studies, but also natural science, the stories of landslides and springs. The times call for more stories, so that we can imagine ourselves into the plight of suffering people and find there the impulse to help.

If you are a teacher, ask your students to climb into an imaginary time machine and take themselves fifty years into the future. What is the world like? What makes people happy? What makes them sad? What message would the people send back with the students into our time? Ask your students to write those messages and bring them to the classroom. Write an article about the project for the local newspaper, giving voice to the children's hopes and fears.

If you are a member of a church, establish a correspondence with a similar church in a part of the world disrupted by the effects of climate change. What are the people's lives like? What are their needs? What is their future? What is the measure of their despair?

If you are a social scholar, study suicides. How are they connected to droughts, floods, ruined crops, dust storms? Map the meteorology of despair.

If you are a writer, tell the stories of hunger and thirst and wandering. Write the new *Odyssey*, the new *Grapes of Wrath*, the new *Jungle*. Write *Exodus II*. While you're at it, write the new songs, the

"Give Peace a Chance" for the times of ecological collapse, the "We Shall Overcome" anthem of the climate crisis. It's music that goes straight to the heart.

If you are a student, learn as much as you can about the effects that climate change is likely to have on people. Study the hydrology of Bangladesh, the topography of Tuvalu, the hunting practices of the far North, the water resources of sub-Saharan Africa. For every numerical chart or graph your professor shows you, ask for a photograph or a story or a song. Commit yourself to a career that does not require you to impose suffering on innocent people. Better yet, commit to lifework that prevents or ameliorates suffering.

If you are a parent, tell your children stories of compassion. And when they are grown, encourage compassion in their lives and their lifework.

10

> *Do we have a moral obligation to take action to protect the future of a planet in peril?*
>
> **Yes, because justice demands it.**

If scientific predictions are right, people of the future will be the ones who suffer most from the effects of environmental disasters created by our comfortable lives. Moreover, people in the poorest situations now, and increasingly in the future, will pay the costs of the profligate use of fossil fuels by people in the richest nations. Both intergenerationally and internationally, those who reap the short-term benefit of abusing the world are not the ones who will pay the price.

This is not fair.

We all seem to be committed to the concept of justice, but we sometimes mean different things. This matters not one bit, because by *any* definition of justice, forcing others to bear the burden of your benefit is unfair.

Consider only one such definition, John Rawls's theory of justice. Rawls says that if you want to know whether a distribution of benefits and burdens is fair, all you have to do is a little thought experiment. Ask yourself, if I didn't know what position I would have in the world (whether I would be a present person or a future person, whether I would be rich or poor, whether I would be African or Inuit), would I choose a situation in which rich, mostly white people of one or two generations reap great benefits and impose the costs on other people, notably future people and the poor? Or do it this way: Would

I approve of this arrangement, if I knew that my worst enemy would have the power to assign me my place in the world?

If the answer is no, then the arrangement in question is unfair. And of course, the answer *is* no. No one would freely choose to pay, in the currency of their suffering and the suffering of their children, in famine and disease and the risk of human life on Earth, the costs of the reckless adventures of the wealthy nations.

We have an obligation to remedy a situation that is currently and patently unjust, and one that will only grow more unjust as time goes on and the future unfolds.

Ethics as if Tomorrow Mattered

Carl Pope

Really, the *new* questions posed by climate change and the global environmental crisis are not ethical at all. The Great Traditions that underlie the world's religious and moral systems already give us the answers to the questions "What is right?" and "What is our duty?" Our existing ethical principles are perfectly adequate.

So why do voices of faith and morals seem so tentative and hesitant about climate? Why do ethicists approach this topic rarely and gingerly?

I'm going to suggest that we hesitate because the scale of the human endeavor—in time and space—has now far outpaced the local context in which ethical principals arose and have been traditionally applied. Global warming means we must put our ethics on steroids—and that makes them much more challenging, threatening, and, yes, subversive and newly controversial.

The ethics of the Great Tradition—whether expressed as the Ten Commandments, the Eightfold Path, or the Golden Rule—were conceived of and applied locally, in time and space. "Love thy neighbor" meant someone you encountered face-to-face. "Thou shalt not kill" referred to murdering someone who was alive when you were.

CARL POPE is executive director of the Sierra Club. A veteran leader in the environmental movement, Pope has been with the Sierra Club for more than thirty years. He is a regular contributor to the Huffington Post and coauthor, with Paul Rauber, of *Strategic Ignorance: Why the Bush Administration Is Recklessly Destroying a Century of Environmental Progress* (2004). Pope has served as a board member for the National Clean Air Coalition, California Common Cause, and Public Interest Economics Inc.

But if the fuels we burn to generate electricity may put villages in Vietnam under water a century from now, this feels like a whole new ballgame. Do I really have to live my life that intentionally? Can I take responsibility for all of the consequences of all of my actions, limited only by my (ever-expanding) knowledge? Is it reasonable, or even possible, to consider all the possible impacts of global warming before I set my thermostat at night?

The great religious traditions teach us how to treat each other. They lay down the spirit with which we should approach the rest of life. But by themselves, they do not provide us with clear guideposts for how technological societies can respect those injunctions.

Science tells us how the world works. It enables us to project from our actions to their distant or future consequences. But it provides no ethical guidance for how we should weigh those consequences.

What the global warming and the environmental crisis have done is raise this question: How do we combine the realms of science and social and religious ethics?

In one sense, the major contribution of modern environmentalism has been to show that these two seemingly disparate disciplines can be combined in a way that makes sense and is consistent—but not very comfortable.

Ethics or religion can tell us that we should not poison others; medicine can tell us that mercury is a neurotoxin; chemistry that burning coal emits mercury; biology that mercury released into the air will be concentrated in the tissues of fish and of the people who eat the fish. Environmentalism's contribution is to bring these insights together and to conclude that we ought to clean up coal-fired power plants because they are poisoning others. Suddenly the ethical injunction "do not kill" has implications for the corporations that operate such power plants and the customers who benefit from the electricity.

The way in which environmentalism melds ethics and science may not be the only possible approach—but I think it is the most fully developed. And applying it to climate creates no new problems not already confronted in asking about deforestation or pollution.

This approach was perhaps summed up by the founder of the Sierra Club, John Muir, when he famously said, "When we try to pull one thing in the universe, we find it is hitched to everything else." It has at its heart the idea that you just take your actions, follow all of their consequences, and then apply traditional ideas of right and wrong to the fruits of those actions.

I'm going to illustrate by exploring three moral questions, each one an expansion of a traditional ethical virtue in time and space. An environmentalist would argue that if you accept the traditional virtues, and now possess the knowledge that extends them globally and into the future, you must apply them in this new and much more demanding way.

And if we do, I would argue, the ethics of climate becomes very simple.

Legacy

Does the future matter?

Environmentalism says, "Oh yes." Indeed, a concern for legacy may be environmentalism's ethical bedrock.

Obviously, there are dissidents. Those who believe that the world will end soon, that there is no long future, see no reason to prepare for it. Nor do those who believe in an invisible economic hand which ensures that the actions of present generations will, without intentionality, automatically be beneficial to future ones. Politically, there are still heirs to the congressman who famously challenged Teddy Roosevelt's emphasis on conserving natural resources by barking, "Posterity. What has posterity ever done for me?" And many economists, in their use of "discount rates" to devalue future outcomes, produce the same results, if not always intentionally.

Where does our obligation to the future come from? Its simplest ethical root is the love of parents for children—and their children. Human beings have for millennia sacrificed for their descendants.

But the principle that conservatives call "piety" links us in a much longer chain. The core idea of piety is that man is not the measure of all things. Pride is the great enemy. There is a contractual relationship between the past, the present, and the future. We—including posterity—are all in this together. As Edmund Burke put it, society "is a partnership not only between those who are living, but between those who are dead, and those who are to be born." Frank Meyer, the seminal figure in the emergence of modern American conservatism, said, "In a deep sense conservatives must have piety towards the constitution of being." Adlai Stevenson captured the same concept in liberal rhetoric when he described us as "passengers on a tiny spaceship, Earth."

Now science provides the long time line of human consequence. We understand the story of our species and our planet far better than we did a century ago, when John Muir was arguing—correctly, but as a dissident—that Yosemite Valley was carved by glaciers.

And that time line tells us that ecological change—not wealth, or power, or even our technological and cultural achievement—is the true inheritance we pass on to our posterity.

We write across the landscape in a sturdy script—ecological devastation. Only a handful of lines from the ancient Greek poetess Sappho survives. But the impoverished ecosystems of Sappho's time and place, the fruits of deforestation and overgrazing, endure.

Even global warming cynics, as the so-called skeptics should be called, concede that *if* the Antarctic ice cap melted, sea levels around Bangladesh and Manhattan Island would rise, flooding would increase, lives would be lost, property destroyed. The life cycles of songbirds of the American Midwest, called "neotropicals" because they spend the winter in Central American rainforests, would be lethally disrupted.

We know that our emissions of greenhouse pollutants may multiply that ecological devastation to an almost incomprehensible degree. Only by shutting our eyes to those consequences can we absolve ourselves of the duty to act—because if we do not, we are, quite literally, stealing from our posterity.

Justice

Questions of justice have become among the most complex and challenging in today's world. The issue of how to balance economic freedom and equality in a complex industrial society is far from resolved. These seem to me honestly difficult problems.

But if we are talking about climate justice, I think it's much simpler. "We are all equal in the eyes of the Lord" is the teaching of the great faiths. "Love thy neighbor as thyself." The air, the water, the Earth belong equally to all of us. Because human beings did not create the biosphere, and because they are dependent for their livelihood on ecological services provided by a myriad of its creatures, including such "lowly" forms as dung beetles, it is hard to argue that the rest of life has no right to make a livelihood off those common ecological services as well.

Take what in other contexts is the great challenge to traditional ethical concepts of social justice and equality—market capitalism. In any true market, you should own what you sell—otherwise it's fencing. You should pay for what you take—otherwise it's shoplifting. Transactions, across the board, should be voluntary and freely chosen on both sides.

But both justice-based and capitalist approaches must concede some basic facts. Anyone who emits carbon dioxide into the atmosphere is appropriating a portion of the world's carbon sinks—that molecule does not remain where we emit it, and it must be taken care of somewhere else. Those, largely in the industrial world, who emit CO_2 do not own those carbon sinks. ExxonMobil does not have the deed to Indonesia's rainforests in its corporate vault in Irvine, Texas. Those who emit greenhouse pollutants are not paying for the damages they cause others. When I pay my electrical bill, no portion of it goes to compensate farmers in India for diminishing rainfall on their fields. And those who suffer from climate change have not voluntarily agreed to this exchange; the Maldive Islands have never sold themselves to be used as a flood storage area for oceans swollen by climate change.

So whether or not we think today's climate is somehow "the best there can be," basic fairness suggests that we have no right to knowingly change it without the consent, and to the detriment, of those who will be losers.

Responsibility

Are we responsible for the foreseeable results of our actions, even the distant and future ones, or only for following the rules of our society?

The Bible and Aesop both suggest that if we have a warning of a famine to come, we are supposed to get ready for it. Historically, limited human knowledge meant that we couldn't foresee distant or remote events. Science now means that we have enormously greater knowledge of the fruits of our deeds. So environmentalism argues that we must be held accountable for the impacts of our economic actions, not just our intentions.

Environmentalism's view draws from the dominant American ethical stream, which says that if actions have predictable consequences, actors should bear those consequences.

We are all taught—on the schoolyard—that our right to swing our fist stops at someone else's chin. Even before the schoolyard our parents made us clean up the milk we spilled—even if we didn't mean to spill it.

Responsibility captures the nexus between our deeds and their results. It says we should respect the integrity of others, pay for what we take, keep our promises, make amends for our transgressions. Without responsibility the core moral concepts of fairness and justice make no sense.

What environmentalism has done is to apply science to extend responsibility across time and space. I could not dump sulfuric acid over my back fence directly into your garden. Why can a copper smelter do the same things from hundreds of miles away, merely because the wind determines whose backyard is fouled? I could not dump organic solvents in your well. Why can a chemical company pollute hundreds of wells by dumping its wastes in a swamp, merely because the poison moves into the drinking water over the course of years or even decades? I could not come onto your property and cut down the trees to improve my view. Why am I permitted to emit so much CO_2 into the atmosphere that a changed climate kills your forest?

Nothing I've said above is very complicated or very new—the logic line is straightforward, the ethics familiar. Our basic moral duties of compassion, justice, and responsibility—outlined by millennia of social and religious ethics—are clear. Science is now showing us the real-world implications of accepting those duties or ignoring them.

Of course the conclusions are highly controversial; they ask of us a moral courage we're not sure we can achieve. Of course we all wish science had got it wrong.

It's uncomfortable, at the least, to recognize our connectedness. We are genetically more than 90 percent chimpanzee; Shakespeare had it only partly right when he said, "What a piece of work is man, how noble in reason." Reason is a gift, but it is also a threat. I really don't want to think mindfully about my daily life. Who does? If we did, we would have to dramatically change our ways of living.

But facts, as Ronald Reagan said, are stubborn things. Many people might want a world in which science cannot predict the results of our consumption of fossil fuels, because the only ethical response to that knowledge is collective, global, and yes, democratic, majoritarian action.

But they don't live in such a world. And neither do the rest of us.

Sustainability as a Founding Principle of the United States

Michael M. Crow

In the summer of 1787, two watershed processes in world history were in their earliest stages of development. First, the Industrial Revolution was gathering momentum in Europe and ultimately exerting its impact on the burgeoning American economy. Second, at this same pivotal moment, the nascent republic known as the United States was just completing its earliest aspirational blueprint, the Constitution. The coincidence of these revolutionary processes and products—one economic and the other political—is significant because, however defining for future generations each may have been, both are in one sense only the result of merely incremental progress in human consciousness. Both represent crude and inchoate forms of social and economic redesign that could have been inestimably more successful had the processes of redesign been undertaken with some awareness of the context and content of the natural world.

The American Constitution is an extraordinary articulation of the design of a state that at once establishes democratic governance, liberty, and justice, as well as other core personal and social aspirations intended to be realized around bedrock political institutions. The Industrial Revolution, resulting from the evolution of fundamental

---

MICHAEL M. CROW is president of Arizona State University. He has served as executive vice provost of Columbia University, where he was also a professor of science and technology policy in the School of International and Public Affairs. He played the lead role in the creation of the Columbia Earth Institute and helped found the Center for Science, Policy, and Outcomes in Washington, DC. He is a Fellow of the National Academy of Public Administration and has authored numerous books and articles relating to the analysis of research organizations, technology transfer, science and technology policy, and public policy.

principles of capitalism and cultural reorganization, consolidated the formats and structures through which society could be reorganized around new kinds of economic institutions.

In neither the Constitution nor the basic principles of capitalism, as best represented by Adam Smith in *The Wealth of Nations,* is there evidence of any meaningful awareness of the fact that the natural systems of the Earth and our constructs and designs as humans must advance in sustainable ways. An appreciation of the interrelationship between natural processes and human design is a prerequisite for any adequate conception of sustainability. This hybrid concept can be summarily defined as the stewardship of natural capital for future generations, but its implications are far broader than any of these terms, embracing not only the environment and economic development, but also health care, urbanization, energy, materials, agriculture, business practices, social services, and government. While sustainable development means balancing wealth generation with continuously enhanced environmental quality and social well-being, it is a concept of a complexity, richness, and significance comparable to other guiding principles of modern societies, such as human rights, justice, liberty, capital, property, governance, and equality.

While even this list of the implications of sustainability is incomplete on its face, any such tally is the product of hindsight derived from our twenty-first-century intellectual culture. Any notion regarding our responsibility to maintain natural capital for future generations or to advance economic and technological progress with a sense of stewardship was not present in the eighteenth-century designs that still drive so much of our economic thinking. While we may parse the deliberations and discussions of the era for evidence of some incipient appreciation of our predicament, we only know with certainty that the understandings we derive from John Muir, Aldo Leopold, and Rachel Carson had yet to be formulated, much less realized. At the time we were still held captive by a millennia-old Malthusian-style constraint model in which each advance in population resulted in a series of negative constraints greatly limiting our collective quality of life by constantly cycling in ways in which personal income could not be enhanced. Not surprisingly, then, the new economic order of the eighteenth century and the new political order being realized in the

United States at the same time were so powerful in their transformative effect that only now can we look back in both awe and fear at what these revolutions have wrought.

Two-hundred-plus years into this new political and economic order, for all its vicissitudes the world has advanced in many positive and constructive ways. The pre–Industrial Revolution economies of subsistence agriculture and the long-term persistence of poverty endured by all but an elite handful have largely passed from the social order. The masses, formerly voiceless and without any political power, now speak loudly and often and can be heard in many new settings. Yet at the same time we sit on the edge of a precipice of a significant failing. Because neither our economic nor our political models have factored in the natural limits of the Earth, and because the Constitution outlines neither aspirations nor outcomes relative to man's relationship with the natural world, we are at this very moment in time on a path toward a condition where the natural rights of man and national laws of economics collide with the natural systems of the Earth, to the ruinous long-term detriment of us all.

As vigorous and dynamic a modern society as we are, and as hard-working and productive as we have been, one would expect our nation to have exerted an impact on the environment. Yet it is surprising to me that in only 250 years we have actually altered the natural patterns of the atmosphere and both land and ocean ecosystems to the extent that future natural capital inputs for our long-term well-being are actually at risk. It is almost beyond comprehension that the political and economic designs that have allowed most of us to leave behind the brutish world of our ancestors are the same designs that have brought us to the brink of environmental collapse.

Both our economic and political designs are at once too limited and too simplistic to address the complex problems intrinsic to the discourse of sustainability, such as intergenerational equity, biodesign, adaptive management, industrial ecology, and natural capital conservation—new principles for organizing knowledge production and application. These inherent limitations are a consequence of not only the relative immaturity of our economic and political tools but also, and more important, the implicit "aspiration of self" that the Constitution endorses. We all operate out of self-interest to some extent, which is entirely rational, but the parameters that our foundational national document establishes in many ways simply constitute a justification for us to indulge in selfish, or let us say at least nakedly self-interested, pursuits and therefore might be just too simplistic to be a completely successful design for long-term societal success. As a consequence of our economic and political system, the individual perspective has inevitably outweighed the collective, with the result that adequate protection for the collective has lost out. In part because of the inevitable limitations of a document drafted in the eighteenth century—however brilliant and visionary it may have been—efforts to advance the long-term interests of the whole by controlling the short-term behavior of the individual are doomed to failure.

While we have pursued our aspirations of self, roughly 20 percent of the planet's bird species have been driven into extinction, 50 percent of all freshwater runoff has come to be consumed, seventy thousand synthetic chemicals have been introduced into the environment, the sediment load of rivers has increased fivefold, and more than two-thirds of the major marine fisheries on the planet have been fully exploited or depleted. What right do we have to eliminate the fishing stock of the oceans for generations or to alter the atmosphere of the planet? What rights of man or pursuit of happiness grants us the power to condemn future generations to the impact of human-induced sea level rise? Of course, the answer is we have no such rights. Likewise, what logic permits the extraction of such quantities of natural capital from the Earth in the ten to fifteen generations that will have presided between 1850 and 2150, leaving future generations with only a diminished basis to use the natural systems from which we have greatly benefited? No such logic can be found.

In an effort to redeem ourselves, let us at last reconsider our design, derived from the framers of the Constitution in the eighteenth century. However belatedly, it is at long last time to add one more value to the concept of the self as expressed by the Constitution. To provide for the common good we cannot only consider justice for those of us present; we must also conceptualize and enact into law provisions for justice for future generations. To ensure the equitable pursuit of happiness we cannot look only at the 40 or 50 years ahead of or behind us; individually we must come to terms with the realization that decisions made during the past 250 years have put humanity during the next several thousand years at potential risk.

It is time for America to take yet another first step, just as we took a first step in 1789. In the twenty-first century we must at last declare sustainability a core aspirational value of the American people, on the same level as liberty and justice and equality. With such a declaration we would see changes in law, changes in behavior, changes in teaching and learning, and, yes, even changes in economics. With such a declaration we would fulfill the expectations of the visionary framers of the Constitution of the United States of America.

Climate Change and Intergenerational Responsibility

Steve Vanderheiden

Two years after the Intergovernmental Panel on Climate Change released its first assessment report, confirming that human activities were deleteriously altering the Earth's climate system, delegates from 172 nations met in Rio de Janeiro for what would become known as the Earth Summit. These negotiations during June 1992 yielded the landmark United Nations Framework Convention on Climate Change, which declared climate change to be a "common concern of mankind" and pledged international cooperation to prevent "dangerous anthropogenic interference with the climate system." The treaty's first principle declares its fundamentally ethical mission: "The Parties should protect the climate system for the benefit of present and future generations of humankind, on the basis of equity and in accordance with their common but differentiated responsibilities and respective capabilities. Accordingly, the developed country Parties should take the lead in combating climate change and the adverse effects thereof."

In these two sentences can be seen the outlines of a moral imperative of unprecedented ambition and scope, addressing a global problem of monumental difficulty. Perhaps most remarkable is its articulation of an ethic of responsibility: responsibility for our increasingly interdependent world. Although responsibility has more often been used in recent political discourse to deny claims of justice than to

---

STEVE VANDERHEIDEN is a professor of political science and environmental studies at the University of Colorado at Boulder. He has written numerous articles and book chapters on environmental political theory. He recently published *Atmospheric Justice: A Political Theory of Climate Change* (2008) and edited *Political Theory and Global Climate Change* (2008).

acknowledge them (where denying aid to the vulnerable is euphemistically described as encouraging them to "take responsibility" for their misfortune), here it serves as the core concept around which a new global climate policy effort would be organized and through which the urgent demands of climate justice could be understood.

But what follows from observing that we all have "common but differentiated responsibilities" for global climate change? We can be more or less responsible (or not responsible at all) for harm that happens to ourselves or to other people, and being among the causes of bad outcomes is not enough to make us responsible for them. No one is responsible for accidental harm, but we can be responsible for harm that we do not foresee or intend if we act recklessly or negligently. And we can be responsible for harm we do not initiate, if it results from our failure to act in cases where we could easily prevent some harm from occurring, as in our decision not to rescue an imperiled swimmer when this could be done with minimal risk to ourselves. Sometimes we share fault and responsibility with others, though we may not be equally at fault, as when some are in a better position to rescue a drowning swimmer than others are, but no one acts to do so. These analyses reveal what persons can be held responsible for in this moral sense: we are responsible for bad outcomes that result from our voluntary acts and omissions, insofar as those outcomes could reasonably have been anticipated and avoided. In legal theory, this is known as "liability based in contributory fault."

Signatories to the Framework Convention not only accepted the scientific findings that people have been and continue to be highly unequal contributors to the problem, but also decided that these were not faultless contributions. Following the 1990 release of the Intergovernmental Panel's first assessment report, detailing the causes and effects of climate change, further greenhouse emissions caused foreseeable (and therefore culpable) rather than accidental harm. No longer could parties plead ignorance concerning the causes and effects of climate change as a tenable strategy for deflecting responsibility for climate-related harm they continued to cause. After 1990, responsibility for climate change was "common" to all, but widely "differentiated" among rich and poor nations and persons. Hence, all would have to cooperate in order to remedy the problem, but the relative burdens that various parties would be assigned were to be based on their ongoing responsibility for it. Such was the judgment of those gathered for the Earth Summit, and rightly so. The world's affluent industrialized nations were primarily responsible for the problem and so would have to be primarily responsible for its remedy. They must, as this first principle declares, "take the lead" in this respect.

Notice what follows from denying that the biggest greenhouse polluters were culpable for their post-1990 contributions to the problem (as the United States steadfastly maintained through 2008 with its rejection of the Kyoto Protocol). If climate change were genuinely a natural disaster, for which none is culpable so that fault-based liability cannot be assigned, the responsibility to remedy would instead be assigned to those in the best position to help. Like the potential rescuer that is most able to assist the imperiled swimmer, the very same nations with the highest post-1990 greenhouse emissions would also be most able to reduce those emissions after learning of the harm that they would continue to cause, even if they consistently denied that anybody could reasonably have known about such causes prior to their initiation of a remedy. Whether or not one accepts "differentiated responsibilities" to remedy on the basis of differentiated culpability for causing the problem or on the basis of varying "respective capabilities" to provide that remedy, the judgment remains the same: the world's affluent must accept responsibility for mitigating the harm of climate change.

To fully comprehend the urgency of this obligation, we must again consider the concept of responsibility, now understood from the perspective of the victim of harm for which others are responsible. Sometimes bad things happen to people without anyone being at fault for that harm. As with the imperiled swimmer, these situations still require a remedy, and potential rescuers can incur a responsibility to come to the aid of those that they can assist without excessive risk to themselves, and the failure to do so under such conditions can be faulted. But cases of accidental harm are the least interesting of those associated with responsibility from the victim's perspective. Persons can be culpable for harming themselves, and while in some such instances there may be good reasons to try and prevent them from doing so—with drug addicts, for example, or those under extreme duress—cases of self-harm can only be unwise or unfortunate, not unjust. Where persons are made to suffer harm for which they are only minimally or not at all responsible, and when some other party is responsible in the moral sense described above, then that harm must be described as unjust. In some cases, justice can be restored if one party culpably harms another but then provides the victim with an adequate remedy, but even where this is possible it is almost always better to avoid the initial injustice than to harm and then restore. Culpable parties acquire a responsibility to right the wrongs they cause when they commit their culpable acts or omissions, but from the victim's perspective one should never be made to suffer harm for which someone else is responsible.

We are now in a position to say what this moral sense of responsibility has to do with climate change, and what climate change in turn has to do with the international and intergenerational justice duties noted by the Framework Convention. Whether or not we intend or are aware that our high rates of greenhouse gas emissions cause the harmful consequences of climate change, we are now responsible for that harm and so acquire two additional kinds of responsibility. First, to the extent that we can avoid continuing to harm, we are morally obligated to do so. To use the language of climate policy, we must first either minimize our contributions to climate change by dramatically reducing our emissions or else insulate current and future others from that harm of climate change we have already caused or continue to cause (or some combination of "mitigation" and "adaptation"). Second, insofar as we fail to do so and thus unjustly cause others to suffer harm for which we are responsible and they are not, we must seek an imperfect remedy to that regrettable injustice. In such cases, we are morally responsible for compensating others for the injury we culpably inflict upon them.

We in the wealthy, industrialized nations are responsible for assuming the burdens associated with reducing our greenhouse emissions (*mitigation*), assisting those victims of climate change that bear far less responsibility for the disproportionate amount of harm they are expected to suffer (*adaptation*), and righting any remaining wrongs that cannot be avoided by either mitigation or adaptation (*compensation*). Poor residents of developing countries are far less responsible for the harmful effects of climate change than are their affluent counterparts, so to allow them to suffer when those culpable for their injury and in a position to assist could avoid or remedy this harm is plainly unjust. Likewise for future generations, who are even less responsible for the harm of climate change they are expected to suffer, even if we take strong proactive measures to curb our future emissions, since they will have in no way contributed to the accumulated greenhouse gases that will continue to wreak their climatic damage for over a century after their initial emission. The future poor will be harmed most of all, since they will inherit a damaged climate system and will be subject to the harm that results on a part of the planet that will have benefited the least of all from the industrial processes that led to this degraded environment.

Unless we *take* responsibility now, we shall certainly fail to do as justice requires, fail to "protect the climate system for the benefit of present and future generations of humankind." Victims will be made to suffer avoidable harm for which they are in no way responsible and for which we will be culpable.

Those for whom such responsibility constitutes a terrible psychological burden frequently meet this analysis with resistance, and they want to know what they can do to opt out of it. Unlike those refusing to accept the economic burdens of their responsibility to others, this response accepts that harming connotes responsibility to remedy but seeks to avoid that harm in order to avoid all consequent responsibility. Such a response may be understandable but is misguided. We should of course always try first to avoid becoming responsible in this way, but in the case of climate change it is impossible to undo our past contributions to the problem or to extricate ourselves from the benefits that accrue to us from living in a society that has become wealthy by unjustly appropriating more than its share of a common resource. By the time we could become cognizant of our responsibilities, as this response is, we are so inextricably bound up in the processes that generate them that we can no longer opt out. Genuine self-reliance—that is, accepting all of our responsibilities but being responsible to none but ourselves—is no longer an available option in the contemporary world. As John Holdren has observed of our alternatives in collectively responding to climate change, "we have only three options: mitigation, adaptation, and suffering." As individuals, we have only two options: accepting our responsibilities to others or acting irresponsibly. But this need not cause dismay. Accepting our responsibility is the most uniquely human action we can take, and we can no more remove ourselves from this responsibility than we can forfeit our humanity or cease to depend on the planet for our welfare. Rich and poor, present and future, in the stewardship of our shared planet and its climate system, we are all responsible to each other. Whatever burden is created by acknowledging our responsibility for the welfare of others is surely outweighed by the very real sense that our personal welfare largely derives from the fact that a great many others have taken and continue to take responsibility for our welfare. With advantage comes responsibility, and we have inherited both from our ancestors. We now have great responsibilities to others as an inherent part of our good fortune, along with the capacity to meet them if we have the integrity and resolve to do so. As persons and people our advantages and responsibilities are intertwined, and from the web of moral obligation that they together create we must bequeath a healthy planet to future others.

Still an American Dilemma

Lauret Savoy

I recently attended a day of talks given by several internationally respected activist-writers working to curb climate change, environmental degradation, and run-amok capitalism. The New England village's lecture hall was filled by a vocal, fairly affluent audience of a few hundred. I was the only brown-skinned person in attendance. Around the room, conversations on a green economy, on 350.org and the Copenhagen climate talks, and on what "we Americans need to do" shimmered with an energy infused by these movers and doers. But the dimensions of class, gender, and race were absent in the lectures and in all of the exchanges I overheard.

A child born today enters a world of rapid and extensive ecological changes. The list is often repeated: Human population continues to grow. Ecosystems globally have never before been so widely fragmented or degraded by human action. (Nearly 1,400 scientists convened by the United Nations cautioned in their 2005 *Millennium Ecosystem Assessment* that "humans have changed ecosystems more rapidly and extensively than in any comparable period of time in human history," resulting "in a substantial and largely irreversible

---

LAURET SAVOY writes across threads of cultural identity to explore their shaping by relationship with and dislocation from the land. A woman of African American, Euro-American, and Native American heritage, she is a professor of environmental studies and geology at Mount Holyoke College. Her books include *Bedrock: Writers on the Wonders of Geology*, *The Colors of Nature: Culture, Identity and the Natural World*, and *Living with the Changing California Coast*.

loss in the diversity of life on Earth.")[1] Coal, petroleum, and other fossil hydrocarbons, once abundant and seemingly cheap "resources," literally fueled industrial revolutions and the mechanization of food production. And because of this fossil-fuel economy, greenhouse gas levels continue to climb, ever exceeding the highest atmospheric concentrations since our species evolved.

While the types, rates, and degrees of environmental change might be unprecedented in human history, the embedded belief and political-economic systems behind them in the United States—the most energy-consumptive nation—are not. Their long, deep roots have allowed and continue to amplify fragmented ways of seeing, valuing, and using "nature" and human beings. The factors and economic frames considered to measure the human (or ecological) footprint on Earth, for example, mask how the exploitations of land and of people are interconnected. Quantifying the area of productive land and water needed to provide ecosystem "services" or resources—like clean water, food, fuel—that are used or consumed, and the wastes then generated, is just a partial measure of the biosphere's regenerative capacity. (And by this measure humanity's footprint already exceeds Earth's ecological limits.)

But American prosperity and progress have come at great human costs, too. Forced removals of the continent's native peoples yielded land to newcomers from Europe and their descendants. The new republic became an economic and political power in large measure from a system of industrial agriculture fueled by enslaved labor. Consuming *other* people's labor, dispossessing *other* people of land and life connection to it, devaluing human rights, and diminishing one's community, autonomy, and health—these are not just events of the past. In this globalizing world, agribusiness giants like Cargill and ConAgra can take advantage of enslaved labor in Brazil at a comfortable moral distance. And far too many degraded environments in America are also citizens' homes: forty out of forty-four states with hazardous waste facilities have disproportionately high percentages of people of color and the economically poor living next to those sites.

---

1. *Millennium Ecosystem Assessment* (2005), www.millenniumassessment.org.

As I listened to conversations that ignored these facts and their consequences for climate change action, *An American Dilemma* came to mind.[2] When the United States entered the Second World War, a social economist named Gunnar Myrdal wrote from Princeton, New Jersey, of his troubled and hopeful impressions of American society. He'd traveled across the nation, particularly in the South, and finally concluded that Americans seemed to be under the spell of a "great national suggestion"—the ideals of an "American Creed of liberty, equality, justice, and fair opportunity for everybody"—that failed many of the nation's citizens in actuality. This failure, Myrdal thought, came from a moral struggle that raged in the minds and hearts of most white Americans as creed clashed with daily life. The inequities that were forced on many of the nation's citizens because of differences in race, class, and gender were "perhaps the most glaring conflict in the American conscience and the greatest unresolved task for American democracy." These words come not from letters to Myrdal's distant home in Stockholm, but from the collective labor of a research team he directed under the auspices of the Carnegie Corporation and published in *An American Dilemma*, a volume of nearly 1,500 pages.

Structural racism and exclusion need not require intentionality so much as ignorance or compromise. People of color were largely left out of the major social policies of the early and mid-1900s that built today's middle class, and current policies largely, even if unintentionally, reinforce disparities created then. The Social Security Administration, which was established as part of FDR's New Deal in the 1930s, long excluded domestic and agricultural workers—more than two-thirds of the African American population at that time—from retirement, disability, and unemployment benefits. The reason: Roosevelt yielded to southern politicians for their votes. Nearly all major New Deal labor laws excluded migrant farm workers, and today's federal laws don't protect them from unfair labor practices. Even an attempt by the Truman administration to nationalize health care in the early 1950s failed in part because the white South feared integration of the health-care system.

---

2. Gunnar Myrdal, *An American Dilemma: The Negro Problem and Modern Democracy* (New York: Harper & Row, 1944).

The unspoken issues in that lecture hall reminded me, just as much as the recent rise in hate crimes, that the American dilemma remains. This society contains many walls, seen and unseen, reflecting patterns of living that dis-member, alienate, and exclude. I've often been asked why people of color don't participate in or contribute to the environmental movement. (We do.) And I've been asked, "Why don't more of 'you' come to the table?" (Well, my preference is that "we" design and build a table together rather assume "others" should come to one already built without their input.) Gated suburban communities—with Keep Out and No Trespassing signs—are paralleled by perhaps less visible but no less powerful borders segregating ideas and senses of who "we" really are. *It doesn't affect me. It's somebody else's problem. It isn't my job.* Those with the wealth and power to distance themselves from the impacts and consequences of environmental contaminants make "not in my back yard" a legal but deadly joke.

How many members of the conservation and environmental movements of the last half century chose to focus within a narrow frame, thinking there was no need to recognize any intersections with other movements like civil rights? People directly experiencing the impacts of contaminated environments or climate change as loss (of home, of food source, health, or livelihood)—and those who are just trying to meet basic human needs—haven't had the opportunity to choose.

This country long ago became accustomed to the unacceptable. A presumption of equality might exist in the minds of many Americans, but huge gaps do exist between expectation and reality. At times I think the anomalous "moral lag in the development of the nation" that Myrdal posited really is a reinforcing symbiosis of embedded otherings (by classism and racism), capitalism's growth-for-profit imperative, and an unexamined mythology of American democracy.

I hesitate writing "we must" or "we should," because such imperatives mean little without both a widespread recognition and moral acceptance of all "we" are. And Aldo Leopold's call to enlarge the moral boundaries of the "community of interdependent parts," and of the social conscience, to the "land" writ large can't be answered if the community's human members deny or avoid our own interconnectedness beyond hierarchies of dominance.

There are no simple answers. Perhaps there are none at all unless each of us chooses to re-member human identity in the largest possible life-sense. Each day, each moment, offers a call to be conscious of divisions and an opportunity to recognize to what and whom we are all connected and thus obligated. The alternative may be a life framed by false problems in false situations.

There Is a Tide

Ismail Serageldin

Half a century ago, half a world away, a youthful President John Kennedy addressed young graduates at American University. Speaking of the need for all the peoples of the world to collaborate to push back the dangers of nuclear war and to promote world peace, he said: "For, in the final analysis, our most basic common link is that we all inhabit this small planet. We all breathe the same air. We all cherish our children's future. And we are all mortal."

Those words are as pertinent today as they were a lifetime ago. In the decades since he spoke those words, we have pushed back the specter of nuclear holocaust, gone to the moon, connected the human race with satellites and fiber, mastered the digital realm of bits and bytes, probed the most distant stars, unlocked the secrets of matter, and tinkered with the very building blocks of life. We have revolutionized our daily lives in myriad ways. Yet our greed and our shortsightedness leave the world a precarious place, almost as precarious as when President Kennedy spoke those fateful worlds.

New challenges loom ahead. These challenges are as serious and as daunting as those President Kennedy spoke of. They are challenges that need our ethical commitment, our solidarity, and a sense of com-

---

ISMAIL SERAGELDIN is director of the Library of Alexandria and chairs the boards of directors for each of the Bibliotheca Alexandrina's affiliated research institutes and museums. He has been awarded twenty-two honorary doctorates and serves as chair and member of a number of advisory committees. Serageldin has also served in a number of capacities at the World Bank, including as vice president for Environmentally and Socially Sustainable Development (1992–1998) and for Special Programs (1998–2000). He has published over sixty books and monographs and over two hundred papers.

mon purpose, for indeed, we are all inhabitants of this small planet, and our greed and our misdeeds are risking past gains and future prospects. We must confront globalization, poverty, and hunger, and above all take to heart the cause of the environment. Yes, the environment. For we are all dependent on the ecosystem that we share in this small planet, which we have put at risk, as never before, by the changes in our climate that we have triggered by our inexorable drive for more, more, and ever more.

No one denies that much has been done to make the world a better place for all. The twentieth century was one of struggle for emancipation. The colonies were liberated; women got the franchise. Totalitarianism has been largely defeated, and democracy is spreading more widely than ever before. Racial, ethnic, and religious minorities and nonconformists were all acknowledged to have political and civil rights that derive from their common humanity. Around the planet, more people than ever enjoy these freedoms. This has not come easily, and the blood of millions was the price that was paid to reach where we have reached today.

On the socioeconomic front, in the last fifty years the developing countries have doubled school enrollments, halved infant mortality and adult illiteracy, and extended life expectancy at birth by an amazing twenty-five years. But despite that, much, so much, remains to be done. A global developmental agenda demands our efforts and our solidarity.

Consider the paradox of our times. We live in a world of plenty, of dazzling scientific advances and technological breakthroughs. Yet our times are marred by conflict, violence, economic uncertainty, and tragic poverty. A sense of insecurity pervades even the most affluent societies. Nations are looking inward, and the rich turn their backs on the poor. The world spends fourteen times more on military spending than on development assistance.

Greed and a lack of ethics have brought the world economic system to a crash of unprecedented magnitude. Bankers piled up credit default swaps that exceed the GDP of the entire world and paid themselves astronomical salaries as they bankrupted their institutions, the companies of the world, and the treasuries of their nations. Now they dodge the consequences of their actions as millions are thrown into unemployment.

Greed and lack of ethics have made a mockery of the commitment that world leaders took in the beginning of this millennium to reduce the number of the hungry from 850 million to 425 million by 2015. Today, in 2009, it is at 950 million and inching toward a billion people! Driven by drought, scarcity of water, and climate change as much as by war, civil strife, and failed socioeconomic policies, a billion people are going hungry.

In the nineteenth century, some people looked at the condition of slavery and said that it was monstrous and unconscionable—that it must be abolished. They were known as the abolitionists. They argued not from economic self-interest, but from moral outrage. Today the condition of hunger in a world of plenty is equally monstrous and unconscionable and must be abolished. We must become the "new abolitionists." We must, with the same zeal and moral outrage, attack the complacency that would turn a blind eye to this silent holocaust, which causes some forty thousand hunger-related deaths every day.

We must, with the same clarity of purpose and determination, devote ourselves to challenging the greed and the environmental destruction that surrounds us.

We, the new abolitionists, must say to our leaders: Enough! We want a new world order, a more equitable and caring world order. We want a more ethical order that looks beyond the immediate and addresses the overarching problems of our time, with climate change at the top of the list. No more denial, no more Band-aids. It is time that we acted as true stewards of the Earth.

Consider for a moment the alternative, the costs of the inaction that threaten the world if we continue with business as usual. For *there is a tide* . . .

There is a tide of carbon and methane emissions that are driving the changes in our climate and bringing storms and destruction, droughts and floods.

A changed climate, where habitats disappear, food production is threatened, and life becomes precarious in vast parts of the world.

The future beckons. We must embrace it and bend it to the pattern of our dreams.

So let us create a united front of caring. To think of the unborn, remember the forgotten, give hope to the forlorn, and reach out to the unreached, and by our actions from this day onward lay the foundation for better tomorrows.

A Fair Deal

Peter Singer

How will we divide up the diminishing capacity of the atmosphere to absorb our greenhouse gases?

If we apply the "You broke it, you fix it" principle, then the developed nations have to take responsibility for our "broken" atmosphere, which can no longer absorb more greenhouse gases without the world's climate changing.[1] According to United Nations figures, in 2002 per capita emissions of greenhouse gases in the United States were 16 times higher than in India, 60 times higher than in Bangladesh, and more than 200 times higher than in Ethiopia, Mali, or Chad. Other developed nations with emissions close to those of the United States include Australia, Canada, and Luxembourg. Russia, Germany, Britain, Italy, France, and Spain all have levels between a half and a quarter that of the United States. This is still significantly above the world average, and more than fifty times that of the poorest nations in which people will die from global warming.

Americans tend to talk a lot about morality and justice. But most Americans still fail to realize that their country's refusal to sign the

---

PETER SINGER is an Australian philosopher who specializes in applied ethics. He is the Ira W. DeCamp Professor of Bioethics at Princeton University and laureate professor at the Centre for Applied Philosophy and Public Ethics at University of Melbourne. He served as chair of philosophy at Monash University and founded its Centre for Human Bioethics. He coauthored his most recent book, *The Ethics of What We Eat: Why Our Food Choices Matter* (2007), with Jim Mason.

---

1. The preceding material is from Peter Singer, "A Fair Deal on Climate Change," June 2007, http://www.project-syndicate.org/commentary/singer24.

Kyoto Protocol, and their subsequent business–as-usual approach to greenhouse gas emissions, is a moral failing of the most serious kind. It is already having harmful consequences for others, and the greatest inequity is that it is the rich who are using most of the energy that leads to the emissions that cause climate change, while it is the poor who will bear most of the costs.

To see the inequity, I merely have to glance up at the air conditioner that is keeping my office bearable. While I've done more than the mayor of New York requested, setting it at 82 degrees F (27 C), I'm still part of a feedback loop. I deal with the heat by using more energy, which leads to burning more fossil fuel, putting more greenhouse gases into the atmosphere, and heating up the planet more. It even happened when I watched *An Inconvenient Truth:* on a warm evening, the cinema was so chilly that I wished I had brought a jacket.

Heat kills. A heat wave in France in 2003 caused an estimated thirty-five thousand deaths, and a hot spell similar to the one Britain had in July 2009 caused more than two thousand, according to official estimates. Although no particular heat wave can be directly attributed to global warming, it will make such events more frequent. Moreover, if global warming continues unchecked, the number of deaths that occur when rainfall becomes more erratic, causing both prolonged droughts and severe floods, will dwarf the death toll from hot weather in Europe. More frequent intense hurricanes will kill many more. Melting polar ice will cause rising seas to inundate low-lying fertile delta regions on which hundreds of millions of people grow their food. Tropical diseases will spread, killing still more people.

Overwhelmingly, the dead will be those who lack the resources to adapt, to find alternative sources of food, and who do not have access to health care. Even in rich countries, it usually isn't the rich who die in natural disasters. When Hurricane Katrina hit New Orleans, those who died were the poor in low-lying areas who lacked cars to escape. If this is true in a country like the United States, with a reasonably efficient infrastructure and the resources to help its citizens in times of crisis, it is even more evident when disasters strike developing countries, because their governments lack the resources needed and because when it comes to foreign assistance, rich nations still do not count all human lives equally.[2]

But there is a solution that is both fair and practical:

- Establish the total amount of greenhouse gases that we can allow to be emitted without causing the Earth's average temperature to rise more than 2 degrees C (3.6 degrees F), the point beyond which climate change could become extremely dangerous.
- Divide that total by the world's population, thus calculating what each person's share of the total is.
- Allocate to each country a greenhouse gas emissions quota equal to the country's population, multiplied by the per person share.
- Finally, allow countries that need a higher quota to buy it from those that emit less than their quota.

The fairness of giving every person on Earth an equal share of the atmosphere's capacity to absorb our greenhouse gas emissions is difficult to deny. Why should anyone have a greater entitlement than others to use the Earth's atmosphere?

But in addition to being fair, this scheme also has practical benefits. It would give developing nations a strong incentive to accept mandatory quotas, because if they can keep their per capita emissions low, they will have excess emissions rights to sell to the industrialized nations. The rich countries will benefit, too, because they will be able to choose their preferred mix of reducing emissions and buying up emissions rights from developing nations.[3]

---

2. The material following note 1 is from Peter Singer, "Will the Polluters Pay for Climate Change?" August 2006, http://www.project-syndicate.org/commentary/singer14.

3. The material following note 2 is from Singer, "A Fair Deal on Climate Change."

The Moral Climate
Carl Safina

What most distresses me is the thirty years we've wasted by asking people to live sustainably. In high school in the early 1970s, having grappled with terrible air pollution, oil embargoes, and the tyranny of Big Energy, we knew we needed more economical and efficient cars, we knew we were vulnerable to unsavory governments, and we knew that *real* free enterprise in the form of energy competition would mean innovative, diverse, agile, and decentralized energy sources. From then to now, as a country, we've sat on our hands.

This, I believe, is our shame. Shame not only because we chose it. Shame because the unborn, who did not choose it, will come saddled with all conceivable consequences. Shame because the poor, who likewise did not choose it, will be hit first and worst.

And because that is not merely "unsustainable" but also *unjust*. It is wrong. And so it crosses a line and becomes not just a matter of "sustainability," but a matter of morality.

Dysfunctional values married to catastrophic leadership have led us to the place you go when you are made to believe that solution is sacrifice and that sacrifice for a just cause is not noble but, rather, out of the question. The moral density of this social climate is wafer thin.

CARL SAFINA is founding president of Blue Ocean Institute. He is a MacArthur Fellow and a Pew Fellow and has received the Lannan Literary Award and the John Burroughs Medal, among other awards. *Audubon* magazine named him among the top one hundred twentieth-century conservation leaders. He has been profiled by the *New York Times*, *Nightline*, and Bill Moyers. His books include *Song for the Blue Ocean* (1998), *Eye of the Albatross* (2002), and *Voyage of the Turtle* (2006), and he has been featured in *National Geographic*.

This refusal to "sacrifice" is actually a pathological refusal to change for the better. *That* is the real sacrifice. That refusal is framed and abetted by the disinformation campaigns of companies that would shrink if we realized we would be better off with fewer of them. Think of ExxonMobil; it's probably the best example. Their fear of us—specifically, that we might accept the consequences of reality—compels them into a rather successful effort to retain power over us by distorting our understanding of what's real.

Nearly every just cause is a struggle between the good of the many and the greed of a few. But because greed has the advertising dollars to make selfishness fashionable, it sustains itself by turning enough people against our own self-interest. Foremost, our interest in hanging on to our own money. Second, our health. Third, the options of our unborn.

Of all the psychopathology in the climate issue, the most counterfunctional thought is that solving the problem will require sacrifice. As though our wastefulness of energy and money is not sacrifice. As though war built around oil is not sacrifice. As though losing polar bears, penguins, coral reefs, and thousands of other living companions is not sacrifice. As though withered cropland is not a sacrifice, or letting the fresh water of cities dry up as glacier-fed rivers shrink. As though risking seawater inundation and the displacement of hundreds of millions of coastal people is not a sacrifice—and a reckless risk. *But don't tell me we need a law mandating more efficient cars; that would be a sacrifice!* We think we don't want to sacrifice, but *sacrifice* is exactly what we're doing by perpetuating problems that only get worse; we're sacrificing our money, sacrificing what is big and permanent, to prolong what is small, temporary, and harmful. We're sacrificing animals, peace, and children to retain wastefulness—while enriching those who disdain us.

When we stop seeing our relationship with the whole living world as a matter of sustainability and realize that it is a matter of morality—of right and wrong—we might make the moment we need.

And when we make that moment, we will begin to loosen the noose we've placed around our children's necks. The world of debt we've doomed them to, the upheaval caused by a destabilized climate and the consequent threats to peace, the draining biodiversity. One of the starkest, most telling contrasts between my generation and the one now in high school is that our parents all thought they could leave their children set up to have a better life. Most people now fear for their kids. And the blame is ours. But since the problems are largely of our making, we have the power to flip them. We just need to create the needed resolve. I know we can, and I think we will.

ETHICAL ACTION

How can we respond to the demands of justice in a time of climate change?

To act justly is to take no more than one's fair share of the Earth's resources, including the capacity of the atmosphere to absorb carbon dioxide without causing major climatic change. Simply take the population of the Earth and divide it into the available resources; the resultant number will tell you the share of resources you can fairly claim. Unfortunately, if you are an average American, this number will be approximately one-fifth of your current use.

What to do? The obvious answer is to get as close to your fair share as you can by making good personal decisions (what to eat, whether to travel, how to generate electricity, etc.) and, where your progress is blocked by institutional inefficiencies (absence of public transportation or even sidewalks, lack of solar power, etc.), agitating for change. Here is where you consult the "change out your light bulbs" to-do lists that you can find on a dozen Web sites. You do all the things on the usual list, of course: insulating your home, switching to efficient appliances and fluorescent bulbs, carpooling or bicycling, buying local food and secondhand clothes, turning off the lights and unplugging the electronics, bringing shopping bags and coffee mugs to the market, washing clothes in cold water, xeriscaping the yard. That might get you a little closer—and you absolutely must do it, all of it.

Next? Think hard about the most important personal decision you will make: Whether to bring children into the world. It's easy to think of overpopulation as a moral failing of other people, other religions, other continents. But if you have more than one or two children, overpopulation is a moral problem in your household. No one "deserves" more children than anyone else; in fact, affluent Americans may have difficulty claiming even equal rights to children, given the global impacts of our lifestyles and life spans. Overpopulation in a time of climate crisis creates terrible problems of distributive justice; overpopulation coupled with overconsumption creates injustice that is exponentially worse.

Even the most conscientious probably can't claim only their just portion of the world's resources right off the bat. At the very least, though, you can *acknowledge* that your life is fundamentally unfair. There will be no arguments that you ought to have more than say, a Bangladeshi, or that you have a right to be richer. And it's worse than that. Even as you lead an unjustifiably consumptive life, it will be gracious to acknowledge that not only are you taking more than your fair share, but your lifelong history of taking more than your fair share has in fact exponentially increased the undeserved suffering of others. So there is a reparations issue to deal with as well.

Next step? At this point, we'd all better tithe, donating 10 percent of our annual incomes to climate action or mitigation of the affects of climate change on the less advantaged. That's 10 percent that goes toward our debt to the world's poor, and it's 10 percent less that we're spending on consumer items that will hurt them even more. That's a minimally decent start.

11

> *Do we have a moral obligation to take action to protect the future of a planet in peril?*
>
> **Yes, because the world is beautiful.**

We can make the argument this way:

Premise 1, an aesthetic affirmation: This world is beautiful.
Premise 2, a moral affirmation: What is beautiful must remain.
Conclusion: Therefore, this world must remain.

Premise 1. It's not just the sun in winter, the salmon sky that lights the snow, or blue rivers through glacial ice. It's the small things, too—the kinglet's golden crown, the lacy skeletons of decaying leaves, and the way all these relate to one another in patterns that are beautiful and wondrous. The timeless unfurling of the universe, or the glory of God, or an unknown mystery, or all of these together have brought the Earth to a glorious richness that awakens in the human heart a sense of joy and wonder.

Premise 2. It is right to protect what is beautiful, wrong to destroy it. This is part of what "right" means—to enhance, rather than diminish, what we value. And this is part of what "beautiful" means—to have value in itself, a quality worthy of continued existence. It would be vandalism to slash Van Gogh's glorious painting of a starry night. What greater moral crime, then, to poison an estuary that sustains a multitude of starry flounder and sea stars? And when something beautiful is destroyed forever, what more appropriate moral response than fury at that terrible wrong and grief at the irretrievable loss? And when the destruction is done knowingly and in exchange for something of far lesser value, words can hardly express the appropriate moral outrage.

Conclusion. Some people say that we live in two worlds—the world of what is and the world of what ought to be. But a full appreciation of the beauty and wonder of the world closes the gap between what is and what ought to be. If this is the way the world is—beautiful, astonishing, wondrous, awe-inspiring—then this is how we ought to act in that world—with respect, with deep caring and fierce protectiveness, and with a full sense of our obligation to the future, that this beauty shall remain.

Our Edens: Ecological Homes

Bernd Heinrich

The Hahnheide in northern Germany is a forest of oaks, beeches, and spruce. At the end of World War II, when I was a small child, our family lived a partial hunter-forager existence in that forest. For six years, my parents gleaned food, warmth, and shelter from the land. My most vivid memories are the magical sensations of finding nests: of winter wrens in the upturned root disks of fallen trees, of the song thrush in a spruce tree, the chaffinches' neat and tidy cup of moss and hair on a beech limb, the long-tailed titmouse's lichen-decorated little nest bag wedged into an alder fork. When I went back to visit the Hahnheide a half century later, I felt that I had returned to a dreamland. The details of the surroundings near these nests took me back to a lost world that resonated with an inner happiness. Overcome by the unbearable joy and sadness in those memories, I fell to my knees and wept.

Near my eleventh birthday, we moved to a real-life other world: we came across the ocean and lived in a run-down farm in a rural section of western Maine. The large hand-hewn timber-framed sheep barn was ready to collapse, and the roof on our long-empty house leaked badly. Every evening in the summer a flock of chimney swifts zoomed over our roof. After making several passes, they descended into our chimney, where they glued their nests of small twigs with

---

BERND HEINRICH is a biologist and professor emeritus of biology at the University of Vermont. He has authored many books and nearly two hundred research papers on animal physiology, behavior, and ecology. His scientific and creative nonfiction articles have appeared in *Science, Scientific American, Smithsonian, Natural History,* and the *New York Times*. His latest book, *The Snoring Bird: My Family's Journey Through a Century of Biology* (2007), is a memoir.

saliva onto the bricks sooted by decades of woodsmoke from hearth fires. The barn was surrounded by thriving patches of lavender fireweed, orange jewelweed, overgrowing fields sprinkled with orange and yellow hawkweed, blue vetch, purple and white clover.

The house was surrounded and almost enveloped by seven ancient sugar maple trees. I used the weathered barn boards to build bird boxes that I hung up in our maple and apple trees. They were soon occupied by pairs of kestrels, tree swallows, starlings, and bluebirds. Phoebes nested in our outhouse, and several pairs of barn swallows twittered in the barn and built their mud nests on the beams. The surrounding fields were vibrant with the song of bobolinks, meadowlarks, savannah sparrows. I sometimes slept in the fields in early spring shortly after the snow melted and while the pussy willows were in bloom, going to sleep and waking the next morning to the sounds of the wildly exuberant sky dance of the woodcock. I was in an ecosystem as varied and vibrant as that of the nearby woods, the taiga forest, and I thought I was in heaven.

Our farm was for me the launching place for a love of the northern forest. It stretches nearly unbroken across the Northern Hemisphere. It is a vast assemblage of different ecosystems, and each of them presents four totally different aspects—spring, summer, fall, and winter. The most long-awaited one every year was for me the spring, when the Earth metamorphoses from quiescence to a bursting. Birds by the billions hurry north through the night skies, each to its tiny plot of land maybe centered around a few spruce trees, an old maple tree, a patch of brush by a blowdown—sibilant voices everywhere. And then, in the first week in October, the leaves turn yellow, magenta, red, and brown. Deer rut, and hunters enter the woods. But soon enough, all the leaves are down and matted into the ground, covering moth pupae and the runnels of the mice, shrews, and moles that hunt them. The trees stand in stark silence, except for the howling wind and the snow sweeping through them. Many of the birds have left to fly to another continent, and much of life retreats underground.

Now I live in Vermont, comfortable in a warm house surrounded by woods. I look through the window and see dark skies, and snow falling on an already deep layer that blankets the ground and leaves thick cushions bending down the branches of the evergreens. Temperatures outside are near zero. Within feet of my window, flocks of chickadees flit through the trees, cheeping and chirping and doing acrobatics while picking at the twigs and looking into every little thing as though endowed with an unquenchable curiosity and unbounded energy. When I look at them from up close—they are very tame—I admire their coal-black bibs and caps. The birds' beauty is more than feather-deep. Their exquisite behaviors and physiology are a marvel of adaptation to their boreal home. They have made themselves over to be at home in the boreal forest, but it has taken them millions of years of evolution to accomplish it.

On this particularly cold and snowy day I spot a Carolina wren, a bird I have never before seen in my twenty-eight years in rural Vermont. This particular individual first appeared around our house at a time when there was much movement of the birds. The Canada geese had been migrating high overhead in their large V formations, all heading south. White-crowned sparrows came and stayed even longer before moving on, and then on the eighth of October I heard the excited chittering calls of a wren. It immediately caught my attention because it had a different ring than those of the winter, house, and marsh wrens with whom I was familiar. And then, in the brush next to my study, I saw the large wren with its unmistakably light eyeline. It stayed, and it is still here even though its normal range would keep it in the Carolinas. It now seems to have moved into my woodpile, and it occasionally feeds on our compost heap.

Today I also hear a crow, and two robins feed on the berries of a viburnum bush. Their species-mates are for the most part long gone to more southern climes. The wren may have decided to stay because he found a special little "niche"—a home in the environment that is different from that of any other bird. The woods around me are full of niches, of potential homes. Niches provide homes to millions of species, dividing the environment up into ever-finer partitions. The ecological niches or homes of some of the birds span large portions of the American continent, but the addresses of the individuals are even more fine-grained than the niche home, because they are adapted through learning and imprinting and experience, not just by genetics and evolutionary memory. The birds have their Edens, and we have ours.

Our Edens are here on Earth, and only here. To suppose they might be elsewhere is blasphemous. I use the plural for Eden deliberately, and also the biblical metaphor and connotation of sin. There can be no greater sin and folly than to endanger an ecosystem, and the one truth that science has bestowed upon us is that all of the Edens—all of the Earth—are interconnected and interdependent. Ironically, each of the millions of species in all of the ecosystems on Earth would destroy its own ecosystem, if it could. That is not because it wants to. It is because of its unquenchable drive to expand and grow until it is constrained by other life doing the same, even as each depends on the other. We are indeed a unique species. We are the only one who harbors the knowledge that there is incomparable complexity to an ecological community where each member has a role to play. Each species is not just a component of a beautiful picture. It is also what makes the canvas a living, breathing whole. The harmony developed in a coevolved process involving the members within it, where each was adapted to and with the others through the processes of competition, mutualism, symbiosis, parasitism, and predation, so that all were balanced for an unending sustainability.

It took some 4.5 billion years for evolution to "make" this fabulous life on Earth. Each of the millions of organisms is arguably as fabulous and unique as we are, and any one ecosystem is composed not just of a handful of species, but of thousands, maybe millions. And we, at the present time, are in the process of a mass destruction. Never before in the Earth's history has one species had the power of ecocide on a global scale.

Our home is the Earth, and all that is in it is part of that home. We do not need to spend one single cent of our precious resources to find out if, when, where, and how much water there might be on Mars or any other of the billions upon billions of other objects out there in the cold distant void of space. Not one penny. Not, at least, unless every ecosystem here is safe, and we and other animals in it can lead secure and sustained existences through time and in the chain of life. We have had a fabulously exciting journey into space. What more do we need from elsewhere if we don't know, appreciate, cherish, and protect what we have here? Let us not for one nanosecond entertain the thought of making another distant globe our home, and solving our ecological problem by moving away. Trying to do so is like banking on "heaven" or "god" to bail us out, and saying to hell with the whole creation.

We need a global perspective of deep time, and the relevant timescale is not that of the quarterly earnings of a corporate business, but one that incorporates consequences generations ahead. Any deviation from a system such as the global ecosystem, whose interconnected life took several billions of years to evolve, will have consequences—now, next year, or a century or more ahead. You can't change one thing without affecting the rest. An immediate apparent advantage now all too often merely delays the cost. Payback will come sometime. It always comes. That is one of the few "laws" of nature.

We are now in a frightening and long-sustained population explosion. It is as "natural" as birth and death. It has happened many times before in our history, but always locally. The civilization collapses, then the people die of wars, hunger, and disease, or move to someplace else. It took a while, but when we left Africa we filled the Eurasian continent. It took the Europeans only an instant to fill America, and Australia. Now the rules have changed. Frighteningly, we have made the whole Earth a commons; we are all interconnected, so if we keep on as we are now doing, the collapse will be global, not just local. This time it would be for all time. You can't resurrect the panoply of an ecosystem, with its thousands of integrated species.

At my window this morning, I gratefully drink hot coffee that has been laboriously harvested by peasants in some distant tropical country and transported here, perhaps over an ocean by ships burning oil from Arabia or coal from Appalachia. I boiled it in water pumped up by power from a to-me-unknown source, from an underground aquifer, and I heated it by turning a dial on a propane burner. A blue flame of combusting gas came from underground from god-knows-where, then heated it. We cannot imagine that this home and these behaviors don't have consequences. They do. They have consequences for the chickadees, the geese, the wren in its unaccustomed cold. They have consequences for the seasons, the return of snow. They have consequences for the future.

We are ecologically bound to the world ecosystem, and collectively we are no different from any other species. We conform to the same laws of supply and demand. We have invented one technological fix after another, all the time unconcerned or unable to do anything about the magnitude of the long-term consequences. It happened with the invention of agriculture; then with the tapping of energy from coal, then oil, then the atom; perhaps now even with the computer and more fuel-efficient cars. But all of that is mere blowing in the wind because it merely supports more people, and so pushes the day of reckoning another day ahead. Eventually, the "future" has a way of becoming the present. The "new" that comes onto the horizon all too often seems like salvation but is in reality a Trojan horse full of ultimate surprises. Disruption is expected in something that took billions of years to build. But sudden, sustained, and unrelenting disruption of this scale has not been seen on Earth before. You can't go back or backward, we are told. Yes, we can. The question is only whether we will. Will we do it slowly and deliberately, getting it right along the way? If not, it will come upon us later, and calamitously.

Wolves, Ravens, and a New Purpose for Science

John A. Vucetich

In language as sterile as it is precise, scientists document the realities and predict the risks that humans are causing catastrophic harm to our environment and, in the process, to ourselves. Scientists cannot figure out whether we ought to be merely the messengers of the facts or whether the ethical implications of these particularly dire facts require us to try to influence policies—and, if so, how.

Technocratic decision makers, those whose work is to "manage nature," vacillate between basing their decisions on "best-available science" or "stakeholder involvement." Unable to sensibly resolve these worthwhile but sometimes conflicting fundamentals, their resulting decisions are often a schizophrenic nightmare.

Of those in the general public who even care, most feel unable to influence the institutions of science and management. Many trade hope for despair; others insist that hope is the only option.

The result has been a disastrous failure in our relationship with nature. Scientists, users of science, administrators of science, lovers of science, anyone in love with and committed to knowledge: What is our responsibility in correcting this mess?

Beating on these problems from two sides—one scientific and one

---

JOHN A. VUCETICH is a professor of animal ecology at Michigan Technological University. He coleads research on the internationally recognized wolves and moose project of Isle Royale, the longest continuous study of a predator-prey system in the world. His work also focuses on the relationship between ecology and philosophy and ethics. He has served as a science or policy adviser for wolf and ungulate management issues in Alaska, Alberta, Ontario, Scandinavia, France, New Mexico, and Michigan.

ethical—seems like a good idea. Except that dire facts are already in abundant supply. And our hearts are as resistant as Teflon to admonitions to care for the environment, no matter how powerful the ethical or scientific reasoning. Hasn't ethics always said we have an obligation to compassion and justice? If so, what more can ethics do? And if the root cause of the environmental emergency is a disastrous lack of compassion and justice, how can science help that?

The answer depends on what we take to be the purpose of science. The purpose of science matters. It affects every aspect of science, from how it is conducted to how it is used. What, then, is science for?

Most ecological science is funded by governments, whose purpose is to serve a public whose predisposed interest is to manage and control nature. Consequently, the operational purpose of ecological science is inevitably to develop the ability to manage and control nature, by explaining the causal mechanisms of ecological systems. Since Galileo, the ultimate evidence for such elucidation has been prediction. We continue to fulfill Francis Bacon's prophecy that science will be able to predict nature, so that we can control nature, all for the sake of "easing man's estate"—which is the fundamental and controlling value. This is the ethical assumption underlying the collaboration between modern technocracy and the purpose of science. It is no small insight that ethical attitudes, not science, are the source of our understanding of science's purpose.

After three hundred years, it is time to overhaul the purpose of science. What is science *good for* in this time, in this world? The way forward is an unfamiliar landscape and needs to be shown as much as explained. And we'll need specific examples to critique and use as resources for emulation or avoidance, whichever is appropriate.

In this spirit, consider how we have been experimenting with the purpose of one scientific project, the wolf and moose research project on Isle Royale, a wilderness island in Lake Superior, North America. On the project's fiftieth anniversary, in 2008, we reflected on the question, "What have been the most important lessons learned from the wolves and moose of Isle Royale?" The exercise was supposed to be a referendum on the virtues of long-term research, and I suppose it was. More important, the reflections expose a scientific purpose that transcends the control of nature and may represent some general insight about what role science should play as we move into an uncertain future.

Five decades is certainly enough time to collect a healthy list of discoveries, but which, if any, represent significant knowledge? Consider these discoveries:

First, from the wolf-moose project, we know what kind of moose wolves kill. Wolves tend to kill the weakest moose, those that are starving, old, injured, or vulnerable to severe winters. This seems like common knowledge among ecologists of today, but Isle Royale is the place where that knowledge was first discovered. Among prime-aged moose, we also know that wolves have a strong tendency to prey on smaller moose. These patterns of wolf predation differ fundamentally from the way humans tend to hunt moose, deer, and elk. Consequently, when wolves are removed from an ecosystem, their effect is not replaced by human hunting. Observations like these tell us how wolves are critically important parts of healthy ecosystems.

Second, the project offers insight about the effects of wolf predation on prey populations. Wolves do not decimate their prey. Nevertheless, we know the effect of wolves is important and complicated. This knowledge is vital for honest dialogue about how we plan to share the land with wolves. And living with wolves certainly requires discussing how much—in the way of deer, elk, and moose—hunters should be willing to share with wolves.

Third, the project provides the world's most detailed understanding of how wolves die when humans are not involved in killing them. As wolves recover in several regions of the world where human-caused mortality is a threat, this knowledge from Isle Royale is valuable for understanding how to restore and conserve wolves. Ironically, this knowledge is also useful to understand the maximal amount of hunting that a wolf population can endure and still remain viable.

As I described these discoveries, I gave some sense of why they may be significant. But shouldn't we ask more precisely what makes any piece of knowledge valuable? The answer begins, I think, by recognizing that there are two kinds of knowledge.

One kind of knowledge helps us do things in the world—helps us conserve nature, restore damage we've caused nature, and live sustainably. However, the knowledge that helps us do good things can also be used for the most disgraceful endeavors—to live unsustainably and to unnecessarily exploit others, human or otherwise. Our attitude determines whether we use knowledge to do right or wrong, good or bad.

Knowledge that can change our attitude about nature is the second, and more important, kind of knowledge.

Think about knowledge that makes you go "Wow!" Wow, that's so beautifully complicated . . . Wow, look how magnificently nuanced . . . Wow, how astonishingly connected. Wow: to be held in a state of wonder about nature. It would seem awfully difficult to intentionally abuse nature while being held by its wonder. How can you do anything but care for nature, while astonished by its beauty, complexity, and interrelatedness?

Recently we discovered a special relationship between wolves and ravens. The presence of ravens influences the size of wolf packs: wolves living in larger packs each get more food because they lose less food to scavenging ravens. They do this by eating a moose so quickly that ravens have little time to scavenge. The details are fantastically complicated, and while wolves in larger packs must share food among their brothers and sisters, parents and offspring, that sharing is not so costly as losing food to scavengers. So ravens have something to do with explaining why wolves live such intensely social lives—a trait that is otherwise rare among carnivores. What an astonishing connection. We also recently described connections whereby ticks, affected by climate, cause moose to flourish or suffer, which in turn has a cascading affect on wolves and forest growth.

These discoveries grabbed much press attention. But why? Certainly not because the knowledge is valuable for doing anything, let alone for controlling nature. The discoveries are appreciated because they reveal astonishing and intricate ecological connections. The value of connections like those lies in their ability to generate wonderment and care for nature.

The wolves and moose of Isle Royale have long been in the business of changing attitudes. The Isle Royale wolf-moose project began fifty years ago during the darkest hour for wolves in North America. The genocidal attitudes we held toward wolves rested on our vilifying wolves. The Isle Royale project was an antidote to the vilification.

The project used knowledge to replace our myths with a sense of wonder for wolves, a sense of wonder reflected each time we say that wolves are an important part of healthy ecosystems.

With each passing year, the most important lessons from Isle Royale have been coming to understand how inadequate our previous understandings had been. Every period in the wolf-moose chronology seems to differ from every other period. Although we can expect the next fifty years to differ substantially from the first fifty years, we are strangely in no position to say how. Our abilities to predict what happens next on Isle Royale are comparable to predictions for weather and financial markets.

Some of our unhealthiest relationships with nature rise from the hubris of overestimating our ability to understand and predict. The collapse of cod fisheries in portions of the North Atlantic and the annihilation of the once-rich Aral Sea represent just two dramatic examples of how damaging such hubris can be. Sadly, much more catastrophe is on the way unless we learn that sustainability has less to do with developing more efficient ways to exploit nature, and more to do with learning to be happy with consuming less—a happiness that would rise from being held by a sense of wonder toward nature.

The lesson Isle Royale offers is a paradox—the paradox of being proud for all the environmental sciences have learned, while being humble in the secure knowledge that what we know is a mere drop in an ocean of what one would need to know in order to say we really understand nature. The purpose of the Isle Royale project is not even aimed in the direction of controlling nature. In fact, the project represents scientific evidence for the limits of prediction. The purpose of the Isle Royale project is instead to generate a sense of wonder toward nature in as many people as possible.

Well, that's the example.

It's not a one-two punch—first science, then ethics. It's science and ethics commingling, stirring wisdom into action that creates a flourishing relationship with nature. Commingling is difficult. It requires sustained collaboration among scientists, ethicists, and artists. Expect a smaller portion of time working on traditional scientific activities. Expect to spend more time communicating (and learning how to communicate) your wonder about nature to a broad audience. Not the job of a scientist? Perhaps not, but it is the responsibility of a citizen.

What about politicians, technocrats, teachers, students, and the general public? What is their role in shaping the purpose of science? As consumers of science, their responsibility is to be open to wonder, to expect science to generate wonderment, and to use scientific knowledge for that purpose.

When we decide that the purpose of science is to generate wonder about nature, rather than to control nature, we will not be far from a relationship with nature that can flourish for all time and generations.

Get Dirty, Get Dizzy

Hank Lentfer

Lying in bed watching the winter sunrise light up the treetops, my daughter said, "Papa, the world is beautiful. All the animals are beautiful. Nature only made one mistake."

"What's that, sweetie?"

"It made us."

"Why do you say that?"

"Because we do things that we know are bad for the Earth."

"Like what?"

"Oh, you know, like drive cars and buy things we don't need."

I try to dissuade her. I tell her we are just as beautiful as the trees, the moose, the sky. But she shakes her head, convinced she is a living mistake.

I wish I could blame the television or the Internet, but Linnea has never seen a TV and is too young to log on to the computer. I wish I could blame it on a concrete jungle that cuts her off from the natural world, but growing up in an Alaskan village, she is surrounded by critters and creeks. In her five short years, Linnea has held the still-beating hearts of salmon and helped pull the hide off deer. She has followed her mother through acres of berry patches and washed hundreds of pounds of garden spuds. Still, Linnea feels separate from the natural world. She remains convinced we are an ugly species. She repeats it at the oddest times, when she's out peeing in the snow or

---

HANK LENTFER has a tight home range centered on a small coastal town in Alaska. When not hunting, gardening, pounding nails, playing with his daughter, or chatting with the neighbors, he enjoys writing and picking his guitar. In 2002 he coedited *Arctic Refuge: A Circle of Testimony* with Carolyn Servid.

riding in the sled, or just before she falls asleep looking through the skylight at the stars.

I cannot speak to my moral obligation to future generations—I have my hands full helping my daughter keep sight of her own beauty. I confess that Linnea's words reflect my own struggle. On my best days, after a morning listening to migrating cranes or an afternoon harvesting carrots, I feel and believe I am as beautiful as any other animal. But then the radio tells of prisoners shackled to floor and ceiling so they cannot sleep and my e-mail warns of polar bears swimming in an ice-free sea and my memory recalls a country founded on genocide, and suddenly beauty becomes a place where no human belongs.

A child must learn about speciesism just as she must learn about racism, and maybe I am my daughter's best teacher. If so, I need a new lesson plan. Seeking that guidance, I recall a dear friend who died years before Linnea's birth. My wife and I nursed Steve through his last days. He died at night, and we kept his body in the house through the next day. His young boys crawled into the bed and washed their father's body. All the grace, tenderness, love, and care with which those boys bathed their dad could not bring life back into a body growing colder by the minute. There was no material justification for cleaning Steve. He was, after all, on the way to the crematorium. Those boys were not looking for promises or hoping for miracles. They were groping for connection to a father who could no longer pitch a baseball. The care guiding those warm sponges over cold flesh provided a direction for the long slow path through grief.

Too often such tenderness is trampled beneath the weight of current events. Conversations about shrinking ice caps, the dangers of perpetual economic growth, and the politics of oil are commonplace in my house. Just because Linnea is quietly playing with her doll does not mean she is not listening. As she grows and takes in more news, the avalanche of doom and danger will only gain momentum. At five she is already talking of saving the planet, already feeling she is the only one who cares, already feeling responsible for the world's ills.

By high school Linnea will learn that humans are responsible for high temperatures and low aquifers, widespread poverty and obscene wealth. In college she may read *The End of Nature* and *The Long*

Emergency. She may return home to find fresh roads cut through her childhood forest. The push of anger and pull of urgency may usher her, like her father, into conservation work. She may work long hours for low pay, pounding against the gears of the Industrial Revolution.

Maybe with the help of friends and the richness of memory, she will also realize that violence toward the Earth, toward each other, grows from isolation. Maybe she'll see that in an effort to inspire immediate action, we too often use fear-mongering rhetoric that feeds our illusion of separateness. We know that you can scare someone into attending church but not into feeling a connection with God; but we forget that while you can scare someone into driving less, you can't scare them into feeling like a beautiful part of nature. Maybe she'll come to understand how toxic it is to feel apart from the very beauty we are striving to protect.

When Linnea was the size of a peanut, twisting and growing in Anya's belly, I was slammed by a father's realization that neither the nine months of pregnancy nor the nine decades of a lifetime are long enough to make the world a sane, peaceful place to raise a child, no matter how many letters I write or carrots I grow. But even if I could not secure a peaceful, abundant future for my child, I wondered, were there things I could do to help her live in a violent, stressful one? The longer I sat with this question, the more the tight grip of my desperation relaxed. The notion of prevention, I realized, leaves me paralyzed and numb. The notion of preparation, on the other hand, fills me with purpose and meaning. Impending chaos and need create untold opportunities for grace and service. What greater gift than a row of turnips in a time of famine? What greater need than a calm soul in a sea of panic? What better balance to horror than humor? What richer response to fear than love?

We activists need to understand that acts of caring are healing in themselves, and healing acts express the deepest care. Whether or not we can save the planet, we need to conserve its gifts. Whether or not we can save the planet, we need to care for it. Just as those boys were searching for a connection that transcends their father's death, we need a connection to the Earth that transcends the success or failure of our efforts.

As a father, I want to grow carrots not because agribusiness is a filthy, greedy, heartless beast, but because rooting in the dirt is fun, worms are groovy creatures, and you can't buy the sweet satisfaction of a fresh carrot at any price. I want to live a simple, rooted life not because a place of privilege feeds on other people's poverty, but because meals of venison, potatoes, and nagoonberry pie fill our kitchen with gratitude-crazed grins. I want to leave the Subaru in the driveway not because the carbon spilling from the exhaust will tip the planet into an inferno, but because a bike ride puts wind in your face and birdsong in your ears. It pumps blood through your veins and reminds us that life is a dizzyingly splendid idea.

The Feasting

Alison Hawthorne Deming

The sun hangs low over Race Point at the far tip of Cape Cod, sunlight thick and yellow turning the rippled sand into a patchwork of slate blue shadows and buttery highlights. Each ankle-high ridge and valley of windblown sand becomes a painterly study in color contrast. There is no gray zone at this late hour of day. The tide is receding. It has left a heap of tiny fish behind, like a row of hay raked up in a farmer's meadow. The fish look like little plastic snakes or gummy worms, their bodies glistening clear as spring water. They look like glass, except that the bodies twitch and flip a moment before falling still. The tiny fish are piled up several inches deep, and the row extends down the beach as far as I can see—at least a quarter mile.

When the hidden struggle of living shows itself this way—the wild ones living their dying in plain sight—the shock to human eyes has two sides. One is abundance and one is grief. The water boils as fish leap for safety, and the huge mouths of striped bass stab the surface in pursuit. Thousands upon thousands of fish are cast up and dying. The alewife have bodies clear all the way through, marked only with one silky black line running the creature's length and a piercing black eye hurled toward the sky. The mackerel have a cord of shiny blue tissue

---

ALISON HAWTHORNE DEMING is a professor of creative writing and former director of the Poetry Center at the University of Arizona. A poetry and prose writer, she received a Wallace Stegner Fellowship at Stanford University, two National Endowment of the Arts Fellowships, a Pushcart Prize, and a fellowship from the Fine Arts Work Center in Provincetown, among other awards. Her most recent book, *Genius Loci* (2005), is a collection of poems about human and nonhuman worlds.

running through the clear flesh, so that it seems the fish skin grows within this cocoon of liquid glass. Some show a dim stain of red near the gills where blood has seeped. Thousands upon thousands lie dead in the waves' wake. They are mostly two inches in length. A few have grown to three-inch length and have the iridescent blue skin of adult mackerel. I scoop up some specimens in my hand, turning them just so in the sunlight, and a lilac-colored patch near the gill illuminates.

The striped bass are making their seasonal journey along the Atlantic shore. Fierce feeders, they come within a foot of the shore, their leaping gape through all of this forage, ecstatic. They jump nearly out of their bodies in the feasting. This is the dance they are made for. Perhaps the bass have schooled this close to shore in flight from larger predators. The spectacle is a feast or a terror, depending on whose story you're telling. They remind me, these thousands of fish dying on the beach, of those mindful suicides—like the poet Anne Sexton with her "awful rowing towards God"—suicides who end it because they cannot bear living another day with the knowledge that they will have to die, curing themselves of death's threat by making a preemptive strike. I know the analogy has nothing to do with the instinct that drives the tiny fish to leap to their deaths, yet the paradox of that desperation, whether a mindless or mindful act, stirs up a sense of tenderness toward life and the difficult terms it sets.

I have faith in natural process, in the intricate systems of reciprocity that keep nature from tilting out of balance. I may belong to the last generation for a very long time to feel this faith. Life is tough and resilient. Life overdoes it. Consider the seeds of the dandelion weed, the tadpoles of a frog, and the sperm production of the human male. There are so many more gametes than needed for a species to survive. The system is biased toward continuity. As for the alewife and mackerel beaching at Race Point on a sunny afternoon in May, this event is a random catastrophe of nature, nothing out of the ordinary. How many more thousands or millions of hatchlings are schooling lustily offshore, escaping the stripers' mouths to fatten into their own maturity? There is no malice and no grief in the actions of predator or prey, only the spectacle of muscular effort, the shine of wet skin.

I have been trying to wrap my head around the scale of violence erupting in human homes, tribes, nations, and against our mothering planet in all its intricate wholeness. No one can explain or justify the breadth and depth of human cruelty. Malice and grief abound. The facts come home to us in the triple crown of climate chaos, crashing biodiversity, and ceaseless genocidal war. How terribly ironic it is that we, the animals who brought ethical principles into the equation of living with others, have turned out to be the most heavy-handed lugs on Earth. There is no point denying this: study air, water, oceans, amphibian or mammal conditions. Memorize the names of places where anthropogenic human suffering makes everyone want to forget: Auschwitz, Darfur, Rwanda, the Lower Ninth Ward, Aceh, Baghdad, Bethlehem. Open the mind to anguish at the start of the twenty-first century, and the tsunami will rise. It's no wonder people cover their ears. Who can bear to carry the weight of so much grief?

I recently dreamed a war scene. Kurds wearing head wraps and draped cloth were tending to the wounded after a battle. The man I love lay on the sand, an ancient place that looked biblical, as I recall the exotic photographs from a children's book of religious stories. The man had been so wounded he could not move. His enemies wrapped him in a papyrus mat and loaded him, like a spool of carpet, onto another Kurd's back, who promised to carry his adversary to refuge. He did so because he knew that if he could not bear the weight, the burden would crush him.

The dream came at a time of many losses in my life. I have been caring for my mother, who at age ninety-nine does not want to live any longer. Each visit to her apartment has exercised my compassion, as, clear-minded, she describes the latest physical indignities, the pain of a collapsing skeleton, the logistics of drug regimens, the boredom of aimless days without a desire even for television, the daily wish to go to sleep and not wake up. "Maybe I can kill myself with boredom," she jokes. But the body has a mind of its own, and hers is not yet finished with life. I have also struggled with losses endured by my partner of the past eight years, a rugged western man who suffered a severe ankle injury while rock climbing as a young man that led to an amputation in his sixties, followed by an ocular stroke that took away the sight in one eye and scattered a cascade of tiny occlusions over his brain. No textbook or essay can offer words adequate to the task of describing what happens to a person whose brain labors to reconfigure after such events. The struggle was more difficult than either of us imagined. It cost us our relationship. The gradual loss of these loved ones is, too, an ordinary catastrophe, though it may sound callous to say so. Death and diminishment are the price of entry. Grief either swallows you whole or spits you out to feel compassion for the grief of others.

For several years I have been writing about animals, looking at how important they have been as characters in the human drama from very early in our history, at how much joy and texture and mystery they bring into our lives. These too have become stories of grief, as the news gets bleaker about the animals' fate in a biologically impoverished world. Fifty percent of the world's animals are in decline. One-quarter of the world's mammals face extinction. That includes the elephant, humpback whale, gorilla, orangutan, spider monkey, cheetah, tiger, and polar bear. Imagine a world in which these creatures are merely imaginary, as the dinosaurs and woolly mammoths are to us today. Imagine the loss of wonder and excitement, the growing fear and sorrow as the continuity between human beings and the others tatters. Some of the threatened ones will survive in captive breeding programs, and for this stark generosity, one must give thanks. But the world we leave to the future will be brutally impoverished. Earth's gorgeous palette is fading, and there is no ark for our fellow creatures but us.

I often question why I am writing about animals at a time when I am suffering much more intimate losses. I'm not sure if studying animals is a way to avoid the pain closer to home or a way to console myself that things are not really that bad for my kind. I suppose both motivations make sense. I need to see the beauty of the lavender stain in a dying mackerel's gills, to feel the lurch of the predator in its raucous feasting joy, to honor the struggle each living creature mounts against the threat of annihilation. I need to write so that words may do their part in building the ark; or, if it is too late for ark building, then at least I can help extend the moral imagination so that it reaches beyond personal loss.

It is difficult to believe that our companion creatures are imperiled. The sensuous thrill of life can lead one to the sense of wonder, but it cannot tell you, for example, that there are dead zones in the oceans, a hole in the sky, and a tincture of industro-agro-pharmaceuticals flowing into rivers and oceans, turning them into chemistry sets brewing a monstrous future. Frogs growing extra legs. Striped bass sipping antidepressants, then hanging vertically in the water like wallpaper. Newborn males of our species arriving with alarming genital abnormalities. To know such bad news takes science. It will take science and public policy and invention to gentle our impact on the planet. It will take poetry, stories, songs, paintings, and theater to bear witness to the unpredictable world we are shaping and its impact on our spirits.

My intention is to live by the doctrine of grief, to savor sadness as its own dark memo of instruction from the moral imagination. "Up again, old heart!—it seems to say—there is victory yet for all justice," as Emerson once urged. Feast your spirit on the beauty that remains.

ETHICAL ACTION

What does the beauty of the world call us to do?

Art is the most powerful expression of human values. No wonder that throughout history, art has moved people to moral action.

Let art celebrate the natural beauty of the world. Paint sea otters or octopi on garbage bins and blank walls. Shut down the elevator music and broadcast whale song. Paint icebergs over the advertisements on the city bus. Amplify the music of barn owls on the nest and let that crooning, not Frank Sinatra's, fill the square.

Dress children as butterflies or wasps and let them parade in the street. Create street puppets—long-legged spiders and floppy squids or grotesque sea clams. These can be agents of rejoicing: parades with bands, a great celebration of life. Or they can be fearsome ghosts, dogging the steps of those who would vandalize natural creation, blocking the paths of the bulldozers, haunting the entrances to buildings where death and destruction are charted on budget projections as ascending lines.

Let all our human constructions be works of art that celebrate and enhance life. Let gardens bloom on rooftops and recycled wood glow on the floors. But more important than adornment, let functional structure itself be beautiful—the cattail marshes that purify water, the merry-go-rounds that pump water in the desert, the windows that warm the children. Ban ugliness. Ugliness dishonors the Earth in spirit and in effect.

Let art chronicle loss. Let no species disappear without public notice. If our ways of life are going to destroy infinitudes of lives, let us at least do so knowingly, and grieve for the terrible loss. Evolution is an infinitely branching tree. Each branch branches again, and branches again, budding and branching, a beautiful and endless complexity. Killing off a species not only wipes that creature off the earth forever; it also lops off the budding branch, eliminating for all time the infinite variation that might have grown from that limb. This is a loss literally beyond imagining. It is also a loss that is largely invisible.

Create art that fills the forests with death notices. Transform every stump in the clearcut into a cross, so no one can drive by a ruined hillside without seeing it for what it is—a graveyard that stretches for miles. Let the roadside bloom with shrines adorned with plastic flowers to mark the extinctions of sparrows. Post "missing persons" notices for the white-headed woodpeckers that used to frequent the ponderosa forest. Send an obituary to the newspaper each spring, when the frogs do not sing. Howl across the lake for the gray wolves that once roamed North America. Assemble the choir and sing hymns as the bulldozers gouge out the last checker-lilies in the valley. Print pictures of ivory-billed woodpeckers on milk cartons. Rent a hearse and follow the truck that sprays poisons in the ditches.

Create art based on a faith in human goodness: the faith that if we could only *see*, if we *only knew*, if we could come face-to-face with the enormity of the killings that are incidental to our casual disregard, we would change our ways.

12

> *Do we have a moral obligation to take action to protect the future of a planet in peril?*
>
> **Yes, because we love the world.**

What does it mean to love a person? To want to be near her, to rejoice in that closeness; to delight in her, body and spirit; to feel whole when you are with her, as if you are only half a person when you are apart—all this, of course, but not only this. Loving is not merely a way of feeling. It is a way of acting in the world. To love a person means to act lovingly toward her—to be kind to her, to protect her, to care about her well-being even more than your own, to devote your life to her thriving.

What, then, does it mean to love a place? We do fall in love with places, there is no doubt. The crest of a mountain, the childhood shelter behind the hedge, the rockwrack shore of the sea, the lilac in bloom by the bus stop—each of us has a place that makes us feel whole and happy and alive. Sometimes we fall in love with the entirety of these, the global home so beautiful from space, with its glowing blue skin and soft clouds. Loving a place is a way of feeling, connected and at peace. But loving a place is also a way of acting. Can we claim to love a place if we skim it for our own gain, or slash it and leave it to die? Just as in loving a person, loving a place means being kind to it, protecting it, caring about its well-being as much as your own.

That analogy is the basis of what has come to be called the "ecological ethic of care." Humans are born to love. When we are most fully blessed, we are born into loving relationships, with people and with places. We have treasured memories of caring and being cared for, of sheltering and being sheltered. Our moral responsibilities grow from those relationships. Because we love the world we were born to, a world now so deeply imperiled, we have a responsibility to come to its defense.

Changing Ethics for a Changing World

J. Baird Callicott

Everyone knows that business-as-usual business in the global human economy is challenged by global climate change. It necessitates a paradigm shift in economic production and consumption from brown to green. And though everyone does not yet seem to know it, we'll be healthier and happier to boot as a result of such a paradigm shift. To realize the paradigm shift in the economy from brown to green, however, will take political will backed by a moral commitment. Business-as-usual thinking in mainstream moral philosophy, however, is also challenged by global climate change. A paradigm shift in ethical thinking will be necessary to motivate the practical steps we need to take to get from the brown to the green economy. Unfortunately, mainstream thinking in moral philosophy is even more dismal than mainstream thinking in the dismal science itself. The paradigm shift in ethics that is demanded by global climate change will also make us happier and healthier.

Environmental philosopher Dale Jamieson well illustrates the classical paradigm of mainstream ethical thinking with six little ditties about Jack and Jill:

J. BAIRD CALLICOTT is chair of the Department of Philosophy and Religion Studies and Regents Professor of Philosophy and Religion Studies in the Institute of Applied Sciences at the University of North Texas. From 1997 to 2000 he was president of the International Society for Environmental Ethics. He has authored more than a hundred book chapters, journal articles, encyclopedia entries, and book reviews. He recently published the two-volume *Encyclopedia of Environmental Ethics and Philosophy* (2008).

Consider Case 1, the case of Jack intentionally stealing Jill's bicycle. One *individual* acting intentionally has harmed another *individual;* the *individuals* and the harm are clearly identifiable; and they are closely related in time and space. If we vary the case on any of these dimensions, we may still see the resulting cases as posing a moral problem, but their claims to be . . . paradigm moral problems will be weaker. . . .

Case 2: Jack is part of an unacquainted group of strangers, each of which, acting independently, takes one part of Jill's bike, resulting in the bike's disappearance.

Case 3: Jack takes one part from each of a large number of bikes, one of which belongs to Jill.

Case 4: Jack and Jill live on different continents, and the loss of Jill's bike is the consequence of a causal chain that begins with Jack ordering a used bike at a shop.

Case 5: Jack lives many centuries before Jill, and consumes materials that are essential to bike manufacturing; as a result, it will not be possible for Jill to have a bicycle.

While it may still seem that moral considerations are at stake in each of these cases, this will be less clear than in Case 1, the paradigm case with which we began. The view that morality is involved will be weaker still, perhaps disappearing altogether, if we vary the case on all these dimensions simultaneously.

Consider Case 6. Acting independently, Jack and a large number of unacquainted people set in motion a chain of events that causes a large number of future people who will live in another part of the world to never having bikes.

Jamieson has accurately captured the standard ethical paradigm in Western moral thought with Case 1. And he has accurately captured the challenge of global climate change to that paradigm in Cases 4, 5, and 6. Most environmental philosophers who have taken on the ethical challenge of global climate change, including Jamieson, have nevertheless come at it armed only with the paradigm represented by Jack-and-Jill Case 1 and the conceptual arsenal developed by eighteenth- and nineteenth-century philosophers and honed by twentieth-century philosophers with that paradigm in mind. As a result, the typical philosophical response to the ethical challenge of global climate change concerns, first, international justice and, second, intergenerational justice.

Those individuals of the present generation most severely affected by global climate change are least responsible for causing it and least able to cope with the consequences. Bangladeshis and Micronesians who never owned a car or used an electric dishwasher will be flooded out of house and home; Inuit and other Arctic peoples will fall through thin sea ice and drown while hunting polar bears—who are themselves drowning and starving because of a warming Arctic.

Second, all individuals living in the mid- to distant future will live in a world of disappearing coastlines, drought, flood, severe and souped-up, and god knows what other unpleasant surprises a more energetic Earth has in store for them.

Can one assign responsibility to contemporary affluent individual Jacks for the suffering, due to climate change, of contemporary individual Jills? Hardly in the same way that it would be possible to do so if individual Jacks harmed individual Jills by, say, being personally involved in sex-slave trafficking or lethal organ harvesting, either as dealer or customer. The contribution to climate change of a single Jack is minuscule. Nor is it conceivable that climate justice could be realized by individual Jacks sending personal checks to individual Jills.

So immediately, we must consider the concept of collective responsibility, which without further ado takes us a step beyond the standard Jack-and-Jill paradigm. Affluent fossil-fuel consumers (and their deceased forebears, one might add, going back to the advent of the Industrial Revolution) are collectively responsible for the suffering due to climate change of poor, vulnerable nonconsumers. And justice demands the former compensate the latter for the harm the former have caused the latter. We soon fall back into the familiar ethical paradigm, however, when we consider how such compensation can be effected. One way would be for the several governments of the world's Jacks to send money to the several governments of the world's Jills. But that would involve some kind of carbon tax imposed on individual Jacks to pay not only for their own Industrial Revolution sins but also for those of their industrializing progenitors. And it would assume that the recipient governments would fairly distribute that money to their individual citizens.

Such is the kind of medicinal, restitutional, zero-sum thinking that currently characterizes international-climate-justice ethics. What about intergenerational-climate-justice ethics? Many individuals living today have children and grandchildren who will live on after them into the future. In 2010 my son is forty and may well live another forty years to midcentury, when many of the untoward effects of climate change will be palpable (if they are not already). His son is ten and has a good chance of living into the last decade or two of the present century. So near-future generations are identifiable individuals—my son and his contemporaries, my grandson and his.

But nothing we do now can benefit the future generations of the present century (my son and his contemporaries, my grandson and his) because, due to lag-timed effects, climate change cannot be turned away on a dime, even if greenhouse gas emissions are reduced to zero next year. Near-future generations will suffer the full effect of the global climate change that greenhouse gases already in the atmosphere will bring about. If we radically reduce our collective carbon footprint now, we will reduce the eventual magnitude of global climate change and the burden of its consequences on individuals living hundreds and thousands of years in the future.

Consider, however, the Parfit paradox, named for philosopher Derrick Parfit. Draconian conservation measures have draconian lifestyle consequences, including reproductive choices and chances. If we do nothing, my grandson and his contemporaries will have one set of children. If we adopt draconian conservation measures, different gametes will meet, and the now-living identifiable individuals who will live on into the near future will give birth to a different set of children. In short, draconian conservation measures will change the individual composition of presently unborn future generations. If we mitigated climate change now, the individual Jacks and Jills we sought to benefit in the distant future would not exist; individual Jims and Janes would exist instead. Presently unborn future Jacks and Jills individually—to imagine a permutation of the Rawlsian "original position" behind a "veil of ignorance"—are presented with the following choice: Is it better to exist in a world of drought, flood, and violent and unpredictable storms, or better to be replaced by Jims and Janes and not to exist at all? An infamous expression came to symbolize the absurdity of the American war on Vietnam: "We had to destroy the village in order to save it." The absurdity of the Jack-and-Jill paradigm of ethics in a time of global climate change may be symbolized in an analogous expression: "We had to deprive those Jacks and Jills, who would otherwise have existed, of their existence in order to benefit other future individuals—Jims and Janes."

Ethical individualism and reductionism, Jack-and-Jill ethics, leads, as we see, to problematic and paradoxical conclusions. The remedy involves an ethical paradigm shift in two conceptual domains: (1) moral psychology and (2) moral ontology.

Homo ethicus is as much a fiction of modern ethicists as *Homo economicus* is a fiction of modern economists. For *Homo ethicus*, that individuals are remote in space (living on another continent) and/or remote in time (living in the distant future) should be as irrelevant to their moral considerability as their race or sex is. Both the standard economic paradigm and the standard ethical paradigm assume that *Homo sapiens* are (in the case of the economists) or ought to be (in the case of the ethicists) purely rational beings. By mainstream economists' lights, actual people behave irrationally—they do not always maximize their own welfare or make the most efficient use of the resources at their disposal. And by mainstream ethicists' lights, actual people irrationally feel greater obligations to their family members, friends, neighbors, fellow citizens, and the present generation of humans than to spatially distant strangers and to temporally distant future humans. Is there something wrong with us or with the economic and ethical paradigms according to which we should, economists and ethicists insist, behave?

Mainstream ethicists think we should give equal consideration to equal interests, no matter whose interests are at stake. Twenty years ago my son had an interest in a college education, and so did an orphan girl in Indonesia. Was I wrong to buy my son a college education at the expense of that Indonesian girl? As a good moral philosopher, suppose I had decided that the Indonesian girl had not just an equal but a greater interest in a college education than my son—who, in any case, educated or not, would have been a relatively affluent male American, with all the advantages pertaining thereto. And suppose I had paid for that girl's education with the limited means at my disposal and not for his. How would I be judged from the point of view of moral common sense? I may have done the rationally right thing, but I would be thought a rational idiot.

Ethics, as the contrarian philosopher David Hume observed, is rooted in the moral sentiments—other-oriented feelings of love, well-wishing, loyalty, patriotism—not in reason alone. And as Charles Darwin argued, the moral sentiments are naturally selected to facilitate the existence of cooperative societies. *The ethical paradigm that meets the challenge of global climate change must shift the emphasis in moral psychology from reason to feeling.*

The moral sentiments may target other-than-individual entities. Patriotism is a good example. To what, then, might our moral sentiments be targeted in the time of global climate change other than spatially and temporally distant individual human beings? To the other *species* (as opposed to individual specimens) with whom we share the planet. To the congeries of biotic communities that these other species compose and the congeries of ecosystems whose functions these species perform. They are our fellow voyagers in the odyssey of evolution. Global climate change exacerbates the threat of extinction already posed by human encroachment. This is our Earth household, our beautiful and serviceable habitat.

Yes, ours—and that of the human generations to come. While we may not be able to muster up any of our moral sentiments for the unknown and unpredictable Jacks and Jills (or Jims and Janes, as the case may be) living in the remote future, we may certainly wish ardently—and unselfishly—that *Homo sapiens* has an ongoing and open-ended future. Only the most embittered misanthrope would wish not. But what kind of future would we wish for an ongoing and open-ended human enterprise? Do nothing now about global climate change and a collapsed human civilization may be humanity's future, a new and possibly terminal Dark Age. No, we hope that human civilization will also have a future, and not just because future individuals—future Jacks and Jills (or Jims and Janes)—will have a better life if civilization does not descend into barbarism. Art, philosophy, science, literature, architecture also stir the moral sentiments, the sentiments of pride and wonder in the achievements of our species.

Aldo Leopold begins his famous chapter "The Land Ethic" with an evocation of Homer. He calls our attention back to the dawn of Western civilization and to its earliest extant literature. Some three thousand years have elapsed, Leopold reminds us, since the days when Odysseus's black-prowed galleys clove the wine-dark seas, headed for beautiful Ithaca, headed home. Must we not do our utmost to sustain that civilization, per se, so that it can endure and evolve for at least another three thousand years? Isn't that an object to which we owe allegiance? Isn't that a living entity that we know and love? Isn't it in our hearts to work to preserve it?

To meet the challenge of global climate change, philosophers need to shift the subjects of ethics from Jack and Jill to entities that themselves exist at proportionate temporal and spatial scales: to species—including but not limited to *Homo sapiens*—not specimens; to the planet's congeries of biotic communities and ecosystems that species compose and in which they function; and to the civilization that is the signal achievement of our own species. And they need to shift the moral sentiments into a more prominent place alongside reason in their moral psychology. Then we might have a coherent, practicable, and inspiring ethics with which cheerfully to confront the moral challenge of global climate change.

Touching the Earth

bell hooks

> I wish to live because life has within it that which is good, that which is beautiful, and that which is love. Therefore, since I have known all these things, I have found them to be reason enough and—I wish to live. Moreover, because this is so, I wish others to live for generations and generations and generations and generations.
>
> LORRAINE HANSBERRY,
> *To Be Young, Gifted, and Black*

When we love the Earth, we are able to love ourselves more fully. I believe this. The ancestors taught me it was so. As a child I loved playing in dirt, in that rich Kentucky soil, that was a source of life. Before I understood anything about the pain and exploitation of the southern system of sharecropping, I understood that grown-up black folks loved the land. I could stand with my grandfather Daddy Jerry and look out at a field of growing vegetables, tomatoes, corn, collards, and know that this was his handiwork. I could see the look of pride on his face as I expressed wonder and awe at the magic of growing things. I knew that my grandmother Baba's backyard garden would

---

BELL HOOKS is an author, feminist, and social activist. She has been a professor at the University of Southern California, Oberlin College, Yale University, and the City College of New York. Her writing focuses on the interconnectivity of race, class, and gender. She has published over thirty books, including *Teaching Community: A Pedagogy of Hope* (2003) and *Outlaw Culture: Resisting Representations* (2006).

yield beans, sweet potatoes, cabbage, and yellow squash, that she too would walk with pride among the rows and rows of growing vegetables showing us what the Earth will give when tended lovingly. From the moment of their first meeting, Native American and African people shared with one another a respect for the life-giving forces of nature, of the Earth. African settlers in Florida taught the Creek Nation runaways, the Seminoles, methods for rice cultivation. Native peoples taught recently arrived black folks all about the many uses of corn. (The hotwater cornbread we grew up eating came to our black southern diet from the world of the Indian.) Sharing the reverence for the Earth, black and red people helped one another remember that, despite the white man's ways, the land belonged to everyone. Listen to these words attributed to Chief Seattle in 1854:

> How can you buy or sell the sky, the warmth of the land? The idea is strange to us. If we do not own the freshness of the air and the sparkle of the water, how can you buy them? Every part of this Earth is sacred to my people. Every shining pine needle, every sandy shore, every mist in the dark woods, every clearing and humming insect is holy in the memory and experience of my people . . . We are part of the Earth and it is part of us. The perfumed flowers are our sisters; the deer, the horse, the great eagle, these are our brothers. The rocky crests, the juices in the meadows, the body heat of the pony, and man—all belong to the same family.

The sense of union and harmony with nature expressed here is echoed in testimony by black people who found that even though life in the new world was "harsh, harsh," in relationship to the Earth one could be at peace. In her oral autobiography, granny midwife Onnie Lee Logan, who lived all her life in Alabama, talks about the richness of farm life—growing vegetables, raising chickens, and smoking meat. She reports:

> We lived a happy, comfortable life to be right outa slavery times. I didn't know nothing else but the farm so it was happy and we was happy. . . . We couldn't do anything else but be happy. We accept the days as they come and as they were. Day by day until you couldn't say there was any great hard time. We overlooked it. We

didn't think nothing about it. We just went along. We had what it takes to make a good livin and go about it.

Living in modern society, without a sense of history, it has been easy for folks to forget that black people were first and foremost a people of the land, farmers. It is easy for folks to forget that at the first part of the twentieth century, the vast majority of black folks in the United States lived in the agrarian South.

Living close to nature, black folks were able to cultivate a spirit of wonder and reverence for life. Growing food to sustain life and flowers to please the soul, they were able to make a connection with the Earth that was ongoing and life-affirming. They were witnesses to beauty. In Wendell Berry's important discussion of the relationship between agriculture and human spiritual well-being, *The Unsettling of America*, he reminds us that working the land provides a location where folks can experience a sense of personal power and well-being: "We are working well when we use ourselves as the fellow creature of the plants, animals, material, and other people we are working with. Such work is unifying, healing. It brings us home from pride and despair, and places us responsibly within the human estate. It defines us as we are: not too good to work without our bodies, but too good to work poorly or joylessly or selfishly or alone."

There has been little or no work done on the psychological impact of the "great migration" of black people from the agrarian South to the industrialized North. Toni Morrison's novel *The Bluest Eye* attempts to fictively document the way moving from the agrarian South to the industrialized North wounded the psyches of black folk. Estranged from a natural world, where there was time for silence and contemplation, one of the "displaced" black folks in Morrison's novel, Miss Pauline, loses her capacity to experience the sensual world around her when she leaves southern soil to live in a northern city. The South is associated in her mind with a world of sensual beauty most deeply expressed in the world of nature. Indeed, when she falls in love for the first time she can name that experience only by evoking images from nature, from an agrarian world and near wilderness of natural splendor:

> When I first seed Cholly, I want you to know it was like all the bits of color from that time down home when all us chil'ren went berry picking after a funeral and I put some in the pocket of my Sunday dress, and they mashed up and stained my hips. My whole dress was messed with purple, and it never did wash out. Not the dress nor me. I could feel that purple deep inside me. And that lemonade Mama used to make when Pap came in out of the fields. It be cool and yellowish, with seeds floating near the bottom. And that streak of green them june bugs made on the tress that night we left from down home. All of them colors was in me. Just sitting there.

Certainly, it must have been a profound blow to the collective psyche of black people to find themselves struggling to make a living in the industrial North away from the land. Industrial capitalism was not simply changing the nature of black work life, it altered the communal practices that were so central to survival in the agrarian south. And it fundamentally altered black people's relationship to the body. It is the loss of any capacity to appreciate her body, despite its flaws, Miss Pauline suffers when she moves north.

The motivation for black folks to leave the South and move north was both material and psychological. Black folks wanted to be free of the overt racial harassment that was a constant in southern life, and they wanted access to material goods—to a level of material well-being that was not available in the agrarian South, where white folks limited access to the spheres of economic power. Of course, they found that life in the North had its own perverse hardships, that racism was just as virulent there, that it was much harder for black people to become landowners. Without the space to grow food, to commune with nature, or to mediate the starkness of poverty with the splendor of nature, black people experienced profound depression. Working in conditions where the body was regarded solely as a tool (as in slavery), a profound estrangement occurred between mind and body. The way the body was represented became more important than the body itself. It did not matter if the body was well, only that it appeared well.

Estrangement from nature and engagement in mind/body splits made it all the more possible for black people to internalize white supremacist assumptions about black identity. Learning contempt for blackness, southerners transplanted to the North suffered both culture shock and soul loss. Contrasting the harshness of city life with an agrarian world, the poet Waring Cuney wrote this popular poem in the 1920s, testifying to lost connection:

> She does not know her beauty
> She thinks her brown body
> has no glory.
> If she could dance naked,
> Under palm trees
> And see her image in the river
> She would know.
> But there are no palm trees on the street,
> And dishwater gives back no images.

For many years, and even now, generations of black folks who migrated north to escape life in the South returned down home in search of a spiritual nourishment, a healing, that was fundamentally connected to reaffirming one's connection to nature, to a contemplative life where one could take time, sit on the porch, walk, fish, and catch lightning bugs. If we think of urban life as a location where black folks learned to accept a mind/body split that made it possible to abuse the body, we can better understand the growth of nihilism and despair in the black psyche. And we can know that when we talk about healing that psyche we must also speak about restoring our connection to the natural world.

Wherever black folks live we can restore our relationship to the natural world by taking the time to commune with nature, to appreciate the other creatures who share this planet with humans. Even in my small New York City apartment I can pause to listen to birds sing, find a tree and watch it. We can grow plants—herbs, flowers, vegetables. Those novels by African American writers (women and men) that talk about black migration from the agrarian South to the industrialized North describe in detail the way folks created space to grow flowers and vegetables. Although I come from country people with serious green thumbs, I have always felt that I could not garden. In the past few years, I have found that I can do it—that many gardens will grow, that I feel connected to my ancestors when I can put a meal on the table from food I grew. I especially love to plant collard greens. They are hardy, and easy to grow.

In modern society, there is also a tendency to see no correlation between the struggle for collective black self-recovery and ecological movements that seek to restore balance to the planet by changing our relationship to nature and to natural resources. Unmindful of our history of living harmoniously on the land, many contemporary black folks see no value in supporting ecological movements, or see ecology and the struggle to end racism as competing concerns. Recalling the legacy of our ancestors who knew that the way we regard land and nature will determine the level of our self-regard, black people must reclaim a spiritual legacy where we connect our well-being to the well-being of the Earth....

Collective black self-recovery takes place when we begin to renew our relationship to the Earth, when we remember the way of our ancestors. When the Earth is sacred to us, our bodies can also be sacred to us.

Love, Grief, and Climate Change

Katie McShane

In countries around the world, conversations are under way about what sacrifices we should be willing to make for the sake of reducing greenhouse gas emissions. Some have argued that we should do nothing—that rather than try to prevent climate change, we should just allow it to happen and try to adapt to it.[1] While the cost-benefit analyses behind this recommendation have been criticized, it's also true that economic efficiency isn't the only value relevant to this decision. We should also think about what it would be like to live in a world where climate change is allowed to progress unabated and whether this is a future that we want to choose for ourselves and our descendants.

Even conservative estimates of the scale and pace of environmental disruption that will take place if greenhouse gas emissions continue at their current rate are stunning. The latest IPCC report estimates that by 2100, climate change "will alter the structure, reduce biodiversity . . . perturb functioning . . . and compromise the services" provided by *most of the ecosystems on Earth*. By then 20–30 percent of all species that currently exist will be at "high risk" of extinc-

---

KATIE McSHANE is assistant professor of philosophy at Colorado State University. In 2002 and 2003 she was a visiting scholar at Harvard University's Center for Ethics and the Professions. She has published papers in journals such as *Environmental Ethics; Environmental Values;* and *Philosophy, Place, and Environment.*

---

1. See, for example, Bjorn Lomborg's *Cool It: The Skeptical Environmentalist's Guide to Global Warming* (New York: Knopf, 2007).

tion.[2] Weather patterns will change, likely bringing severe drought to tropical regions and flooding to temperate regions. Changes in weather patterns will alter what kinds of agriculture various regions can support; species' migratory patterns will change; invasive species will become more widespread and further disrupt ecosystemic relationships; and infectious diseases will take over areas made newly hospitable to them.

What will life be like for human communities living in a world where all this is taking place? One likely consequence of such rapid and widespread ecological change is that much of the carefully accumulated knowledge of local environments that human communities have relied on for generations will be rendered useless. Which things are edible, when to plant and harvest, how to manage pests, how to treat diseases—all of this knowledge relies on a level of climatic and ecosystemic stability that will no longer exist. We humans often credit our own ingenuity for the success we have had in the understanding, prediction, and control of the natural world. But we need to take seriously the likelihood that much of our success is due to the fact that human civilizations have developed within a period of unusual climatic stability. Our talent for prediction is due in part to the fact that the world we live in has so far been relatively predictable. How our methods will fare in a much less predictable world is anybody's guess.

Widespread changes in the nature and makeup of the Earth's ecosystems will likely affect human settlement and land-use patterns, agricultural practices, disease rates, and population growth/decline, and those changes will in turn have serious economic and political consequences. People are likely to feel much more insecure about their own welfare in such a world. An increase in the stresses placed on natural systems is likely to result in greater scarcity of the basic necessities of human life: food, water, and shelter. Increased scarcity often leads to increased hostility among those competing for resources; war may become increasingly common. These are threats that people

---

2. A. Fischlin, et al., "Ecosystems, Their Properties, Goods, and Services," in *Climate Change 2007: Impacts, Adaptation and Vulnerability. Fourth Assessment Report of the Intergovernmental Panel on Climate Change*, ed. M. L. Parry, et al., 211–272 (Cambridge: Cambridge University Press, 2007).

who work in national security are already concerned about and taking very seriously.

But in addition to these problems, my own guess is that this will also be a time of profound sadness and disarray in human cultures. Many of the stories that we tell, the songs that we sing, the artworks that we create and preserve are about familiar parts of the natural world. Even those of us who live in modern cities still teach our children the names of farm animals, cook the foods that we grew up eating, dress for the seasons as we expect them to be, and cheer for sports teams named after local fauna. Our sense of our own national and regional identities—our descriptions of what "home" is like and what makes home a special place—often depend on the presence of other species in ways that we don't recognize. Imagine Greece without olives, India without mangoes, or Ireland without grass. How would Americans feel about a world in which California has no more redwoods, Minnesota doesn't get snow, Kansas can't grow wheat, and Louisiana has no more crawfish? Much of our sense of our own identities—our regional affiliations, our personal histories, our stories about what the world is like—have been shaped by living in natural environments that are familiar to us and for which we have come to feel affection. A rapid change in these environments is likely to have a deeply unsettling effect on our sense of ourselves and our place in the world.

To think that the things we love—including the plants, animals, places, and systems that make up our ecological communities—won't ever change is clearly naive; to hope that they won't change is unwise. Most of the things we love do change over time; this is part of what keeps our emotional lives interesting. But in our emotional attachments as well as our economic and ecological contexts, it is the rate of change that makes all the difference. The right combination of stability and change keeps life interesting; too much change and too little stability leave us feeling disoriented, fearful, uncertain, and insecure. A world in which we watch as a quarter of the Earth's species go extinct over the course of a century, and in which we live amid the political and economic turmoil brought about by dramatic ecosystemic changes, is a world in which we will feel adrift, untethered from the familiar, worried about our own security, and unable to be helped by the knowledge passed down to us from our elders and our cultural traditions.

This might strike some readers as a sentimental description of our connections to the natural world, and one might think that sentimental considerations have no place in public policy decisions. But I think we underestimate the importance of our sentimental attachments at our own peril. It is through its connections with our sentiments that the world becomes a place where we feel at home, a place that we can come to love, and where we feel a sense of belonging. These are goods that aren't obviously quantifiable in economic terms, but they're a crucial part of what makes human lives go well.

None of this is to say that arguments from economic efficiency have no place in our current discussions. Dollars aren't just dollars; they can represent diseases treated (or not), lives saved (or not), and people fed (or not). But it is also important to think about those values not so easily quantified. In deciding what sacrifices, if any, we are willing to make to prevent climate change, we need to think about what it would be like to watch as so much of what we have come to love about this world passes away. The moral question we face isn't just "How can we most efficiently spend today's dollars?" but, rather, "What kind of a future do we want for ourselves and our descendants?" A future of relatively peaceful, stable communities with rich and flourishing cultures might well be worth significant economic sacrifices.

For the Love and Beauty of Nature

Stephen R. Kellert

Modern humans have in many ways lost their bearings as biological beings, as just another animal and species in the firmament of creation. Despite a contemporary emphasis on mitigating climate change, on protecting biological diversity, on preventing pollution, on sustainable development, on a green economy, we are even more separated, if not alienated, from our biological roots. In our heart of hearts, most of us view progress and civilization by how much we have seemingly transcended the constraints of our biology, become in effect a new and emergent being, one who through technology, engineering, mass production, construction, and urbanization has become uniquely created through the transformation of physical nature and the suppression of other organisms. Even our efforts at mitigating climate change, preventing pollution, achieving energy efficiency, and fostering biological conservation often promote the very technological tools and econometric models that created the disconnect from nature in the first place.

Yet until we first address how we are in the fullest sense a formative reflection of our relationship with the nonhuman world, we will

---

STEPHEN R. KELLERT is the Tweedy/Ordway Professor of Social Ecology and codirector of the Hixon Center for Urban Ecology at the Yale School of Forestry and Environmental Studies. His work focuses on understanding the connection between human and natural systems. He has served on committees for the National Academy of Sciences and is a member of the International Union for Conservation of Nature Species Survival Commission Groups. He has authored more than one hundred publications, including *Building for Life: Understanding and Designing the Human-Nature Connection* (2005).

not easily resolve the linked environmental, social, and spiritual crises of our time, or avert the catastrophe that could result as much from what Richard Nelson called our "imperiling loneliness ... that isolates us from the natural community" as from the more obvious forms of environmental destruction.[1] In effect, I am asserting that our environmental challenges cannot be addressed solely or perhaps even initially through the likes of climate treaties, cap-and-trade systems, regulating toxins, commodifying ecosystem services, new modes of global environmental governance, or other techno-policy fixes, as important and difficult as these may be. These well-intentioned acts of political engineering largely constitute Band-aids that may bring symptom relief but do not confront the underlying disease, a modern humanity fundamentally at odds with its experiential dependence on the biological world not only for material sustenance but, just as critically, for a host of emotional, intellectual, moral, and spiritual needs as well.

This lack of recognition of the necessity of emotionally and intellectually reconnecting with nature is so widespread that it often prevails even among those who profess an environmental ethic. This was dramatically illustrated to me by participating in a conference in memory of Aldo Leopold, the famous advocate of a "land ethic." Leopold described this moral imperative in this way: "An ethic to supplement and guide the economic relation to the land presupposes the existence of some mental image of land as a biotic mechanism. We can be ethical only in relation to something we can see, feel, understand, love, or otherwise have faith in. ... A thing is right when it tends to preserve the integrity ... and beauty of the biotic enterprise. It is wrong when it tends otherwise."[2] In effect, Leopold's land ethic advanced the idea that our conservation objectives must derive from a fundamental moral affinity based on understanding, appreciating, and recognizing the natural world's beauty, on loving and even spiritually connecting with that world. Despite this famous

---

1. Richard Nelson, "Searching for the Lost Arrow: Physical and Spiritual Ecology in the Hunter's World," in *The Biophilic Hypothesis,* ed. S. Kellert and E.O. Wilson (Washington, DC: Island Press, 1993), 223.

2. Aldo Leopold, *A Sand County Almanac, with Other Essays on Conservation from Round River* (New York: Oxford University Press, 1966), 230.

dictum, the conversation that day focused almost entirely on legal, financial, and political mechanisms for controlling climate change, the loss of biological diversity, correcting market failure by trading in nature's goods and services, and other forms of economic, engineering, and regulatory intervention. It reminded me that Leopold in his time had lamented equally well-intentioned market-oriented technical and policy fixes that attempted to make conservation "easy" but thereby risked making it "trivial."

An analogous danger exists today of believing that technological innovation, economic incentives, and legal-regulatory controls are all that matter and that emotional, cognitive, or values-based considerations focusing on our relationship to the natural world are either unimportant or secondary as a necessary step in securing a sustainable and meaningful future. I believe this perspective is not only misguided, but a self-defeating folly. We require a renewed realization of an ancient awareness that as individuals and social beings, like any species, we are formed by the quality of our relationship with the natural environment. It may help to introduce the concept of biophilia,[3] which, stated simply, asserts that humans possess an inherent biological affinity for the nonhuman world that is instrumental to their health, productivity, and well-being. The more complicated aspect of biophilia is the many ways this affinity manifests itself, from the widely recognized inclinations to utilize and exploit nature to equally important attraction and curiosity, affection and kinship, knowledge and understanding, mastery and control, moral and spiritual relation, even fear of and aversion to the natural world. All these physical, emotional, and intellectual affinities for nature are universal, genetically encoded tendencies because they confer fitness in the ancient struggle to survive and thrive. These inclinations represent a spectrum of self-interest contingent on the quality of our experiential ties to the nonhuman environment. Whether we live off the land or become investment bankers, our capacities for physical health, critical thinking, problem solving, social bonding, emotional

---

3. E. O. Wilson, *The Human Bond with Other Species* (Cambridge: Harvard University Press, 1984); Kellert and Wilson, *The Biophilic Hypothesis*; S. Kellert, *Kinship to Mastery: Biophilia in Human Evolution and Development* (Washington, DC: Island Press, 1997).

attachment, coping and mastery, even moral reasoning rely on a vast matrix of contacts and connections with natural process, especially during the formative years of childhood.

All these physical, emotional, and intellectual affinities, when adaptively expressed, give rise to an ethic of appreciation, love, and respect for nature that motivates us to sustain and protect it for reasons of personal and collective self-interest. As Wilson remarked in the case of biodiversity, but applicable to any environmental challenge: "A robust, richly textured, anthropocentric ethic can . . . be made based on the hereditary needs of our species, for the diversity of life based on aesthetic, emotional, and spiritual grounds."[4] If we lack a diverse base of physical and mental connection to nature, we rarely strive after its conservation, no matter the technological, economic, or policy mandate to do so. Loving nature, delighting in its beauty, extolling its exquisite complexity, and finding meaning from connecting with creation remain no less critical as motivators to sustain natural systems as reaping material gain from utilizing its goods and services or being legally exhorted to prevent its injury.

There is, nonetheless, this cant of mind that regards emotional and valuational concerns about the human relationship to nature as impractical and romantic preoccupations lacking the realism of the only real motivators of action—desire for material gain or fear of environmental catastrophe. There is also the simplistic notion that calling for a renewed relation to nature focuses only on the outdoors in typically undisturbed surroundings lacking humans. On the contrary, the promotion of emotional, intellectual, and spiritual affinities for the natural world is all about self-interest and practicality, and these experiential ties can and must occur in representational ways, indoors, and in the urban built environment as much as through direct experience of the wild.

I return to the Leopold conference by way of illustration. The meeting occurred in the recently completed LEED Platinum–designated building of the Yale School of Forestry and Environmental Studies, the building a new standard of "greenness" measured by such standards as

---

4. E. O. Wilson, "Biophilia and the Conservation Ethic," in Kellert and Wilson, *The Biophilic Hypothesis*, 38.

energy efficiency, carbon neutrality, recycled and recyclable products, nontoxic materials, renewable power, and other low-environmental-impact features. Yet the building also incorporated many *biophilic* design features that sought a positive connection to nature in the built environment.[5] The meeting room, for example, was distinguished by its great spacious vaulted space, its curved ceiling, the extraordinary degree of natural lighting even on that dark rainy day, its extensive natural materials, the fractal geometry of its oak-paneled walls and laminated Douglas fir arches, and other biophilic design features. I asked the audience whether they liked the room and whether they thought Leopold would have enjoyed being educated there. All answered in the affirmative. I asked how they might respond to the room if it had equivalent energy efficiency and carbon-mitigating features but instead was a box, artificially lit, served by processed air, with no windows and no good views. How would they experience that room after two, four, six, eight hours? Might they become restless, fatigued, afflicted (as has been found elsewhere in such rooms) by itchy skin and sore, scratchy eyes?[6] Would they choose to be there the next day or in the future? If they were permanent occupants of the building, would they be motivated to sustain the structure and recycle it generation after generation once the low-environmental-impact features had been surpassed by new innovations?

My point was that the room's biophilic design features, its deliberate attempt to forge a positive connection to nature, was all about practicality and self-interest. The building was a modern structure in an urban setting, but it remained a reminder, whether the occupants recognized it or not, of our evolutionary biology in a natural, not artificial world. Being sustainable means keeping something in existence, and we only sustain those things we feel a deep affection and attachment for because we perceive that their special qualities convey enduring meaning and value. Long before the lawsuits are filed, the regulatory standards are promulgated, the global policies are forged, a necessary step is the commitment that derives from a sense of con-

---

5. S. Kellert, J. Heerwagen, and M. Mador, eds., *Biophilic Design: The Theory, Science, and Practice of Bringing Buildings to Life* (New York: John Wiley, 2008).

6. S. Kellert, *Building for Life: Designing and Understanding the Human-Nature Connection* (Washington, DC: Island Press, 2005).

nection to the spaces and places we inhabit as organisms. As Leopold admonished:

> There must be some force behind conservation, more universal than profit, less awkward than government, less ephemeral than sport, something that reaches into all times and places . . . something that brackets everything from rivers to raindrops, from whales to hummingbirds, from land-estates to window-boxes. . . . I can see only one such force: a respect for land as an organism . . . out of love for and obligation to that great biota.[7]

The conference participants nodded their heads in polite recognition of what I had tried to express. Soon enough, however, they returned to the "hard" stuff of whether to impose a cap-and-trade system, establish regulatory mandates to correct environmental injustice, implement more punitive punishments to correct environmental misdeeds, or create a new system of global environmental governance. They seemed certain that only in these ways could they realize Aldo Leopold's vision of a more environmentally just and harmonious world.

7. Leopold, *A Sand County Almanac*, 198.

Earth Religion and Radical Religious Reformation

Bron Taylor

What is our place in the universe? To whom are we related and upon whom do we depend? How should we live? The answers to these questions may determine the fate of life on Earth.

It is, of course, a contestable premise that our perceptions and thoughts, including those we consider spiritual or ethical, might affect human behavior *to such an extent* that we would address climate change and halt the destruction of the Earth's biological diversity. It may be that impulses such as the sex drive and the need to consume calories so powerfully programs human behavior that cultural mores have little influence. Perhaps these impulses, which so effectively promote human population growth and the diffusion of our species into nearly every earthly habitat, lead inexorably to the tragedy of carrying capacity overshoot, ecosystem collapse, societal breakdown, and mass extinction.[1]

BRON TAYLOR is a professor of religion and nature at the University of Florida and the author of *Dark Green Religion: Nature Spirituality and the Planetary Future* (University of California Press, 2010). He is also the founding president of the International Society for the Study of Religion, Nature and Culture, and the editor of the *Journal for the Study of Religion Nature and Culture* (from 2007), *The Encyclopedia of Religion and Nature* (2005), and *Ecological Resistance Movements: The Global Emergence of Radical and Popular Environmentalism* (2005). See www.brontaylor.com for an unabridged version of this article.

---

1. The term "carrying capacity" refers to "the maximum population of a given species which a particular habitat can support indefinitely" given the available and needed habitat, calories, water, and other necessities. Ecologists variously refer to as "collapse" or "die-off" a sharp decline in numbers when a population of organisms "has exceeded the carrying capacity of its habitat." See William Catton, *Overshoot: The Ecological Basis of Revolutionary Change* (Urbana: University of Illinois Press, 1980), 272, 273.

This is not an idle worry. Research published in *The Limits to Growth* (1972) contended that if trends present then were to continue (especially increasing human numbers and per capita consumption), there would be widespread breakdown of the world's environmental and social systems in the twenty-first century. These trends *have* continued and are increasingly obvious.[2] There are, however, many examples where environmentally friendly values and lifeways have coevolved. It *is* possible for us to live in ways that do not degrade the environmental systems we depend on. But more realistic prescriptions are needed than the ones most commonly offered, such as, that all but the poor should dramatically reduce their levels of consumption. Few have followed or will follow this ascetic prescription, however virtuous it may be to do so. Nor will technological innovation prevent ecosocial collapse. Although some technological innovation reduces environmental deterioration, overall, it is a significant driver of biological simplification.

To transcend facile answers and prescriptions, we must begin the quest for sustainability by establishing taboo-free zones where every premise is examined with an eye toward whether an idea or practice promotes or erodes Earth's genetic and species variety. In this spirit I offer what I think are relevant and important contentions related to knowing our place in the world, our kin, and how we should live.

Scientific understandings about the explosive beginning of the cosmos and the theory of biotic evolution provide the best basis for understanding the origins and diversity of life on Earth. This does not mean that science can fully explain the existence of the universe. As the anthropologist Loren Eiseley eloquently put it, "I am an evolutionist . . . [but] in the world there is nothing to explain the world. Nothing to explain the necessity of life, nothing to explain the hunger of the elements to become life, nothing to explain why the stolid

---

2. Donella Meadows, Jørgen Randers, and Dennis L. Meadows, *Limits to Growth: A Report for the Club of Rome's Project on the Predicament of Mankind* (New York: Universe, 1972); Graham M. Turner, "A Comparison of the Limits to Growth with 30 Years of Reality," *Global Environmental Change* 18 (2008), 397–411; Donella Meadows, Jørgen Randers, and Dennis L. Meadows, *The Limits to Growth: The 30-Year Update* (White River Junction, VT: Chelsea Green, 2004). Many other studies reinforce the thesis in *Limits to Growth*.

realm of rock and soil and mineral should diversify itself into beauty, terror, and uncertainty."[3] Although there are too many limits on knowledge for us to fully comprehend the universe, there are some things we know with reasonable certainty, including:

We belong to the world. The Earth is our home, our place in the universe. Although on rare occasions we shoot a few organisms into space, to remain alive, they must return to Earth. The only place we know for sure that living things exist is here. While there may be life elsewhere, we know that complex life depends on conditions so uncommon that it is rare in the universe; indeed, Earth may be the only place such life exists.[4]

All earthly life shares a common ancestor; therefore, living things are kin, related in a familial sense. The evidence for this kinship is overwhelming, from the genetic structure shared by all organisms to the ways species change, sometimes evolving even into entirely new species. Another way we can perceive kinship, a common bond with other organisms, is by noting that *all* life came to be through exactly the same processes, which include a striving to survive and reproduce. As Charles Darwin once put it, "If we choose to let conjecture run wild, then animals, our fellow brethren in pain, diseases, death, suffering, and famine—our slaves in the most laborious works, our companions in our amusements—they may partake [of] our origin in one common ancestor—we may be all netted together."[5] The final passage in this quote introduces the next, critical insight:

Life exists in complex, interdependent webs. All life-forms are absolutely dependent on other organisms that create and sustain their necessary habitats, as when bacteria recycle waste and plants produce oxygen.

---

3. Loren Eiseley, *All the Strange Hours* (Lincoln: University of Nebraska Press, 2000), 242.

4. Peter D. Ward and Donald Brownlee, *Rare Earth: Why Complex Life Is Uncommon in the Universe* (New York: Springer-Verlag, 2000).

5. Charles Darwin in his "Notebooks on Transmutation," cited in Donald Worster, *Nature's Economy: A History of Ecological Ideas,* 2nd ed. (New York: Cambridge University Press, 1994), 180.

The above-mentioned facts are more *obviously* true than are beliefs about invisible, immaterial forces, worlds, or beings. This is not to say it is impossible that invisible things or forces exist. It is instead to assert that rational, well-informed individuals, even those who believe in the existence of spiritual realities, will acknowledge the rational grounds on which agnostics or atheists base their doubt or disbelief. All the above propositions speak to questions regarding our place in the universe and to whom we are related. Along with our uncertainties, they also illuminate questions pertaining to how we should live.

The recognition of biotic kinship, combined with our moral imagination, leads to identification with and felt empathy for all other life-forms. Such kinship ethics involve a desire to treat all earthly organisms with respect and to protect the biological processes on which all life depends. This does not mean that we can treat equally or avoid killing every organism, for all life depends on the death of other living things. It means that we understand that there are *natural laws* that must be respected for the Earth's living systems to flourish.

Many human cultures have kinship ethics and are based in an understanding of nature's laws. These understandings and ethics are often encoded in myths, beliefs, and practices that may not at first glance have an environmental dimension. On close scrutiny, we can discern the ways some cultural narratives and practices promote the flourishing of ecological communities. But today, these examples are small in scale and found among relatively homogenous groups, especially in the few remaining enclaves indigenous peoples inhabit, and have been little impacted by the global market's voracious appetite for resources. Is it possible for large human societies, and international bodies such as the United Nations, to promote an environmentally sustainable world? Is there anything that might unify today's contentious political, economic, and governmental actors?

As unlikely as this may seem, what might unify our species is an accurate understanding of our place in the universe and the nature of our earthly relations, our kinship with other organisms, and the recognition of our absolute dependence on the biosphere and its ecosystems. There are signs that these sorts of understandings, although fledgling, are growing globally. Indeed, I believe a new religious form is evolving—a naturalistic nature religion, or, at least, a religion-resembling nature spirituality. The adherents to such nature spirituality, which in recent work I have called "dark green religion," generally consider nature to be sacred, even though they rarely explain the term. This sort of spirituality is spreading, especially where an evolutionary-ecological worldview has taken root.[6] Evidence of the rapid growth and influence of this spirituality suggests that we may be witnessing the emergence of a global, civic Earth religion, which the political theorist Daniel Deudney aptly labeled "terrapolitan Earth religion."[7]

To understand terrapolitan Earth religion, in which loyalty and felt citizenship is to the Earth itself, we must understand the idea of civic (or civil) religion. This term refers to cultures in which a nation is invested with transcendent meaning and sacred purpose, thereby promoting group identity and a willingness to sacrifice for the good of the whole. The overall message is that God is responsible for establishing the nation and securing its future. Such religious nationalism consecrates the nation through myths and speeches about its sacred origins and mission, and national rituals during holidays and inaugurations, and at memorials.

An important aspect of civil religion is that it is inculcated through nonspecific and nonsectarian references to the divine. In this way, religious references do not hinder the "we feeling" needed for shared identity and citizen-embracing loyalty, even when people have different religious perceptions and beliefs. While civil religion often supports the status quo, it can also have a prophetic dimension, teaching

---

6. Bron Taylor, *Dark Green Religion: Nature Spirituality and the Planetary Future* (Berkeley and Los Angeles: University of California Press, 2010).

7. Deudney first discussed terrapolitan Earth religion in "Global Village Sovereignty: Intergenerational Sovereign Publics, Federal-Republican Earth Constitutions, and Planetary Identities," in *The Greening of Sovereignty in World Politics*, ed. Karen Litfin (Cambridge, MA: MIT Press, 1998). See also "In Search of Gaian Politics: Earth Religion's Challenge to Modern Western Civilization," in *Ecological Resistance Movements: The Global Emergence of Radical and Popular Environmentalism*, ed. Bron Taylor (Albany: State University of New York Press, 1995), and "Ground Identity: Nature, Place, and Space in Nationalism," in *The Return of Culture and Identity in IR Theory*, ed. Yosef Lapid and Friedrich Kratochwil (Boulder: Lynne Rienner, 1996).

that if the people do not fulfill their religious duties, divine blessing might be withdrawn. With civil religion there are, therefore, both positive reasons for ethical behavior (the joys of belonging), and negative ones (avoiding misfortune). In sum, with civil religion, identity and loyalty are "based upon the experiences and feelings of connectedness to a particular place or area."[8] With terrapolitan Earth religion, these feelings and the corresponding ethical obligations are to the Earth and the planet's diverse life-forms and ecosystems. It yields what some call natural piety, biophilia, or religious naturalism, among other terms. It could provide, Deudney believes, a unifying Earth identity and a cultural basis for international environmental cooperation, even a federal republican Earth constitution. Moreover, Deudney argues, its potential is partly because such religion provides "a scientifically credible cosmology" that coheres with an evolutionary-ecological worldview, unlike most long-standing religions. This is one reason for the growing influence of dark green or terrapolitan Earth religion, and the possibility that in a salutary way, such religion could become decisive in the human and planetary future. Such religion self-consciously seeks to promote cultures well adapted to their regional and global habitats.

These positive developments pose a most troubling question: Are the world's long-standing religions fundamentally maladaptive? The late environmental anthropologist Roy Rappaport was one of many scholars who thought so, asserting that the world's major religious traditions, largely because they were written down and thus had become inflexible, were "adaptively false." In other words, they are ecologically maladaptive. He and a number of other eminent scholars who have been promoting this "dark green religion" believe the world's predominant religions should be jettisoned in favor of new spiritual forms that cohere with scientific cosmologies promoting reverential behaviors toward the entire natural world. Are they right? Will maladaptive religions eventually die out because they lead to the destruction of the habitats of their carriers? If the answers to these questions are affirmative, then no time should be wasted in replacing

---

8. Deudney, "Global Village Sovereignty," 313, which also contains his review of other writers and thinkers who have articulated such views. See 317 for the subsequent Deudney quote about credible cosmology.

the old, maladaptive forms with nature spiritualities grounded in an evolutionary-ecological worldview.

More time will be needed to judge whether any long-standing religious traditions will prove malleable enough to be adaptive long term; certainly many within the world's religions are laboring to make them environmentally responsible. Nevertheless, religions with ancient roots have more historical and conceptual obstacles to overcome before they can promote comprehensive green behavior than do post-Darwinian forms of nature spirituality. This is why very little of the energy expended by participants in the world's religions is currently going toward the protection and restoration of the world's ecosystems.[9] Conversely, participants in nature spiritualities steeped in an evolutionary-ecological worldview appear to be more likely to work ardently in environmental causes than those in religious traditions with longer pedigrees.[10] I would be delighted if decisive majorities in mainstream religious traditions were to become more environmentally engaged than individuals with other backgrounds and worldviews, but there is little evidence of such a trend.[11] This sad fact casts doubt on the hope that the greening of conventional religions will lead the way toward the urgently needed changes. If worldviews

---

9. For the most comprehensive compilation of the available evidence, see B. Taylor, ed., *Encyclopedia of Religion and Nature,* and for subsequent scholarly analysis, the *Journal for the Study of Religion, Nature and Culture* (see www.religionandnature.com).

10. A recent study of college students found a significant correlation between "biospheric altruism" among college students and environmentally beneficent behavior; see Thomas Dietz, Amy Fitzgerald, and Rachael Shwom, "Environmental Values," *Annual Review of Environmental Resources* 30 (2005): 335–372. Environmental sociologist Bernard Zaleha accurately summarized the parts of this study that are especially pertinent here: "Those who ascribe to some type of nature-venerating religion probably can be expected to have a higher rate of pro-environment behaviors" than those with a more anthropocentric attitude toward nature. Bernard Zaleha, "Our Only Heaven: An Investigation of the Global Spread and Significance of Nature-Venerating Religion," unpublished paper, University of California, Santa Cruz, April 2009.

11. Survey researchers in 2006 concluded, based on research assuming and focused on religion as conventionally defined in the United States, that it was "not a major influence on environmental views." Pew Research Center report, "Americans uneasy with mix of religion and politics," 24 August 2006, http://pewforum.org/publications/surveys/religion-politics-06.pdf. I have found no little empirical data to suggest that participants in the world's predominant religious traditions are dramatically more environmentally active than other citizens.

matter, a much more profound worldview change may be needed than many assume. It may be that reforming long-standing religions is another form of incrementalism that the planet can ill afford. For this reason, I will put my energies into promoting a more radical religious reformation.

I believe that our greatest hope resides not in invisible beings or the reformation of traditions believing in them, but in the unfolding evolutionary-ecological worldview, which teaches interdependence *and* mutual dependence, evokes humility and felt kinship with other organisms, and imparts a feeling of belonging and connection to our biosphere. With such nature spirituality, we understand that all life got here in the same way, that we are all subject to the same laws, and that although we have unique talents, we have no greater right to be here than any other living thing. With such perception, we might agree to shrink our numbers to ensure the planet's other species have the habitats they need. With a clear understanding of the limits of human knowledge, even though we might disagree about ultimate causes, we can agree that we are lucky to belong to this wondrous and mysterious planet and cosmos. We could also, with such shared understandings, learn to participate self-consciously and responsibly in the Earth's ongoing process of biocultural evolution. This, I think, is our beautiful and daunting challenge.

Coda

Humble yourself as you face the ocean, for it is the source of life.
Humble yourself as you explore the terrestrial world, for it is the expression of the life force.
Humble yourself in the midst of both watery and earthly worlds, for you are utterly dependent on them.
Humble yourself as you contemplate the awesome universe.
Humble yourself as you ponder the mysteries of life, for you will never fully understand them.
Then, celebrate these mysteries, joyous at your good fortune.
Rejoice that you are alive and belong to the only place we know life exists.
Evolve in ways that respect life's diversity.
And, if you feel an urge to worship, worship life.

A Promise Made in Love, Awe, and Fear

Wendell Berry

As industrial technology advances and enlarges, and in the process assumes greater social, economic, and political force, it carries people away from where they belong by history, culture, deeds, association, and affection. And it destroys the landmarks by which they might return. Often it destroys the nature or the character of the places they have left. The very possibility of a practical connection between thought and the world is thus destroyed. The little that survives is attenuated—without practical force. That is why the Jews, in Babylon, wept when they remembered Zion. The mere memory of a place cannot preserve it, nor apart from the place itself can it long survive in the mind. "How shall we sing the Lord's song in a strange land?"

The enlargement of industrial technology is thus analogous to war. It continually requires the movement of knowledge and responsibility away from home. It thrives on the disintegration of homes, the subjugation of homelands. It requires that people cease to cooperate directly to fulfill local needs from local sources and begin instead to deal with each other always across the rift that divides producer and

---

WENDELL BERRY is a poet, essayist, farmer, and novelist. He has taught at Stanford University, the University of Kentucky, Georgetown College, and Bucknell University, among others. Berry has received numerous awards and honors, including a Guggenheim Foundation Fellowship, a Rockefeller Foundation Fellowship, a National Institute of Arts and Letters award for writing, the Lyndhurst Prize, and the Aitken-Taylor Award for Poetry. He has written more than forty works of fiction, nonfiction, and poetry, including *The Mad Farmer Poems* (2008) and *Whitefoot* (2009).

consumer, and always competitively. The idea of the independence of individual farms, shops, communities, and household is anathema to industrial technologists. The rush to nuclear energy is powered by the industrial will to cut off the possibility of a small-scale energy technology—which is to say the possibility of small-scale personal and community acts. The corporate producers and their sycophants in the universities and the government will do virtually anything (or so they have obliged us to assume) to keep people from acquiring necessities in any way except by *buying* them.

People who are willing to follow technology wherever it leads are necessarily willing to follow it away from home, off the Earth, and outside the sphere of human definition, meaning, and responsibility. One has to suppose that this would be all right if they did it only for themselves and if they accepted the terms of their technological romanticism absolutely—that is, if they would depart absolutely from all that they propose to supersede, never to return. But past a certain scale, as C. S. Lewis wrote,[1] the person who makes a technological choice does not choose for himself alone, but for others; past a certain scale he chooses for *all* others. If the effects are lasting enough, he chooses for the future. He makes, then, a choice that can neither be chosen against nor unchosen. Past a certain scale, there is no dissent from a technological choice.

All the grand and perfect dreams of the technologists are happening in the future, but nobody is there.

What can turn us from this deserted future, back into the sphere of our being, the great dance that joins us to our home, to each other and to other creatures, to the dead and the unborn? I think it is love. I am perforce aware how badly and embarrassingly that word now lies on the page—for we have learned at once to overuse it, abuse it, and hold it in suspicion. But I do not mean any kind of abstract love, which is probably a contradiction in terms, but particular love for particular things, places, creatures, and people, requiring stands and acts, showing its successes or failures in practical or tangible effects. And it implies a responsibility just as particular, not grim or merely dutiful, but rising out of generosity. I think that this sort of love

---

1. C. S. Lewis, *The Abolition of Man* (New York: Macmillan, 1975), 70–71.

defines the effective range of human intelligence, the range within which its works can be dependably beneficent. Only the action that is moved by love for the good at hand has the hope of being responsible and generous. One cannot love the future or anything in it, for nothing is known there. And one cannot unselfishly make a future for someone else. Love for the future is self-love—love for the present self, projected and magnified into the future—and it is an irremediable loneliness.

Because love is not abstract, it does not lead to trends or percentages or general behavior. It leads, on the contrary, to the perception that there is no such thing as general behavior. There is no abstract action. Love proposes the work of settled households and communities, whose innovations come about in response to immediate needs and immediate conditions, as opposed to the work of governments and corporations, whose innovations are produced out of the implicitly limitless desire for future power or profit.

I come, in conclusion, to the difference between "projecting" the future and making a promise. The "projecting" of "futurologists" *uses* the future as the safest possible context for whatever is desired; it binds one only to selfish interest. But making a promise binds one *to someone else's future*. If the promise is serious enough, one is brought to it by love, and in awe and fear. Fear, awe, and love bind us to no selfish aims, but to each other. For when we promise in love and awe and fear, there is a certain kind of mobility that we give up. We give up the romanticism of progress, which is always shifting its terms to fit its occasions. We are speaking where we stand, and afterward we shall stand in the presence of what we have said.

The Call to Forgiveness at the End of the Day

Kathleen Dean Moore

May 25, 2025

All those years, the Swainson's thrushes were the first to call in the mornings. Their songs spiraled like mist from the swale to the pink sky. That's when I would take a cup of tea and walk into the meadow. Swallows sat on the highest perches, whispering as they waited for light to stream onto the pond. Then they sailed through the midges, scattering motes of wing-light. Chipping sparrows buzzed like sewing machines as soon as the sun lit the Douglas firs. If I kissed the knuckle of my thumb, they came closer and trilled again.

For years there were flocks of goldfinches. After my husband and I poisoned the bull thistles on the far side of the pond, the goldfinches perched in the willows. When they landed there, dew shook from the branches into the pond, throwing light into new leaves where chickadees chirped. The garbage truck backed down the lane, beeping its backup call, making the frogs sing, even in the day.

Oh, there was music in the mornings, all those years. In the overture to the day, each bird added its call until the morning was an ecstasy of music that faded only when the diesel pumps kicked on to pull water from the stream to the neighbor's bing cherry trees.

Evenings were glorious too. Just as the sun set, little brown bats began to fly. If a bat swooped close, I heard its tiny sonar chirps, just

---

KATHLEEN DEAN MOORE (coeditor) is Distinguished Professor of Philosophy at Oregon State University, where she directs the Spring Creek Project for Ideas, Nature, and the Written Word. She is an environmental philosopher and essayist whose most recent books are *The Pine Island Paradox* (2005), *Rachel Carson: Legacy and Challenge* (2008, with Lisa Sideris), and *Wild Comfort* (2010).

at the highest reach of my hearing. Each downward flitter of its wings squeezed its lungs and pumped out another chirp, the way a pump-organ exhales Bach. Frogs sang and sang, but not like bats or birds. Like violins, violin strings just touched by the bow, the bow touching and withdrawing. They sang all evening, thousands of violins, and into the night. They sang while crows flew into the oaks and settled their wings, while garter snakes, their stomachs extended with frogs, crawled finally under the fallen bark of the oaks and stretched their lengths against cold ground.

I don't know how many frogs there were in the pond then. Thousands. Tens of thousands. Clumps of eggs like eyeballs in aspic. Neighborhood children poked them with sticks to watch their jelly shake. When the eggs hatched, there were tadpoles. I have seen the shallow edge of the pond black with wiggling tadpoles. There were that many, each with a song growing inside it and tiny black legs poking out behind. Just at dusk, a hooded merganser would sweep over the water, or a pair of geese, silencing the frogs. Then it was the violins again, and geese muttering.

In the years when the frog choruses began to fade, scientists said it was a fungus, or maybe bullfrogs were eating the tadpoles. No one knew what to do about the fungus, but people tried to stop the bullfrogs. Standing on the dike, my neighbor shot frogs with a pellet gun, embedding silver BBs in their heads, a dozen holes, until she said *How many holes can I make in a frog's face before it dies? Give me something more powerful.* So she took a shotgun and filled the bullfrogs with buckshot until, legs snapped, faces caved in, they slowly sank away. Ravens belled from the top of the oak.

When the bats stopped coming, they said that was a fungus too. When the goldfinches came in pairs, not flocks, we told each other the flocks must be feeding in a neighbor's field. No one could guess where the thrushes had gone.

Two springs later, there were drifts of tiny white skins scattered in the shallows like dustrags in the dusk. I scooped one up with a stick. It was a frog skin, a perfect empty sack, white, intact, but with no frog inside—cleaned, I supposed, by snails or winter—and not just one. Empty frogs scattered on the muddy bottom of the pond. They were as empty as the perfect emptiness of a bell, the perfectly shaped absence ringing the angelus, the evening song, the call for forgiveness at the end of the day.

As it happened, that was the spring when our granddaughter was born. I brought her to the pond so she could feel the comfort I had known there for so many years. Killdeer waddled in the mud by the shore, but even then, not so many as before. By then, the pond had sunk into its warm, weedy places, leaving an expanse of cracked earth. Ahead of the coming heat, butterflies fed in the mud between the cracks, unrolling their tongues to touch salty soil.

I held my granddaughter in my arms and sang to her then, an old lullaby that made her soften like wax in a flame, molding her little body to my bones. *Hush-a-bye, don't you cry. Go to sleep, you little baby. Birds and the butterflies, fly through the land.* I held her close, weighing the chances of the birds and the butterflies. She fell asleep in my arms, unafraid.

I will tell you, I was so afraid.

Poets warned us, writing of *the heartbreaking beauty that will remain when there is no heart to break for it.* But what if it is worse than that? What if it's the heartbroken children who remain in a world without beauty? How will they find solace in a world without wild music? How will they thrive without green hills edged with oaks? How will they forgive us for letting frog-song slip away? When my granddaughter looks back at me, I will be on my knees, begging her to say I did all I could.

I didn't do all I could have done.

It isn't enough to love a child and wish her well. It isn't enough to open my heart to a bird-graced morning. Can I claim to love a morning if I don't protect what creates its beauty? Can I claim to love a child if I don't use all the power of my beating heart to preserve a world that nourishes children's joy? Loving is not a kind of *la-de-da*. Loving is a sacred trust. To love is to affirm the absolute worth of what you love and to pledge your life to its thriving—to protect it fiercely and faithfully, for all time.

My husband and I were there when the last salmon died in the stream. When we came upon her in the creek, her flank was torn and moldy. She had already poured the rich, red life from her muscles into her hopeless eggs. She floated downstream with the current, twitching when I pushed her with a stick to turn her upstream again. Sometimes her jaws gaped, still trying to move water over her gills. Sometimes she tried to swim. But she bumped against rocks, spilling eggs onto the stones. Without reason, she pushed her head into the air and gasped. We waded beside her until she died. When she was dead, she floated with her tail just above the surface, washing downstream until she lodged on a gravel bar. The music she made was the riffle of rib-bones raking water, then no sound at all as her body settled to the bottom of the pool.

I buried my face in my hands, even as I stood in the water with the current shining against my shins. Oh, we had known the music of salmon moving upstream. When the streams were full of salmon, crows called again and again, and seagulls coughed on the gravel bars. Orioles sang, their heads thrown back with singing. Eagles clattered. Wading upstream, we walked through waves of carrion flies that lifted off the carcasses to swarm in our faces, buzzing like electrical current. Water lifted and splashed, swept by strong gray tails, and pebbles rolled downstream. It was a crashing coda, the slam and the buzz and the gull-scream.

Ring the angelus for the salmon and the swallows. Ring the bells for frogs floating in bent reeds. Ring the bells for all of us who did not save the songs. Holy Mary mother of god, ring the bells for every sacred emptiness. Let them echo in the silence at the end of the day. Forgiveness is too much to ask. I would pray for only this: that our granddaughter would hear again the little lick of music, that grace note toward the end of a meadowlark's song.

Meadowlarks. There were meadowlarks. They sang like angels in the morning.

ETHICAL ACTION

How can we express our love for the world?

No letters to your congressional representatives tonight. No second cup of shade-grown coffee. No Web search for rates of ice-cap melting or declining numbers of polar bears.

Turn off the lights. Go outside. Shut the door behind you.

Maybe rain has fallen all evening and the moon, when it emerges between the clouds, glows on the flooded streets and silhouettes leafless maple trees lining the curb. Maybe the tide is low under the docks and warehouses, and the air is briny with kelp. Maybe cold air is sinking off the mountain, following the river wall into town, bringing smells of snow and damp pines. Starlings roost in a row on the rim of the supermarket, their wet backs blinking red and yellow as neon lights flash behind them. In the gutter, the same lights redden small pressure waves that build and break against crescents of fallen leaves.

Let the reliable rhythms of the moon and the tides reassure you. Let the smells return memories of other seas and times. Let the reflecting light magnify your perception. Let the rhythm of the rushing water flood your spirit. Walk and walk until your heart is full.

Then you will remember why you try so hard to protect this beloved world, and why you must.

13

> *Do we have a moral obligation to take action to protect the future of a planet in peril?*
>
> **Yes, to honor and celebrate the Earth and Earth systems.**

Begin with the Big Bang. Imagine the swirling gas, the accreting matter. Imagine lightning at the edge of water. Imagine first life.

Or imagine there is no beginning. Imagine the People hoeing corn forever under sweet rain that returns and returns.

However we got to this day, this decade, whatever unfolding of time and distance, here we are, in the time of rain, the time of spring flowers, sage, salt water, fertile soil that brings forth gardens choked with vegetables, family and friends healthy and happy. The astonishing fact of this moment shakes us to the core—its contingency, its blessing.

We are the beings through which the universe becomes aware of itself and celebrates itself. This is our duty.

That we might, through carelessness or callousness, be the ones, after billions of years, to undermine this unfolding, to cut off unfurling forms of life, to fail in our praise, is unspeakable.

The Great Work

Thomas Berry

The most basic issue of our time is human-Earth relations. We have disturbed the geological structure, the chemical composition, and the biological forms of the planet in a disastrous manner with our population explosion and technological power. We have closed down the creativity of the Cenozoic era (the last 65 million years) and are ending a chapter of the geobiological history of Earth. Earth is now in a state of recession; its basic life systems have become disturbed, toxic, or are extinguished. The tragedy is that the Cenozoic has been a lyrical period in Earth's history. This was the period when Earth came to its full florescence. Trees, songbirds, flowering plants, marine life, tropical rainforests: all of these and more came into being during this era.

It was in such a setting, amid such awesome surroundings, that we as humans came into being and had our primordial experience of existence, an experience so overwhelming that our awakening into conscious awareness may well have been simultaneously an awakening to the divine.

My generation has done what no previous generation could do, because they lacked the technological power, and what no future gen-

---

THOMAS BERRY (1914–2009) was a writer, cultural historian, and eco-theologian. His work was widely recognized as creating a new vision of the human-Earth relationship. He received eight honorary doctorate degrees and was recognized by numerous organizations, including the United Nations, the Center for Respect of Life and Environment, and the Catholic Church, for his outstanding contributions to peace and justice in the world. His books include *The Great Work: Our Way into the Future* (2000) and *The Dream of the Earth* (2006).

eration will be able to do, because the planet will never again be so beautiful or abundant.[1]

The current extinction is being caused by human action within a cultural tradition shaped in a biblical-Christian and classical-humanist matrix. The tragic flaw in both traditions seems to be an anthropocentrism that has turned into a profound cultural pathology. The biblical story, however valid, however unique in what it offers, no longer seems sufficient to address the issues before us. We also need the story of our past and our dream of the future Ecozoic era, for this coming era must first be dreamed. Through the dream comes the guidance, the energy, and the endurance we will need. The transition that is before us will cost an immense effort and require a wisdom beyond anything that we have known before.[2]

To recover a situation where humans would be present to the Earth in a mutually enhancing manner, I believe we must return to a sense of intimacy with the Earth akin to that experienced by many indigenous peoples of earlier times. This can be done through our new story of the universe, which is now available to us through empirical inquiry into the origin, structure, and sequences of transformations through which the Earth has come to its expression at the end of the twentieth century.

Articulating this story fully would be the supreme achievement of modern intelligence. Then we can see that this story of the universe is in a special manner our sacred story, a story that reveals the divine particularly to ourselves, in our times. We will also be able to appreciate the primordial unity of origin of every other being in the universe. This is especially true of living beings of Earth, all of which have descended through the same life processes. Through this sharing in a common story, we come to recognize our total intimacy with the entire natural world. An impenetrable psychic barrier is removed. We are no longer alienated objects but communing subjects. We recognize that in every aspect of our being, we are a subsystem of the universe system.

We discover the Earth in the depths of our being through partici-

---

1. Excerpts from Thomas Berry, "Religion in the Ecozoic Era," in *The Sacred Universe*, ed. Mary Evelyn Tucker, 89–91 (New York: Columbia University Press, 2009).
2. The material following the previous note is from Berry, *The Sacred Universe*, 98.

pation, not through isolation or exploitation. We are most ourselves when we are most intimate with the rivers and mountains and woodlands, with the sun and the moon and the stars in the heavens, when we are most intimate with the air we breathe, the Earth that supports us, the soil that grows our food, with the meadows in bloom. We belong here. Our home is here. The excitement and fulfillment of our lives is here. However we think of eternity, it can only be an aspect of the present. The urgency of this psychic identity with the larger universe about us can hardly be exaggerated. Just as we are fulfilled in our communion with the larger community to which we belong, so too the universe itself and every being in the universe is fulfilled in us.

We might say that the universe, through its vast extent in space and its long sequence of transformations in time, is a single multiform celebratory event. The human might be described as that being in whom the universe reflects on and celebrates itself and the deep mysteries of existence in a special mode of conscious self-awareness. Within this larger universe, the planet Earth constitutes a single integral community. It lives or dies, is honored or degraded, as a single interrelated reality.

As regards the future, it can be said quite simply that the human community and the natural world will go into the future as a single sacred community, or we will both experience disaster on the way.[3]

Perhaps the most valuable heritage we can provide for future generations is some sense of the Great Work that is before them of moving the human project from its devastating exploitation to a benign presence. We need to give them some indication of how the next generation can fulfill this work in an effective manner. For the success or failure of any historical age is the extent to which those living at that time have fulfilled the special role that history has imposed upon them. No age lives completely unto itself. Each age has only what it receives from the prior generation. Just now we have abundant evidence that the various species of life, the mountains and rivers, and even the vast ocean itself, which once we thought beyond serious impact from humans, will survive only in their damaged integrity.

3. The material following the previous note is from Berry, *The Sacred Universe*, 94–96.

The Great Work before us, the task of moving modern industrial civilization from its present devastating influence on the Earth to a more benign mode of presence, is not a role that we have chosen. It is a role given to us, beyond any consultation with ourselves. We did not choose. We were chosen by some power beyond ourselves for this historical task. We do not choose the moment of our birth, who our parents will be, our particular culture, or the historical moment when we will be born. We do not choose the status of spiritual insight or political or economic conditions that will be the context of our lives. We are, as it were, thrown into existence with a challenge and a role that is beyond any personal choice. The nobility of our lives, however, depends upon the manner in which we come to understand and fulfill our assigned role.

Our own special role, which we will hand on to our children, is that of managing the arduous transition from the terminal Cenozoic to the emerging Ecozoic era, the period when humans will be present to the planet as participating members of the comprehensive Earth community. This is our Great Work and the work of our children.[4]

---

4. From *The Great Work,* by Thomas Berry, copyright © 1999 by Thomas Berry. Used by permission of Bell Tower, a division of Random House, Inc.

An Ethic of the Earth

N. Scott Momaday

Once in his life a man ought to concentrate his mind upon the remembered Earth, I believe. He ought to give himself up to a particular landscape in his experience, to look at it from as many angles as he can, to wonder about it, to dwell upon it. He ought to imagine that he touches it with his hands at every season and listens to the sounds that are made upon it. He ought to imagine the creatures there and all the faintest motions of the wind. He ought to recollect the glare of noon and all the colors of the dawn and dusk.

I am interested in the way that a man looks at a given landscape and takes possession of it in his blood and brain. For this happens, I am certain, in the ordinary motion of life. None of us lives apart from the land entirely; such an isolation is unimaginable. We have sooner or later to come to terms with the world around us—and I mean especially the physical world, not only as it is revealed to us immediately through our senses, but also as it is perceived more truly in the long turn of seasons and of years. And we must come to moral terms.

N. SCOTT MOMADAY (NAVARRO SCOTT MAMMEDATY), a Kiowa Indian, was born in Lawton, Oklahoma, in 1934 and grew up in close contact with the Navajo and San Carlos Apache communities. He received his BA in political science in 1958 from the University of New Mexico. At Stanford University he received his MA and PhD in English. His books of poetry include *In the Bear's House* (St. Martin's Press, 1999), *In the Presence of the Sun: Stories and Poems, 1961–1991* (1992), and *The Gourd Dancer* (1976). His first novel, *House Made of Dawn* (1969), won the Pulitzer Prize for Fiction. He is author of several other novels, prose collections, the children's book *Circle of Wonder* (1994), and the play *The Indolent Boys*. He is also the editor of various anthologies and collections.

There is no alternative, I believe, if we are to realize and maintain our humanity, for our humanity must consist in part in the ethical as well as in the practical ideal of preservation. And particularly here and now is that true. We Americans need now more than ever before—and indeed more than we know—to imagine who and what we are with respect to the Earth and sky. I am talking about an act of the imagination, essentially, and the concept of an American land ethic.

It is no doubt more difficult to imagine the landscape of America now than it was in, say, 1900. Our whole experience as a nation in this century has been a repudiation of the pastoral ideal which informs so much of the art and literature of the nineteenth century. One effect of the technological revolution has been to uproot us from the soil. We have become disoriented, I believe; we have suffered a kind of psychic dislocation of ourselves in time and space. We may be perfectly sure of where we are in relation to the supermarket and the next coffee break, but I doubt that any of us knows where he is in relation to the stars and to the solstices. Our sense of the natural order has become dull and unreliable. Like the wilderness itself, our sphere of instinct has diminished in proportion as we have failed to imagine truly what it is. And yet I believe that it is possible to formulate an ethical idea of the land—a notion of what it is and must be in our daily lives—and I believe moreover that it is absolutely necessary to do so.

It would seem on the surface of things that a land ethic is something that is alien to, or at least dormant in, most Americans. Most of us have developed an attitude of indifference toward the land. In terms of my own experience, it is difficult to see how such an attitude could ever have come about.

Ko-sahn could remember where my grandmother was born. "It was just there," she said, pointing to a tree, and the tree was like a hundred others that grew up in the broad depression of the Washita River. I could see nothing to indicate that anyone had ever been there, spoken so much as a word, or touched the tips of his fingers to the tree. But in her memory Ko-sahn could see the child. I think she must have remembered my grandmother's voice, for she seemed for a long moment to listen and to hear. There was a still, heavy heat upon that place; I had the sense that ghosts were gathering there.

And in the racial memory, Ko-sahn had seen the falling stars. For her there was no distinction between the individual and the racial experience, even as there was none between the mythical and the historical. Both were realized for her in the one memory, and that was of the land. This landscape, in which she had lived for a hundred years, was the common denominator of everything that she knew and would ever know—and her knowledge was profound. Her roots ran deep into the Earth, and from those depths she drew strength enough to hold still against all the forces of chance and disorder. And she drew from the sustenance of meaning and of mystery as well. The falling stars were not for Ko-sahn an isolated or accidental phenomenon. She had a great personal investment in that awful commotion of light in the night sky. For it remained to be imagined. She must at last deal with it in words; she must appropriate it to her understanding of the whole universe. And, again, when she spoke of the Sun Dance, it was an essential expression, something of her relationship to the life of the Earth and to the sun and moon.

In Ko-sahn and in her people we have always had the example of a deep, ethical regard for the land. We had better learn from it. Surely that ethic is merely latent in ourselves. It must now be activated, I believe. We Americans must come again to a moral comprehension of the Earth and air. We must live according to the principle of a land ethic. The alternative is that we shall not live at all.

Spring's Hopes Eternal

Curt Meine

The time of renewal has arrived again in the American Midwest. In the wetlands, migrating ducks and geese find open water, silver maples bud out, marsh marigolds explode with sun-yellow blooms, muskrats jostle for territory, chorus frogs click, and spring peepers testify to their name. In the woodlands, ephemerals appear. A month ago, sugar maples sprouted sap buckets. Now the anemones and Dutchman's breeches and spring beauties emerge from below, and transitory warblers arrive overhead. Ferns unfurl. The invasive garlic mustard seizes the daylight and overtakes entire woodlots. In the prairies, pasque flowers launch the season, soils and grasses dry, and human beings with drip torches, matches, and water cans train fires to restore the land's vitality. In gardens and barnyards and fields and orchards and pastures, we mingle again with soil and plants and animals in the year's opening acts of production; vegetables, greens, grains, fruit, milk, and meat will follow. In towns and cities, we shed, layer by layer, our winter pelage and torpor. We all come to that instant when we are struck by the lengthening and warming of the days—and take the moment to congratulate ourselves for *making it* through the months of cold and darkness.

Every spring in Wisconsin our topography recapitulates our glacial geology. Winter gives way to spring as the Pleistocene gave way to

---

CURT MEINE is director of the Center for Humans and Nature, a Senior Fellow with the Aldo Leopold Foundation in Baraboo, Wisconsin, and a research associate with the International Crane Foundation. He is a conservation biologist, historian, and writer who has edited and authored several books, including *Correction Lines: Essays on Land, Leopold, and Conservation* (2004).

the Holocene. The ice melts back, south to north. Snow piles persist late on the north slopes, in the deeper valleys, under the conifers. Thawing snowbanks leave behind mini-moraines of sand and gravel and grit, lost gloves, beer cans, and candy wrappers. People venture forth again into reopened territory, and plants and animals reclaim the land. Pools of meltwater linger in fields, holding the tractors back and the ducks on for at least a few weeks. When the water table is high, as it has been lately, and the pools stay, cattails and arrowheads sprout amazingly from the over-corned, mucky seed banks.

Maybe it was this annual recapitulation that informed the flood and origin stories of the Midwest's Native people. Maybe it inspired early naturalists and geologists to see so well the region's glacial past and to understand the reality of geologic time and climatic change. In his fresh Wisconsin youth in the 1850s, John Muir imprinted on the region's glaciated lands and waters, developing the imagination that later allowed him to see the way of the ice through Yosemite's high cirques and valleys. In the 1870s, geologist T. C. Chamberlin began his field studies of the moraines of southeastern Wisconsin. This work allowed him to piece together the epic story of North America's glacial past, to track the pulsing advances and melt-backs that shaped the lay of the land. Chamberlin's expertise led him to literally define the terms of the Pleistocene. In an 1896 contribution to the *Journal of Geology*, he provided the first classification and names for North America's glacial stages. Chamberlin called the most recent period the Wisconsin stage, reflecting the prominence of glacial features in his home landscape—moraines, boulder fields, basins, drumlins, eskers, kames, lakes, ponds, beach ridges. In speculating on the causes and effects of glaciation, Chamberlin was among the first scientists to consider the Earth as an entire, complex, dynamic system. He was also among the first to identify the critical role that long-term carbon cycling played in influencing climatic conditions.

The work of Chamberlin and other early Earth scientists pulled back the curtain on Earth's environmental past, revealing its incredible panoply of climatic change. Older stories of creation, change, and human origins took on new meaning. In the process, we took a bite out of Eden's apple, gained self-awareness, came to new understandings of our earthly reality. But in the bargain we also lost the pure innocence of spring and the easiness of hope.

Once upon a time, we knew nothing about carbon dioxide and methane concentrations, Milankovich cycles and paleoclimates, albedo effects and thermohaline circulation. We could appreciate the spring for what we experienced it to be: the return of annual warmth and light and life. The hope we found in the spring could be wildly and wonderfully *ungrounded*—effortless, irrational, even unrealistic. We could simply feel the season's sensations and processes: anticipation, energy, promise, renewal, revival, rebirth.

Then we gained critical understanding of the seasons. We came to know the return of spring as a contingent phenomenon, an expression of complex and interacting natural forces changing ceaselessly over space and through time. We grew to understand how the creatures of the air and land and sea responded to such changes. How, as the continental glaciers melted back over the millennia, species ranges expanded and shrank, populations swelled and ebbed, migration routes stretched out, anatomies adapted, life cycles shifted, ecological relationships reordered themselves. And how, toward the end of the Pleistocene, human beings began walking out of Africa.

Then we gained critical understanding of our contribution to a changing spring. How we came out of the ice ages with new ways to both exploit and symbolize the living world around us, intensifying our relationships within it and our impacts on it. How we built human civilization over twelve postglacial millennia by drawing upon the energy-rich carbon held in the soils, forests, coal beds, oil and natural gas fields, using it to fuel our Agricultural, Neolithic, Industrial, and Information Revolutions. How through systematic exploitation of those pools of energy we changed the land, with consequences both intended and unintended. How those consequences included the further buildup of greenhouse gases, the scrambling of ecosystems, the altering of soils and hydrologies and chemistries, and the diminishing of biological diversity at accelerated rates and on expanded scales. How we changed ourselves and our communities in the process. We now track the oncoming spring not just as an astronomical or meteorological phenomenon, but as in part a *cultural* phenomenon. Through phenological studies, we chronicle the timing of changes in frosts and thaws, bloomings and callings, migrations and hatchings—but now we calculate degrees of human influence on those changes.

At the beginning of this century, scientists coined a new term—the Anthropocene—to distinguish the current geologic era of unprecedented human impacts on the Earth, its systems, and its other lifeforms. As our actions as *Homo sapiens* have changed the spring, so have they changed the very geography and phenology of hope. How can one *hope* when spring now signifies such drastic change—when it does not just warm our chilled bodies, but liquefies the polar ice sheets? How can one *hope* when facing an increasingly uncertain future?

And so, in the Anthropocene, hope too becomes an increasingly human artifact. From time immemorial hope has been a joyful human *response* to geological, orbital, and environmental flux. Now hope must become more and more a human *creation*. In a spring that is several degrees more humanized, we cannot just rely on the Earth's circling of the sun to provide our free supply of hope. We will need to do more than just gather the wild expectations and aspirations that arise spontaneously from the Earth and sun. We will need to cultivate it, to *make it*. We will need to generate hope out of the human heart, expanding the circle of our concern and compassion.

We confront a great challenge and a daunting obligation: to ensure that, for our children and grandchildren, spring remains a season of hope and renewal rather than of concern and fear. To do so we will need to ground our hope, redefine it, put it to work. In its innocent, pre-Anthropocene incarnation, hope was (in Emily Dickinson's timeless rendering) a thing with feathers that perched in the soul and sang its wordless tune. In its newly sobered form, hope must become a thing of encouragement, attentiveness, and mindfulness; of competence and commitment; of confidence in our ability to perform wisely and well as human beings. If our naive hope has flown off, our mature hope must gain the self-assurance of the spring migrant, ever alert to change, yet able and determined to wing its way onward. We are bound now to renew the experience of renewal itself.

Dawn for All Time

Linda Hogan

The moon is filling, a bowl of earthlight remaining in the first of dawn. Venus is near Earth in its orbit behind the black branches of a tree where a large bird sits. Morning arriving.

Standing with others, we smell pine smoke, hear the whispers, and gaze toward the mountains.

And then it is blue dawn. At the top of the mountain is a deer. No, he of the antlers is a man. No, the man is a deer; they are one, standing majestic and powerful. Smoke rises from behind the mountain in dark gray clouds. It smells like the history of pines. It is the odor of ancient places, old trees the deer and the gatherers walked beneath. At the crest of the next mountain, another deer with antlers wide and great as old branches. It takes my breath away.

For thousands of years this has been a moment of awe, this holy beauty. They come down the mountain, sticks in their front hands used as forelegs of the deer, walking in graceful, animal movement. As one comes to a patch of snow, he moves to the side, around sage, mesquite, and walks beautifully through the chalmisa plants. The other animals come from behind the mountains, crying out with all their life. Drums begin as a heartbeat, and the old men sing, wrapped in woven blankets. From an old adobe building, the sacred deer moth-

---

LINDA HOGAN is a Chickasaw essayist, poet, and novelist whose writings explore the connection between humans and animals. She has served on the National Endowment for the Arts Poetry Panel and recently retired from teaching creative writing at the University of Colorado. She has published sixteen books, including the Pulitzer Prize–winning novel *Mean Spirit* (1990), and her most recent book, *People of the Whale: A Novel* (2008).

ers are brought out, untouchable by human hands. They take their place, solemn and with heavy grace. The dance that has been here in the long past begins once again, at one more dawn. The animals and the hunter dance with reverence and awe on their faces, elders watching on the sides to be certain all is correct, and the Pendleton and Navajo blankets are beautiful standing with human beings inside them. The people have done this for thousands of years, since before any written history, and they will do this for all the tomorrows.

I am a Native woman, a Chickasaw from the Southeast, people who were removed from our homelands during the Trail of Tears. We dwell now in the place once called Indian Territory. It was then created for all the indigenous peoples of this continent by the American government. At the time, all tribes were going to be placed here, in Oklahoma, and a wall built around the borders so no Indians could escape. Black Kettle's band was pursued for so many years that their route eroded the land near Goodland, Kansas.

Nevertheless, in spite of attempts at acculturation, we have maintained our language, and some "outlaws" retained the songs and dances, so some tradition remains, even if it is not like the Pueblo dances, which have been on the same lands for thousands of years. Still, I look at the language and find in it the way we understood our world in the original homelands, the way a people's view was at work before the tear in our lives. A word for animal, *Nan okcha,* means "all alive." It means more than just that which is animated. Embedded in the language, it says that the animals have lives and being and are sentient, a significant part of a whole. They have relationship and connection with other lives and the world at large, an animate world. We have an awe of them and an obligation to keep *all alive*. That is our purpose here. Our ancestors survived in order for us to be here, and we have a debt to them as well. It is an ethic, a way of being with this Earth and its inhabitants, all sacred.

The long histories of invasion and "conquest" changed cultures, destroyed languages and their peoples, cleared the forests of medicines that might have been remembered if the people hadn't been broken. Still, these cultures remain in regions of the world once thought to be places without people, with nothing of value. Now we face other challenges shared by all people as polar ice melts and villages are flooded, as the mines use up the aquifers, as forests are cut and the water leaves, taking over lands in other places as lethal floods, hurricanes, and tsunamis, as droughts occur in places where rainforests were cut. Lives are disappearing through the hole we have created that grows over our shared world. What we don't know is large, and it is leaving us, as the regard for life is now often missing from other humans.

When I consider how we have unbalanced our world, changed its tilt, created the opposite of creation, I know we have to remember. Writer Meridel LeSeuer called it re-membering the dismembered. Over here is a stone with lichens older than we can remember. My great-grandfather walked by those same ancient lichens. Re-member. I want to remember this land. We need to remember the plants are alive and speak with each other. They tell one another when they are under attack by using a language of hormones. They recall their first language. We must remember songs the plants taught us, about when the corn grew its first tassels, its silken hair with the odor of fresh earth. Remember the green cedar when it first made an opening to rise through earth, the birth of an oak breaking open the acorn.

I remember when we had dreamers and they knew the water and its first songs, and I remember that the dreamers found water and medicine for the people. *Nan okcha.* All alive. Remember.

So here it is, the remembered dance at the Pueblo. The dancer lifts his antlers with eagle-down feathers in them, and the sunlight is full now, as full as the remembering in the blood what is left to offer this Earth, animal, all alive.

He turns like a deer.

We are participants in the world, with the universe. What we do changes things, and we need to remember this as much as we need to persevere this way, not just as islands in our cultures but with all people knowing that they are part of it and somewhere in their past are the deep channels of memory, the dream water, the tender shoots of green, and the welcome magic of continuing for the tall grasses, our grandchildren, the unborn infants turning like the deer in a mother, a healthy future, right and good, knowing that if we live well, it will welcome us.

The Universe Story and Planetary Civilization

Mary Evelyn Tucker and Brian Swimme

As we see our present interconnected global challenges of widespread environmental degradation, climate change, crippling poverty, social inequities, and unrestrained militarism, we know that the obstacles to the flourishing of life's ecosystems and to genuine sustainable development are considerable.

In the midst of these formidable challenges, we are being called to the next stage of evolutionary history. This requires a change of

---

BRIAN SWIMME received his PhD in mathematics from the University of Oregon and is on the faculty at the California Institute of Integral Studies, in San Francisco, where he teaches courses on evolutionary cosmology to graduate students in the humanities. He is the author of *The Universe Is a Green Dragon* and *The Hidden Heart of the Cosmos*, coauthor with Thomas Berry of *The Universe Story*, and creator of the DVD series "The Powers of the Universe" (brianswimme.org).

MARY EVELYN TUCKER is a Senior Lecturer in Religion and the Environment at Yale University, holding joint appointments as a research scholar in the Divinity School, the School of Forestry & Environmental Studies, and the Department of Religious Studies. With John Grim, she cofounded the Forum on Religion and Ecology. Tucker and Grim also coordinated a ten-conference series titled "World Religions and Ecology" at Harvard's Center for the Study of World Religions. Tucker has been a committee member of the Interfaith Partnership for the Environment at the United Nations Environment Programme (UNEP) since 1986 and is vice president of the American Teilhard Association. Author of many books on religion and ecology, she has recently published *Worldly Wonder: Religions Enter Their Ecological Phase* (Open Court Press, 2003). She is the coeditor of books on ecological views of Buddhism, Confucianism, and Hinduism. She has published *Confucian Spirituality*, coedited with Tu Weiming, and *The Record of Great Doubts: The Philosophy of Ch'i*.

consciousness and values, an expansion of our worldviews and ethics, for the evolutionary life impulse moves us forward from viewing ourselves as isolated individuals and competing nation-states, to realizing our collective presence as a species with a common origin story and shared destiny. The human community now has the capacity to realize our intrinsic unity in the midst of enormous diversity. Most especially, it has the opportunity to see that this unity arises from the dynamics of the evolutionary process itself. For the first time we have a scientific story of the evolution of the universe and Earth that shows us our profound connectedness to this process. We are still discovering the larger meaning of the story.

Our sense of the whole is emerging in a fresh way as we feel ourselves embraced by the evolutionary powers unfolding over time into forms of ever-greater complexity and consciousness. We are realizing too that evolution moves forward with transitions such as the movement from inorganic matter to organic life and from single-celled organisms to plants and animals, transformations that sweep through the evolutionary unfolding of the universe, the Earth, and the human. All such transitions come at times of crisis. They involve tremendous cost, and they result in new forms of creativity. The central reality of our times is that we are in such a transition moment. This is not an easy moment, as human suffering and environmental loss are already widespread. It is not a guaranteed transition, as it will require tremendous human creativity, emotional intelligence, and spiritual strength.

Surrounding this moment is an awakening to a new consciousness that challenges older paradigms of the human as an isolated being in a random, purposeless universe. Peter Raskin called this the Great Transition, while Joanna Macy named it the Great Turning. From valuing hyperindividualism and independence, our consciousness is shifting to embrace interdependence and kinship on a vast scale. The Enlightenment values of life, liberty, and the pursuit of happiness are being reconfigured. Thus life now includes the larger life of the Earth, individual freedom requires responsibility to community, and happiness involves more than material goods. A sense of a larger common good is emerging—the future of the planet and its fragile biosphere.

In this spirit, we are in a transition from an era dominated by competing nation-states to one that is giving birth to a sustainable multicultural planetary civilization. This birth is occurring within the context of our emerging understanding of the universe story.

Over the past century, science has begun to weave together the story of a historical cosmos that emerged some 13.7 billion years ago. The magnitude of this universe story is beginning to dawn on humans as we awaken to a new realization of the vastness and complexity of this unfolding process.

Just as we begin to understand the vast and complex history of the universe, we are becoming conscious of the multidimensional environmental crisis and of the rapid destruction of species and habitat that is taking place around the planet. Just as we are realizing the vast expanse of time that distinguishes the evolution of the universe over some 13.7 billion years, we are recognizing how late is our arrival in this stupendous process. Just as we are becoming conscious that Earth took more than 4 billion years to bring forth this abundance of life, it is dawning on us how quickly we are foreshortening its future flourishing.

We need, then, to step back, to assimilate our cosmological context. If scientific cosmology gives us an understanding of the origins and unfolding of the universe, philosophical reflection on scientific cosmology gives us a sense of our place in the universe. And if we are so radically affecting the story by extinguishing other life-forms and destroying our own nest, what does this imply about our ethical sensibilities or our sense of the sacred? As science reveals to us the particular intricacy of the web of life, we realize we are unraveling it, although in part unwittingly. Only recently have we become fully conscious of the deleterious consequences of our drive toward economic progress and rapid industrialization.

As we begin to glimpse how deeply embedded we are in complex ecosystems and how dependent on other life-forms, we see we are destroying the very basis of our continuity as a species. As biology demonstrates a fuller picture of the unfolding of diverse species in evolution and the distinctive niche of species in ecosystems, we are questioning our own niche in the evolutionary process. As the size and scale of the environmental crisis is more widely grasped, we are seeing our own connection to this destruction. We have become a planetary presence that is not always benign.

The simultaneous recognition of our cosmological context and our environmental crisis is clearly demonstrated by two major permanent exhibits at the American Museum of Natural History in New York. One is the Rose Center, which houses the Hall of the Universe and the Hall of Planet Earth. The other exhibit is the Hall of Biodiversity.

The Hall of the Universe is architecturally striking. It is housed in a monumental glass cube, in the center of which is a globe containing the planetarium. Suspended in space around the globe are the planets of our solar system. In a fascinating mingling of inner and outer worlds, our solar system is viewed against the backdrop of garden plazas and street scenes visible through the soaring glass panels of the cube. After first passing through a simulation of the originating fireball, visitors move onto a spiral pathway that descends through time, tracing the 12-billion-year cosmic journey, from the great flaring forth in the fireball, through the formation of galaxies, and finally to the emergence of our solar system and planet. It ends with the evolution of life in the Cenozoic period of the last 65 million years. Under a circle of glass, the breadth of a single human hair represents all of human history. The dramatic effect is stunning, as we are called to reimage the human in the midst of such unfathomable immensities.

The Hall of Planet Earth continues this evocation of wonder, revealing the remarkable processes of the birth of Earth, the evolution of the supercontinent Pangaea, the formation of the individual continents, and the eventual emergence of life. It demonstrates the intricacy of plate tectonics, a theory that has been accepted for fewer than fifty years, and it displays geothermal life-forms around deep-sea vents discovered only a decade ago. The exhibit illustrates how new our knowledge of the evolution of the Earth is and how much has been discovered within the last century.

In contrast to the vast scope of evolutionary processes evident in the Hall of the Universe and the Hall of Planet Earth, the Hall of Biodiversity displays the extraordinary range of life-forms that the planet has birthed. A panoply of animals, fish, birds, reptiles, and insects engages the visitor. A plaque in the exhibit observes that we are now living in the midst of a sixth extinction period due to the current massive loss of species. It notes that while the five earlier periods of extinction were caused by a variety of factors, including meteor collisions and climate change, the cause of this present spasm of extinction is, in large part, human activity.

This realization calls into question not only our role as a species, but our ultimate viability. Along with those who recognized the enormity of the explosion of the atomic bombs in Japan, we are the first generations of humans to actually imagine our own destruction as a species. And while this may be extreme, some pessimists suggest this may not be such a regrettable price to pay for the survival of other life-forms.

The exhibition notes, however, that we can stem this tide of loss of species and habitat. The visitor walks through an arresting series of pictures and statistics where current destruction is recorded on one side and restoration processes are highlighted on the other. The contrasting displays suggest that the choice is ours—to become a healing or a deleterious presence on the planet.

These powerful exhibits on cosmic evolution and on species extinction illustrate how science is helping us enter into a macrophase understanding of the universe and of ourselves as a species among other species on a finite planet. As the Rose Center exhibits demonstrate, the evolution of the universe and the Earth is an unfolding story in which humans participate. Indeed, the introductory video to the Hall of the Universe observes that we are "citizens of the universe," born out of stardust and the evolution of galaxies, and that we are now responsible for its continuity. Humans can assist in stemming the current extinction spasm, the Hall of Biodiversity suggests—a bold position for an "objective" and "unbiased" science-based museum.

Scientists no longer stand completely apart from what they study. They are helping us witness the ineffable beauty and complexity of life and its emergence over billions of years. They point toward a more integrative understanding of the role of the human in the midst of an extinction spasm. Some of this shift in the museum's perspective arose in the late 1990s, when the curators were searching for an ornithologist. Of the final six candidates, four studied bird species that vanished into extinction during the course of their research. This alarmed the museum curators, who realized they could not simply stand by with a disinterested objectivity and witness extinction. The Hall of Biodiversity emerged from this realization.

It can be said, then, that the new moment for science involves three intersecting dimensions: to understand the evolution of the universe and the Earth with the best scientific methods, to integrate the evolutionary narrative as a whole (cosmic, Earth, human), and to reflect on the story with a sense of our responsibility for its continuity. From this emerges a new integration of scientific facts, story, and meaning.

Environmental ethicists and scholars of the world's religions are also called to contribute to this understanding of the universe story. The challenge for religion and ethics is both to reenvision our role as citizens of the universe and to reinvent our niche as members of the Earth community. This requires reexamining such cosmological questions as where we have come from and where we are going. In other words, it necessitates rethinking our role as humans within the larger context of universe evolution, as well as in the closer context of natural processes of life on Earth.

What is humankind in relation to 13.7 billion years of universe history? What is our place in the framework of 4.6 billion years of Earth history? How can we foster the stability and integrity of life processes? These are critical questions underlying the new consciousness of the universe story. This is not simply a dynamic narrative of evolution; it is a transformative cosmological story, which engages human energy for a future that is sustaining and sustainable.

The goal of the universe story, then, is to tell the story of cosmic and Earth evolution by drawing on the latest scientific knowledge in a way that makes it both relevant and moving. What emerges is an intensely poetic story that evokes emotions of awe and excitement, fear and joy, belonging and responsibility.

The universe story is a dramatic one. Throughout billions of years of evolution, triumph and disaster have been only a hair's breadth apart. Violence and creativity are pervasive. The ability of matter to organize and reorganize itself is remarkable—from the formation of the first atoms to the emergence of life. Simple hydrogen has become a vibrant living planet, with beings that now are able to investigate how this has happened and imagine a life-sustaining future. We are coming to realize that the energy released at the very beginning has finally become capable, in the human, of reflecting on and exploring its own journey of change.

Waking up to our fundamental relationship with the cosmos will be a way to reengage with life. The universe story enables us to connect more deeply with the universe and the Earth of which we are a part. In doing this, we will appreciate the need for a sustainable human presence on the planet.

Thus the integrated story of the origin and development of the universe, of Earth, and of humans could become an inspiring vision for our time. This is because the story gives us a sense of common evolutionary heritage and shared genetic lineage. This new understanding of kinship of humans with each other and with all life could establish the foundations for rediscovering our past and sustaining the future. Carl Anthony, one of the leaders of the environmental justice movement, has said this perspective has been profoundly transforming for his life and work.

We can indeed be inspired by this view of nested interdependence—from galaxies and stars to planets and ecosystems—so that we sense how personally we are woven into the fabric of life. We are part of this ongoing journey. From this perspective, we can see that our current destructive habits toward the environment are unsustainable. In an evolutionary framework the damage we are causing is immense—indeed, cataclysmic. We can thus recognize ecological, economic, and social change as not only necessary but inevitable.

A great transition is indeed upon us. Beyond world wars and the cold war, there beckons the sense of a larger planetary whole—an emerging, multiform, planetary civilization. It is in participating in this transition moment that we will fulfill our role as humans on behalf of future generations. It requires a profound change of consciousness and values—both an expanded worldview of the universe story and a comprehensive global ethics that embraces the Earth community. In this way we will indeed become citizens, not simply of nation-states but of a planet that beckons us with beauty, mystery, and grace.

ETHICAL ACTION

How can we honor and celebrate the Earth's systems?

Study the stars. How far away are they? How ancient? How long does it take their news of the past to reach our eyes? How can they burn with that brightness, and where do stars go when they die?

Study decomposition. How does it happen that death hurries back into life, even in the course of a season? Why is the skeleton of an alder leaf so beautiful, when its life is gone and scattered? What would the noise sound like if we could hear all the chewing underfoot?

Study pollination.

Study predation.

Study rain.

Study imagination, the biochemistry of new ideas, the frazzle-ended nerve cells reaching out to each other or the divine wind whispering in the soul. How can we think of what does not exist? How can we hope for what we have never seen, never in the history of the world?

Study the beginning and the end. How does it come to be that we are here, humans on the green Earth, precariously perched in the middle of the perfect cosmic explosion?—not too weak or all would collapse back into a single point; not too strong or all would disperse, every particle equidistant in space extending infinitely in all directions.

Study celebration. How is it that humans sing in harmony, bringing different voices and different tones together into chords that vibrate stained-glass windows and shake the leaves on the trees? Where does dancing begin? Why does a child's laughter make us cry with happiness? Can the universe celebrate its splendor without us, or does it need human joy?

These are things we will never completely understand. But trying makes all the difference in the world. The wonder of the ongoing processes of creation astonishes us. Their beauty makes us humble. Their power scorches every bit of arrogance from our raised heads. Their mystery dazzles and confounds us and invites us into the sacred.

14

> *Do we have a moral obligation to take action to protect the future of a planet in peril?*
>
> **Yes, because our moral integrity requires us to do what is right.**

Integrity: oneness, wholeness. Think *integer,* "a positive whole number." Think *integrate,* "to bring together into a whole." People of moral integrity are whole in just this way: there is no division between what they believe is right and how they act in the world. People of integrity are the opposite of hypocrites, who say one thing and do another.

People of integrity simply refuse, like conscientious objectors in wartime, to do what they believe is wrong. Rather, their lives are shaped by their values. The decisions they make about how to live—what work they do, or refuse to do; what they buy, or refuse to buy; the children they bear, and what they raise them to care about most deeply—all are expressions of their moral beliefs about what is right and good. There is rare joy and satisfaction in such a life entire.

So much of human unhappiness is caused by the fracturing of the self. We all know the curses of a fractured life—the dishonest stories we tell ourselves about the harms we do or allow to be done, the disempowerment we feel when we allow ourselves to be manipulated by corporate decisions, the anger at ourselves, the consequent frantic striving after something (anything!) that would make us worthy in our own minds.

On this argument, much of environmental degradation is caused by hypocrisy. It's not that we are moral idiots. Our values are sound: we care about the well-being of our families and neighborhoods; we truly believe in justice; we value meaningful connection to the natural world and to powers greater than ourselves. But we don't act as if we did. As consumers and employees, we allow ourselves to be agents of powerful destructive forces. Here lies the intractability of the environmental disasters—and their solution.

The environmental emergencies call us to live lives that honor our values. If we think it's wrong to poison children, then why would we buy poisons? If we think it's wrong to poison the air, then shouldn't we ride bicycles? If we object to habitat destruction, why would we ever shop at that mall? The suggestion that individual choices make no real difference—that they won't impact the world, help the future, stop global climate change—is entirely irrelevant. The suggestion that such a life requires renouncing wealth or fame or power? Irrelevant. The suggestion that it is challenging, maybe impossible, to live a life at right angles to the structures of society? Irrelevant. Our own integrity requires us to strive to do what is right, because that is what we believe is right.

Moral Responsibility Is the Price of Progress

Ernest Partridge

> The earth is in much more danger from human action than from natural disasters. This is not a prediction of doom but a wake-up call. We have to recognize the dangers and control them. I am an optimist, and I believe we can.
> STEPHEN HAWKING

Responsibility: The Elements

In his inaugural address, U.S. President Barack Obama spoke of "a new era of responsibility—a recognition . . . that we have duties to ourselves, our nation, and the world, duties that we do not grudgingly accept but rather seize gladly."

It was especially fitting that the president should proclaim this "era of responsibility" at this time and occasion, for one might argue (as I shall) that at no time in history has a nation and a generation borne a heavier burden of moral responsibility.

This bold pronouncement follows from an analysis of the concept

---

ERNEST PARTRIDGE is a researcher, lecturer, and consultant in applied ethics (especially environmental ethics) and has taught moral philosophy and environmental ethics at three University of California campuses and Northland College in Wisconsin. He coedits the progressive political Web site The Crisis Papers and has published hundreds of scholarly papers, Web articles, and encyclopedia entries dealing with moral philosophy, moral psychology, politics, policy analysis, and environmental ethics.

of "moral responsibility." That concept entails four essential elements: *knowledge, capacity, choice,* and *value significance*. To elaborate:

Knowledge

To an unprecedented degree, the sciences have provided us with knowledge of the consequences of our actions and policies. This condition includes a capacity to acquire knowledge that is both available and relevant. Willful and deliberate ignorance, the infamous "plausible deniability" of the politician and publicist, provides no moral excuse. Climate scientists are in near unanimous agreement that industrial civilization, and in particular the consumption of fossil fuels, will bring about catastrophic changes in the global climate. We know full well that the actions we are taking today are harmful to both the present and the future.

Capacity

Similarly, today's *technology*, the application of these sciences, has created a historically unprecedented capacity of industrial societies to affect the future, proximate and remote, deliberately and accidentally, for better or for worse. So while the technology of the Industrial Revolution is largely responsible for the release of greenhouse gases that cause global warming, we are not powerless in the face of this crisis. Applied science indicates it may be possible to develop and employ benign and renewable sources of energy, thus mitigating some of the damaging effects of climate change.

Choice

We have the opportunity to choose among several options to deal with anticipated consequences, including the option to do nothing. The capacity and choice conditions together indicate that moral responsibility resides in the continuum between the polar opposites of *inevitability* and *impossibility*. There is no moral credit for doing that which is inevitable, and no moral blame for failing to do that which is impossible. While it might be possible, through new technologies, to avoid the more dire consequences of global climate change, humanity collectively might instead opt for "business as usual." Formidable financial and media resources are today invested toward precisely that result.

Value Significance

These consequences affect the rights, welfare, and life quality of "morally considerable" individuals now and into the future. If climate scientists are to be believed, the moral stakes in the climate emergency are unprecedented. As physicist Michio Kaku observes, "The generation now alive . . . is a generation that controls the destiny of the planet itself. And unless we are careful, our children and grandchildren may say to us that we ruined everything."

If this analysis of the concept of moral responsibility is correct, then we find justification for the claim that the burden of responsibility on the present generations is unprecedented. Before the middle of the twentieth century, the very idea that human activities might seriously and permanently affect the global atmosphere and oceans, or the gene pool or our species and others, seemed preposterous. We believed we were too puny and the planet too vast for such consequences. Now the sciences have disabused us of such assurances. Examples of the Earth's vulnerability to human impacts are both numerous and familiar. These include the introduction of chemicals and radioactive substances unknown to nature into the seas, the atmosphere, and the global ecosystem. These substances cause ozone depletion, global warming, the acidification of the oceans, the contamination of aquifers, and the deposition of radioactive waste that will remain toxic for hundreds of thousands of years. While these are the by-products of benefits to the present generation, they all exact postponed costs to remote generations. Paradoxically, while technology, thoughtlessly applied to our immediate needs and interests, is the cause of our planetary crises, we must look to science and technology to help lead us past these emergencies.

To summarize: the accelerating advances of science and technology have made it compellingly clear that future generations are vulnerable to the consequences of our acts and policies. Furthermore, through science we have come to understand the long-term consequences of these policies, and through technology we have acquired the capacity to affect these consequences, either through deliberate application of technology or through a careless disregard of the unintended side effects thereof. This *knowledge, capacity,* and *choice* together entail a burden of responsibility that we cannot escape, so long as we willingly accept the enlightenment of science and the capacities of our technology. "To do nothing is to do something"—namely, to assent to existing trends and entailments.

The Ignorance Excuse

A prominent criticism of the contention that the present generation has a profound burden of responsibility toward the future is based on the knowledge condition of responsibility. This criticism claims that we do not and in principle cannot know what future generations will need or value and thus can make no provision for them. How, for example, could previous generations have known of our need for rare semiconducting elements such as germanium? And conversely, what if they had needlessly sacrificed by storing up vast quantities of whale oil, with no anticipation of the coming ages of petroleum and electricity? When we examine the faulty predictions of fifty and one hundred years ago regarding life at the close of the twentieth century, how can we with any confidence forecast conditions of life in the remote future?

Granting all this, there are, nonetheless, some fundamental facts that we can know about future generations:

- First of all, they will be humans, with well-known biotic requirements necessary to sustain their health.
- Second, future persons for whom we are responsible will be moral agents, which means they will be sentient and self-conscious, having a sense of themselves and other persons as continuing beings with the capacity to choose among alternative futures, and with the capacity to reason abstractly and thus to act on principle. All this entails that these future persons will be bound by familiar moral categories of rights, responsibilities, and the demands of justice.
- Third, if these future persons are to live and flourish, they must be sustained by a functioning ecosystem, and they will require the same global climate from which the human species, and then human civilization, evolved and thrived.

- And finally, they will require stable social institutions and a body of knowledge and skills that will allow them to meet and overcome cultural and natural crises that may occur during their lifetimes.

Clearly, we now have the scientific knowledge and the technological capacity to profoundly affect these conditions and these resources that we will bestow upon our successor generations.

Policy Guidelines

Assuming that we know enough about the welfare of future persons to act responsibly on their behalf, what guidelines might direct our policies toward future generations? Prominent among those proposed by philosophers and others are the following:

First of all, do no harm. Because the "ignorance excuse" is not without some merit, an insight from the utilitarians would be very helpful to the policymakers: namely, we should favor policies that mitigate evil over policies that promote good. This precept is supported by commonsense considerations. First, avoidable or treatable pain demands the moral attention of everyone, while the "pursuit of happiness" is the appropriate concern of the individual. Furthermore, it is much easier to identify and address the causes of misery than it is to promote the wellsprings of happiness. This is especially so with regard to the future. The pains and tribulations of future persons, like those of ourselves, can often be traced to disruptions in the fundamental biotic, ecosystemic, climatological, psychological, and institutional conditions listed above. Their pleasures and satisfactions will come from a future evolution of culture, taste, and technology we cannot even imagine.

The "critical Lockean proviso." According to John Locke, it is morally permissible to "take from nature," mix one's labor with the taking, and claim the result as one's private property, so long as one leaves "as much and as good for others." While this may have been true in a world of frontiers and homesteads, it is no longer possible. Once a barrel of petroleum is extracted from the Earth and consumed, there is no longer "as much and as good" remaining for our successors. But if we were to share equally our petroleum resources with all generations far into the future, we would be allocated a cupful each. So we must instead adopt a "critical Lockean proviso," whereby we leave for the future not the very resource that we deplete, but the opportunity to obtain whatever it was for which the original resource was utilized. Thus, while future generations may not need petroleum (just as we no longer need whale oil), they will need what petroleum provides, namely energy. And need it or not, they will not have much petroleum, since "peak oil" is now upon us. Thus, it is our responsibility to find replaceable energy sources that are also sustainable. The proviso also entails that we utilize recycling technologies and "interest-bearing" (i.e., renewable) resources, such as sustained yield forestry and fisheries. And this in turn validates the need to preserve natural ecosystems.

Preserve the options. This rule is clearly entailed by the previous two. While we cannot predict the technological solutions to future resource scarcity, we owe future generations a full range of options and opportunities for research and development of these technologies. This in turn entails a continuing investment in scientific and technical education and research. Happily, such an investment benefits our own generation and that of our immediate successors, as it also benefits the remote future.

Anticipation and prevention is preferable to cure. We should therefore keep an informed eye on impending impacts upon the future. "Earlier" is easier and cheaper than "later." Accordingly, our responsibility to future generations must include technological and environmental impact studies that will foresee, and expand the capacity to foresee, developing crises and the consequences of our projects and policies far into the remote future. Obvious examples of this "duty of anticipation" include studies of stratospheric ozone depletion, global warming, chemical hormone disruption, and nuclear waste disposal.

Just forbearance. This dimension of the duty to posterity clearly follows from the previous, for once we have determined through scientific research how our actions might affect the remote future, we may face a clear duty to forgo advantages for the sake of future generations. To cite our examples once again, studies of atmospheric physics and chemistry may determine that we face a choice between having our grandchildren protected from ultraviolet radiation or having our generation enjoy the convenience of aerosol sprays and supersonic aircraft. Similarly, due to the so-called "greenhouse effect," our voracious appetite for fossil fuel energy may be inconsistent with a tolerable climate for our successors. Accordingly, a decision to favor future generations would, in these instances, require just forbearances on the part of those now living. A policy of "just forbearance" is a conservative approach to provision for the future, which is often favored by environmentalists. The ecosystem, they argue, is a network of complex and subtle interrelationships, the intricacies and ramifications of which we can never fully comprehend. Rather than carelessly toss aside components of this system (e.g., species and nutrients), we should approach the planetary life community with humility and care. If our information is incomplete, it is better to postpone, or even to abandon, projects that threaten the integrity of the system.

Doing well by doing good. We should favor policies that work to the advantage of both ourselves and our successors—policies that, other factors being roughly equal, are least burdensome to the present generation. This rule is responsive to the constant political problem of convincing the public to accept sacrifices now to bring about benefits that they will never see. On reflection, it seems that a significant number of our "duties to the future" also benefit us and those we directly care about: our children and grandchildren. Among these benefits are the control of pollution and population. Furthermore, research and development of alternative energy sources and methods to mitigate global warming, while they improve the life prospects of succeeding generations, will also coincidentally stimulate scientific, technological, and economic advancement for the present.

The urgency of the posterity issue is new to the literature and debates of moral philosophy and of policy studies. But now, having made its appearance, the question of our responsibility to future generations cannot be returned to obscurity. For if our analysis of "moral responsibility" (as *knowledge, capacity, choice,* and *value significance*) is correct, then an attempt to escape from this responsibility through a disavowal of the knowledge provided by our sciences, and an abandonment of the capacity and choice bestowed by our technology, would itself be a profoundly immoral course of action. If, on the other hand, we continue to support the advancement of science and technology and yet ignore the long-term consequences thereof, we will not avoid our moral responsibility; we will be in default of it. We will thus be properly condemned by the generations that succeed us.

The wisdom of the myth of Eden resounds through history to speak to us today. We have nourished ourselves from the tree of knowledge, and by so doing we learned of good and evil and thus have become inalienably responsible for our use of that knowledge. There is no turning back. Abraham Lincoln spoke the enduring truth of our burden and our opportunity: "We cannot escape history. We will be remembered in spite of ourselves. . . . The fiery trial through which we pass will light us down in honor or dishonor to the latest generations. . . . We . . . hold the power and bear the responsibility."

Climate Change: What Is Required of Us?

Terry Tempest Williams

On 8 March 2003, approximately two weeks before the United States invaded Iraq, thousands of individuals, largely women and children, gathered in Washington, DC, for a Code Pink rally in the name of peace. It was also the International Day of Women.

We walked from Martin Luther King Park through the streets of the nation's capital to Lafayette Park, located directly in front of the White House. When we arrived, we were met by a wall of Washington police outfitted with black combat gear, bulletproof vests, and rifles. We were not allowed to proceed on to the public park.

Medea Benjamin, one of the organizers of Code Pink, began to negotiate with the police captain. While these negotiations were under way, Rachel Bagby, an African American poet and musician, stood directly across from a policeman and focused her attention on one officer in particular, also African American. She began singing, with all the power of her God-given voice, "All we are saying is give peace a chance." Over and over she kept singing, "All we are saying is give peace a chance."

Other women began to join her. She never took her eyes off that man, but just kept singing to him in her low, dignified voice. In that

---

TERRY TEMPEST WILLIAMS is the Annie Clark Tanner Scholar in Environmental Humanities at the University of Utah. She is the author of thirteen books, including *Refuge: An Unnatural History of Family and Place* (1991) and *Finding Beauty in a Broken World* (2008). Her writing has appeared in the *New Yorker*, the *New York Times*, *Orion*, and numerous anthologies. She has been honored with a Robert Marshall Award from The Wilderness Society, the Wallace Stegner Award from The Center for the American West, and the John Simon Guggenheim Fellowship in creative nonfiction, among many other awards.

moment, it was clear neither one of them would be who they are, or where they are, without the voices of dissent uttered by their parents, without the literal acts of civil disobedience practiced by their parents' parents and their parents' parents before them.

The African American policeman quietly stepped aside, creating the opening we walked through.

This is what the Open Space of Democracy looks like.

"Disobedience to be civil has to be open and nonviolent. . . . Disobedience to be civil implies discipline, thought, care, attention. . . . Disobedience that is wholly civil should never provoke retaliation . . . but love." So speaks Gandhi.

Henry David Thoreau understood this when he chose to spend a night in jail for refusing to pay his taxes because of his opposition to the Mexican-American War. "Cast your whole vote, not a strip of paper merely, but your whole influence," wrote Thoreau in his essay "Civil Disobedience."

Martin Luther King Jr. understood this when he chose to spend time in the Birmingham jail and wrote: "I am cognizant of the interrelatedness of all communities and states. I cannot sit idly by in Atlanta and not be concerned about what happens in Birmingham. Injustice anywhere is a threat to justice everywhere. We are caught in an inescapable network of mutuality, tied in a single garment of destiny. Whatever affects one directly, affects all indirectly. Never again can we afford to live with the narrow, provincial 'outside agitator' idea. Anyone who lives inside the United States can never be considered an outsider anywhere within its bounds.

"For years now I have heard the word 'Wait!' It rings in the ear of every Negro with piercing familiarity. This 'Wait!' has almost always meant 'Never.' We must come to see . . . that 'justice too long delayed is justice denied.'"

We have a history of bravery in this nation, and we must call it forward now. Our future is guaranteed only by the degree of our personal involvement and commitment to an inclusive justice.

To engage in responsive citizenship, we must become citizens who respond. Passionately. This is how we can make a difference.

This is, ultimately, about climate change: Climate Change as the heating up of the planet; Climate Change as a change of minds and hearts, inspiring direct action, both politically and personally.

Thomas Jefferson said, "I believe in perilous liberty over quiet servitude."

May we commit ourselves to "perilous servitude."

Call it a "peaceful uprising." Call it the open space of democracy.

It is easy to believe that we the people have no say, that the powers in Washington will roll over our local, on-the-ground concerns with their corporate energy ties and thumper trucks. It is easy to believe that the American will is focused only on how to get rich, how to be entertained, and how to distract itself from the hard choices we have before us as a nation.

I refuse to believe this. The only space I see truly capable of being closed is not the land or our civil liberties but our own hearts.

The human heart is the first home of democracy. It is where we embrace our questions. Can we be equitable? Can we be generous? Can we listen with our whole beings, not just our minds, and offer our attention rather than our opinions? And do we have enough resolve in our hearts to act courageously, relentlessly, without giving up—ever—trusting our fellow citizens to join with us in our determined pursuit of a living democracy?

The heart is the house of empathy whose door opens when we receive the pain of others. This is where bravery lives, where we find our mettle to give and receive, to love and be loved, to stand in the center of uncertainty with strength, not fear, understanding this is all there is. The heart is the path to wisdom because it dares to be vulnerable in the presence of power. Our power lies in our love of our homelands.

It is time to ask, when will our national culture of self-interest stop cutting the bonds of community to shore up individual gain and instead begin to nourish communal life through acts of giving, not taking? It is time to acknowledge the violence rendered to our souls each time a mountaintop is removed to expose a coal vein in Appalachia or when a wetland is drained, dredged, and filled for a strip mall. And the time has come to demand an end to the wholesale dismissal of the sacredness of life in all its variety and forms, as we witness the repeated breaking of laws, the relaxing of laws in the sole name of growth and greed.

A wild salmon is not the same as a salmon raised in a hatchery. And a prairie dog colony is not a shooting gallery for rifle recreationists, but a culture that has evolved with the prairie since the Pleistocene. At what point do we finally lay our bodies down to say this blatant disregard for biology and wild lives is no longer acceptable?

Twelve thousand students gathered together in Washington in March 2009 to say, "Enough!" We marched together, physically, symbolically, to ask the United States Congress to think about alternatives to coal-fired power plants, to mountaintop removal, to exploiting our fragile wildlands for oil and gas development that commits us to a future that is based on shortsighted energy retrieval, not a vision toward energy renewal.

Climate Change. What is required is to create a climate of consciousness.

We have made the mistake of confusing democracy with capitalism and have mistaken political engagement with a political machinery we all understand to be corrupt. It is time to resist the simplistic, utilitarian view that what is good for business is good for humanity in all its complex web of relationships. A spiritual democracy is inspired by our own sense of what we can accomplish together, honoring an integrated society where the social, intellectual, physical, and economic well-being of all is considered, not just the wealth and health of the corporate few.

Climate Change. What is required is a change of consciousness.

I do not believe we can look for leadership beyond ourselves. I do not believe we can wait for someone or something to save us from our global predicaments and obligations. I need to look in the mirror and ask this of myself: If I am committed to seeing the direction of our country change, how must I change myself?

We are in need of a reflective activism born out of humility, not arrogance. Reflection, with deep time spent in the consideration of others, opens the door to becoming a compassionate participant in the world.

"To care is neither conservative nor radical," writes John Ralston Saul. "It is a form of consciousness."

Climate Change.

Are we ready for the next evolutionary, revolutionary step? May we act on behalf of the restoration of justice for all species, not just our own. And may our collective acts of disobedience be civil and spirited and beautiful. This is the Open Space of Democracy.

Being Cool in the Face of Global Warming

David James Duncan

What kind of stories best serve in a crisis? We've all heard and read global warming stories ranging from outraged denial to pure grief and fear to effective analysis. Which touch us most deeply? Which might motivate people here in the northern Rockies? Which might paralyze or polarize the same people? What are the differences between an empowering narrative and a paralyzing one? What are the mechanics and secret fuel of a story that directs us toward acts of courage, good sense, good humor, and hope?

Pico Iyer, in a recent *Orion* magazine piece, writes, "Our outer environment can only begin to be healed by our inner life. . . . I'm not sure we can ever truly tend to our polluted waters, our shrinking forests, the madness we've loosed on the air, until we begin to try to clean up the inner waters and attend to the embattled wild spaces within us. Action without reflection got us into this mess. . . . The answer is not more action, but clearer reflection."

Reading Iyer, I am drawn to one of the most popular words in our language: the slang word "cool." No other word means what "cool" means. Cool's staying power and international reach are inar-

---

DAVID JAMES DUNCAN is a novelist and essayist whose writing deals with environmental issues, spirituality, and fly-fishing. His books *The River Why* (1983) and *The Brothers K* (1992) received the Pacific Northwest Booksellers award, and *The Brothers K* was a New York Times Notable Book and won a Best Books Award from the American Library Association. His recent books are *God Laughs and Plays: Churchless Sermons in Response to the Preachments of the Fundamentalist Right* and *My Story as Told by Water*.

guable. Translate "cool" literally into French and you get *frais*, "kind of cold." But the French themselves know this is no description of James Dean's acting or Miles Davis's jazz, so they pepper their proud tongue with our American "cool," untranslated and unexplained. So do the Germans. And Mandarin has *kù*, Russian *coolnah*, and Swedish *coolt*, all trying for *cool*'s effect.

Dictionaries will tell you "cool" means "great," "fine," "excellent," "fashionable," or "the opposite of sucks." But each of these meanings is just a shard of the greater (and cooler) meaning the slang term has come to own. "Cool" has been cool since at least 1933, thanks to black American English and to black jazz musicians. But I've got a Buddhist scholar friend who believes the term was born a few decades before that.

The five-thousand-page Buddhist bible known as the Pali Canon was translated into English by T. W. Rhys David in 1917 and soon became modestly popular among an intelligentsia that almost certainly included a few African American jazz players. In his translations, Rhys David applies the quality of "cool" to the most evolved human souls, who are known in Pali as *Arahants*. The word translates "saint," "wise one," "illumined one." So in the Brahmin Sutta, for instance, we find "The Arahants, the Saintly Cool, whose work is perfected, and they sane and immune [to illusion]." Or, in the Dhatuvihanga Sutta: "He [the Arahant] comprehends that . . . as the soul's trajectory nears its end, all experiences here on Earth become cool."

"The term *siti-bhuta*, which means to *become cool*," Rhys David adds, "is often combined with *nibbuta*, gone out, extinguished . . . [because] the fires of . . . lust, greed, and anger no longer burn in one who is *nibbuta siti-bhuta*. It is in respect of these terms that a man is extinguished and cooled." The Yakkha Sutras support this understanding of "cool":

> Surely at all times happily doth rest
> The Arahant in whom all fire is extinct,
> Who cleaves not to sensuous cravings
> [Who has] cooled all his being and is rid of [Illusion's] germ.

The pre-English roots of "cool" dovetail with the translated Buddhist meaning. The *Oxford English Dictionary* gives the Old English figurative meaning as "to lose the heat or excitement or passion; to become less zealous."

Why does the Buddhist meaning of "cool" strike me as important? From 2000 through 2008, a zealous American administration purchased whatever "facts" that prostituted science and PR trickery could provide, and whatever "faith" that theological demagoguery could bring, in defense of the position that global warming is a hoax. During those same eight years I stood in abject gratitude beside some of the coolest stream restoration and wild salmon and trout rehabilitation projects in the same misguided nation. It struck me, in every such situation, that the people who were the driving force behind the life-giving projects were never zealous in the Old English or American political sense. They did not take heated political and rhetorical positions. Most took pains, on the contrary, to reach across demographic, religious, and party lines so that diverse humans could work together toward practical goals.

One example is a Madison River Valley rancher named Jeff Laszlo, who inherited a thirteen-thousand-acre cattle ranch that includes the headwaters of Odell Creek, a spring creek tributary that joins the Madison at Ennis, Montana. In 1950, before Jeff was born, his ranch's five-mile portion of wetlands and spring creek were "improved" by one of the post–World War II engineering schemes ubiquitously practiced by agencies like the Army Corps of Engineers and Bureau of Reclamation. The "improvements" at Jeff's place turned a wildlife haven and wild trout stream into a twenty-foot-wide, four-inch-deep, fishless ditch that cattle soon trampled into clay pavement. When Jeff decided, with the help of a few revamped public and private agencies, to turn his cow pavement back into a living spring creek, he figured it'd take a couple of years. Turns out it'll take decades: true restoration is a slow, careful, *reciprocal* conversation with nature. But even given the brevity of Jeff's effort (four years, counting the gleam in his eye) there are three miles of beautiful meandering spring creek now flowing in place of his inherited lifeless ditches.

There was a crew working a variety of heavy equipment the day I walked the project with Jeff. Some of the backhoe or crane operators had on headphones. For all I know, some were listening to talk radio's heated rhetoric as they labored. But the work of their hands and machines showed no sign of heat. That work was restoring meanders and cutbanks, installing woody debris, planting willows and cottonwoods; that work had literally raised the water table by four to five feet and lowered the water temperature by ten degrees. And when human nature achieves both psychological and biological "cool," wild nature responds: herons, cranes, ducks, shorebirds, deer, and moose have returned to the spring-fed pools the equipment operators carefully crafted. As I walked the reborn stream with Jeff, a Pale Morning Dun hatch began, and wild trout, who love cool as much as Buddhists do, began rising. Many had migrated up from the Madison twenty miles below. They now thrive and multiply in the 56-degree water. Upper Odell, four years ago fishless, now supports 1,760 trout per mile.

The cool this projects demonstrates is not just a Fahrenheit lowering of degrees: it is a willingness on the part of many to "lose heat or excitement and become less zealous"; it is a move in the direction of the "saintly cool."

Restoration work is cosmological work. Restoration work is wild nature, abstract thought, and human desire engaged in creative negotiation. Global warming has presented every sentient being with new sets of dire facts that most people have never before heard and that many people would love to deny. The negotiations between the possessors of hard facts, and humanity at large, must therefore be truly creative. Negotiation requires constant intuition and constant adjustment of our narratives, not just data-based policies poured down from on high.

While a certain AM chatterbox makes $32 million a year to pour down the notion that human-caused global warming is a hoax perpetrated by Al Gore, I've been looking for a simple guiding sentence that might cool as many Americans as AM radio manages to overheat. My search for this sentence has not taken place in environmental literature. Words like "environment," I'm sorry, are not lovely enough to inspire the needed change. Say it aloud. *Environment.* Hear the technoid ring? If the book of Genesis began "In the beginning, God created the environment" instead of "In the beginning, God created the heavens and the Earth," the Bible might be out of print. The word's lack of musicality prevents those who lack tin ears from rallying round "the environment" with sufficient love and passion.

What we seek to defend is a holiness. What we're dying to implement is what the Great Plains contemplative Kathleen Norris calls "a way of surrendering to reduced circumstances in a manner that enhances the whole." This is why the guiding sentence I keep liking invokes humanity's own hidden holiness. Although I'm not a Christian, one of my guiding sentences, in the face of global warming, is this famous statement of Jesus: "The kingdom of heaven is within you." I've never recovered from the wondrous confusion I felt when I first heard these words as a child. I pray I never do. Here's one thing I love about the statement: Jesus didn't say the kingdom is within us only in times of tranquility, or that it rises up only in times of crisis. He said the kingdom is *always* within us.

Our overwhelmed, news-blasted, alternately depressed and skeptical minds want to know, "Why should we seek an inner kingdom when our outer kingdom is going to hell in a handbasket?" I return them to Pico Iyer: "Our outer environment can only begin to be healed by our inner life. . . . Attend to the embattled wild spaces within us. Action without reflection got us into this mess."

At the end of *Dakota*, Kathleen Norris's monk friend tells how he attends to the wild spaces within himself. He says, "You have only to let the place happen to you: the loneliness, the silence, the poverty, the futility, indeed the silliness of your life."

When I bring my loneliness, futility, and silliness to the half-wild places of my corner of the world, my loneliness, futility, and silliness are enveloped in an unceasing flow of natural power and pattern. Ever watch a Canada goose *ménage à trois* in late March? I see them every spring on Lolo Creek. Letting the place happen to me, I share loneliness with the goose who loses the contest and ends up being the odd male out. And in sharing, my loneliness is reflected in a handsome goose who has acquired a comic honk. So is it any longer even loneliness? Ever see a magpie playing with a half-grown colt, or a crow teasing a dog that's trying to catch it, or a pair of coyotes chasing carp through shallow water on a warm day, or a bear lying on its back, oblivious to you, playing with its toes? Letting such creatures and their places happen to me, I've shared such exquisite silliness with magpie, crow, coyote, and bear that the wild places within me began somehow to thrive.

When the natural patterns, mysteries, and beauty of the wild interior align with the same pattern, mystery, beauty in the wild outside, it's effortless to realize we're not the droning of our brain, the cravings of our ego, or the consumer impulses of our bodies. We are infinitely and mysteriously more. And so, materially, we require much, much less to thrive.

What can we do to cool our selves, our planet, or one another? If Norris's monk and I might make a few suggestions: *visit* your local fields of lilies, *watch* the fowls of the air around you, *consider* the blue gentians of the ridgelines and the paintbrush and asters of the meadows, *contemplate* the crossbills, swallows, and ravens crisscrossing your local lake-shine, *behold* the movement of mist on the face of erased waters, and let every other special place you inhabit happen to you till it pierces your loneliness, futility, and silliness and shows you a little kingdom. The wild without and the wild within are as reciprocal as reflections on water. And such reflections, witnessed live, are not only cooling and clarifying, they're dirt cheap. They're cheap because the treasure of the entire interior kingdom is already within you. And to shop till you drop in *that* kingdom creates no carbon footprint at all.

"In times so dire," the troubled mind asks and asks, "what practical purpose does even the most vivid experience of a merely interior kingdom serve?"

Here's the reply of another monk, St. Isaac of Syria: "Make peace with yourself, and heaven and Earth make peace with you. Take pains to enter your innermost chamber and you will see the chamber of heaven, for they are one and the same, and in entering one you behold them both."

Everything Must Go

Paul B. Thompson

The human species has made a mess of creation. That is the starting point for any environmental philosophy of the twenty-first century. There are still many people who need to be brought to this starting point, so we really do need an environmental journalism dedicated to the "We're screwed!" message. But my question is, What happens after the moment when someone realizes that the lifestyles of middle-class aspirants to a lifestyle of the rich and famous are, along with the lifestyles of the anonymous poor, turning the Earth to shit? How does one respond to the inconvenient truths of pollution, habitat degradation, resource depletion, and climate change?

Environmental philosophy might grapple with this question in a number of ways, but none of the reasonable ways to grapple with it involves saying, "So what?" I want to contest that thought, but I want to contest it in a way that (I hope) remains consistent with the general spirit of environmental philosophy, a spirit bequeathed in Arne Naess's injunction to embrace a deep ecology, one that involves a commitment to nature that goes far beyond a self-serving understanding of the "We're screwed!" message. A deep ecology is not content with an environmentalism focused simply on saving creation for the use that future human beings might want to make of it.

PAUL B. THOMPSON is the W. K. Kellogg Chair in Agricultural, Food, and Community Ethics in the Department of Philosophy at Michigan State University. He has served on many national and international committees on agricultural biotechnology and contributed to the National Research Council report *The Environmental Effects of Transgenic Plants* (2002). The second edition of his book *Food Biotechnology in Ethical Perspective* was published in 2007.

So admittedly, some ways of saying "So what?" lack a sufficiently well developed sense of natural piety: where others see a mess, some see mere untidiness or, worse, they see *in* these others a kind of obsessive fussiness. Smoke coming from the factory stacks looks and smells like jobs to them, and a cheap steak is a gift that should not be examined too closely. This way of saying "So what?" suggests that the speaker simply has not really been brought to the starting point after all. Yet although it's possible that people who say "So what?" don't know what they are at risk of losing, it's also possible that they really wouldn't care too much about losing it even if they knew. Unpacking this kind of "So what?" response can go on indefinitely, and the way that one unpacks it sets many agendas for environmental writing, filmmaking, and political activism. One agenda aims to teach the facts of risk and loss, while another aims to teach the value that explains this loss *as* loss. But the question of what one does once one sees that humans really have made a mess of creation still remains. Can someone who has seen this situation for what it is still say, "So what?"

Here is a more interesting possibility, one that actually embraces risk and loss. British political theorist John N. Gray is known for his withering critiques of the liberal presumption that problems can be solved, that genuine progress is achievable. He has recently advanced a kind of environmental politics that draws darkly on the Gaia hypothesis of James Lovelock. In his book *Straw Dogs*, Gray argues that the evolutionary history of the human species provides absolutely no resources for the solidarity or political community that would be required for "solving" the environmental problem.[1] But this is (as Gray sees it) actually the good news for environmentalists. Gray believes that Gaia will heal herself when religious warfare, climate change, and resource depletion have conspired to bring human population levels down to the roughly half-billion level the planet's carrying capacity might reasonably be expected to support. Gray's dark vision turns the "So what?" response away from the familiar fixation on jobs and wealth accumulation, and toward resigned acceptance of our inevitable fate.

1. John Gray, *Straw Dogs: Thoughts on Humans and Other Animals* (London: Granta Books, 2002).

I would actually go further than Gray. One can also be philosophical in this way about Gaia herself. Heraclitus is credited with the aphorism "All is flux," or, as I prefer, "Everything flows."[2] Woody Allen allegedly gave up philosophy when his realization that the death of the sun would eventually result in the end of life on Earth drove him into a state of depression. Unable to work, he reports, he was eventually expelled from New York University for "looking into the soul of his neighbor" on a metaphysics test.[3] This is not actually quite the sense in which Heraclitus is usually understood, however. In characterizing the constant transformation and recombination of the four classic elements (earth, air, fire, and water) into one another as ebb and flow, Heraclitus's fragment suggests that of the four elements, water most closely models the underlying form of being itself.

This tantalizing doctrine is open to the interpretation that while everything visible vanishes, a kind of balance or equilibrium endures. Can people who get to the starting point for environmental philosophy take any comfort here? Can they, with knowing smiles of resignation, reply, "So what?" Can we be content with Gray's thought that after humanity is gone, creation will heal itself? I suppose that people are entitled to take comfort wherever they find it, but I am reminded of the closing sections from *Walden,* where Thoreau's discussion of soil and rock formations on a bank spark the Heraclitean speculation that even in this seeming solidity the close observer of nature detects flow. These comments counter Thoreau's earlier stated ambition to "catch the nick of time, and notch it on my stick."[4] Stanley Cavell has argued that we should understand the "stick" passage as indicative of Thoreau's view that writing has the power to crystallize human experience and to preserve its inherent quality in timeless form.[5]

If so, we might attend to what Henry David has to tell us. The "flow" sections from *Walden* oppose the "stick" passage more as

---

2. Heraclitus, from C. H. Kahn, *The Art and Thought of Heraclitus* (Cambridge: Cambridge University Press, 1979).

3. Woody Allen, *Without Feathers* (New York: Ballantine, 1972).

4. Henry David Thoreau, *Walden: Or, Life in the Woods* (Boston: Ticknor and Fields, 1854).

5. Stanley Cavell, *The Senses of Walden: An Expanded Edition* (San Francisco: North Point Press, 1981).

counterbalance than as contradiction. Everything flows; everything must go. Eventually even the protons from which matter is composed decay. Though Gaia might heal herself after humanity, even Gaia is not "for all time." What is more, the stick on which Thoreau notches the nick of time is, after all, only *his* stick. If the passages of *Walden* do seem to many readers to have crystallized a bit of Thoreau's spirit, creating for us an opening to the flowing, perishing experience of that too-quickly-extinguished Concord flame, that is for us, and not forever.

So what if we're screwed? Seize the day. Thoreau might have lived a self-reliant and resigned life on a then badly degraded Walden Pond in utter simplicity without having to write about it, yet that was never his project. His commitment to wildness was to save the world—the presence of being, the nick of time—not the planet. We need no faith in our ability to save the planet to realize that passage and vanishing is vindicated by (and only in) its very wildness, its lack of larger purpose. Thoreau aspired to live so as to become native to this place, this Earth, knowing that, as John J. McDermott writes, any place seized is a place denied.[6] Everything flows, everything must go. All the more reason to cherish and serve the Earth while it lasts.

---

6. John J. McDermott, *The Drama of Possibility: Experience as Philosophy of Culture*, ed. Douglas R. Anderson (New York: Fordham University Press, 2007).

The No-Man's-Land of Ethics

Joerg Chet Tremmel

In his epoch-making book *The Imperative of Responsibility,* philosopher Hans Jonas argues that modern technology increases the potential to irreversibly impair the future fate of humankind and nature by actions and omissions. Jonas works out what held true throughout the ages up to the twentieth century:

> For all his boundless resourcefulness, man is still small by the measure of the elements: precisely this makes his sallies into them so daring. . . . Making free with the denizens of land and sea and air, he yet leaves the encompassing nature of those elements unchanged, and their generative powers undiminished. . . . Much as he harries Earth, the greatest of gods, year after year with his plough—she is ageless and unwearied; her enduring patience he must and can trust, and to her cycle he must conform.[1]

Jonas can be criticized for considering nature stable and indestructible. Such a concept is surely one-sided and is no longer advocated by ecologists in such general terms. In view of the five geological phases of global extinction of animal species and the cycles of ice and warm ages, nature must be seen as far more vulnerable to catastrophes.

Jonas's decisive and indisputable point, however, is that humans

---

JOERG CHET TREMMEL is a research fellow at the Centre for Philosophy of Natural and Social Science at the London School of Economics and Political Science. His research interests include applied ethics, especially intergenerational justice and climate ethics, the "short-termism" of political and business systems, and epistemology.

---

1. Hans Jonas, *The Imperative for Responsibility: In Search of an Ethics for the Technological Age* (University of Chicago Press, 1985), 3.

Relevant time scales for humankind and the environment

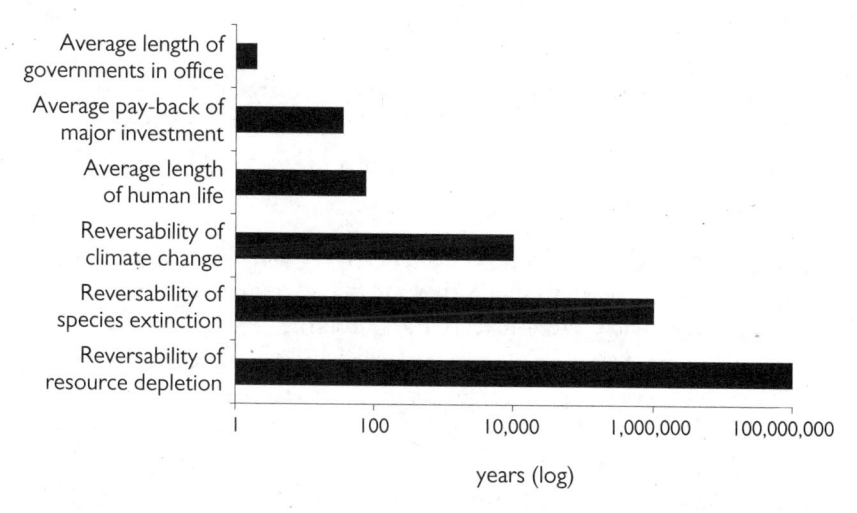

Average length of governments in office	▮
Average pay-back of major investment	▬▬
Average length of human life	▬▬▬
Reversability of climate change	▬▬▬▬▬▬
Reversability of species extinction	▬▬▬▬▬▬▬▬▬
Reversability of resource depletion	▬▬▬▬▬▬▬▬▬▬

1　　100　　10,000　　1,000,000　　100,000,000

years (log)

had relatively little influence on global, supra-regional nature before the modern age. They were not able to throw the ecosystem they lived in off balance. They did not adapt to nature on grounds of reason but simply because they had no choice. Under these circumstances, there was no need for an ethics of responsibility to nature. Rather, humans were well advised to approach nature with as much cleverness and efficiency as possible to sufficiently benefit from its seemingly boundless resources.

But gradually, things humans had once accepted as their fate came within their scope of influence in the twentieth century. The long-term effects of nuclear energy were not conceivable in the past, except in utopias with a science fiction character. The same applies to the magnitude of climate changes—which are, after all, an influence on the basic biophysical conditions of our planet itself.

As the comparison of relevant time scales of humankind and nature shows in the figure above, we are now able to shape the future by intervening in the household of nature to a great extent, for millions of years.

Let us take nuclear waste as an example. A country like Germany, which is not a nuclear power and thus does not reuse plutonium for military purposes, has produced in its nuclear power plants 118 tons of plutonium (Pu-239) as a waste substance. According to our present state of knowledge, one gram of our present waste plutonium will still be left in 310,608 years, and one gram can be lethal for a human being.

If we consider that our written history is only 10,000 years old, the permanence of the burden present generations are placing on the shoulders of future generations becomes quite clear. The semiotician Roland Posner explains:

> In all three fields [nuclear, genetic, and space engineering], we are dealing with time spans that go beyond that of human history up to now. The science, literature, and art of earlier centuries become unintelligible if they are not reinterpreted and translated into new languages every few generations. In the same way, state institutions have rarely existed for more than a few centuries, and they are constantly threatened by war and subversive movements. Even present-day religions are not much older than a few thousand years, and they have not primarily handed down scientific information to us, but myths and rituals.[2]

The human-made climate change, the depletion of the oceans through unsustainable fishing, the clearing of the wilderness for plantations, and the loss of the biodiversity are by no means new phenomena. In earlier times, however, they were limited to specific areas, whereas now they take place on a global scale and at a far greater pace.

The enormous increase in humankind's powers in the twentieth century explains why even the most important moral philosophers of the past paid almost no attention to our responsibility to posterity. Kant, for instance, writes:

> What is disconcerting here, however, is that previous generations seem to have pursued their arduous endeavors only for the sake of later ones, in order to prepare for them a level from which they can raise even higher the structure that nature intended; and that nevertheless only the later generations should have the fortune to dwell in the building which was the work of a long series of earlier gen-

---

2. Roland Posner, "Atommüll als Kommunikationsproblem," in *Warnungen an die ferne Zukunft. Atommüll als Kommunikationsproblem*, ed. Roland Posner (Munich: Raben-Verlag, 1990), 7–15.

erations (albeit without this being their intention), without themselves being able to share in the fortune that they themselves have worked toward.[3]

John Rawls—much like Kant a few hundred years before him—takes an "autonomous social savings rate"[4] for granted, a type of natural law by which the living conditions of future generations will continuously improve. But the image of the "spoiled heir" has now been replaced by the concern that future generations might become the ecological, economic, and social victims of today's generation's shortsighted politics. Still, many moral philosophers believe that every important moral principle has essentially been brought forth and discussed in the long history of ethics, so there can be no fundamental changes. Here, for example, is Robert Spaemann's witticism: "In questions of how to live life rightly, only wrong ideas can truly be new."

For millennia, ethics have dealt with future generations with the confidence that the future is likely to resemble the past. The universe of traditional ethics is limited to contemporaries—that is, to their expected life span. This "neighbor ethics" is no longer sufficient. A certain model of justification of moral standards, namely that of a reciprocal consideration of interests for egoistic reasons, has been particularly popular since Thomas Hobbes. It should be discontinued. Those living a hundred years from now can hardly impose sanctions on us for the harm we are doing to them today.

Up to now, moral philosophy has lived in the Newtonian age. The new territory humankind has conquered with high technology is still a no-man's-land for ethical theory. It is high time to explore this land.

---

3. Immanuel Kant, *Toward Perpetual Peace and Other Writings on Politics, Peace, and History*, ed. Pauline Kleingeld (New Haven: Yale University Press), 6.
4. John Rawls, *A Theory of Justice* (Cambridge, MA: Belknap Press, 2005).

The Advocacy Responsibility of the Scientist

José Galizia Tundisi

We all agree that scientists from all walks have the task of promoting and increasing knowledge through their professional activities. Their findings are then permanently added to the cacophony of the world's ideas with the expectation that someday, in the near or distant future, those findings can be useful to humankind. It was in this way, at least in part, that enormous progress was achieved in medicine, agriculture, and engineering in the twentieth century.

In the twenty-first century, however, we scientists face a very different problem, and the world needs science and scientists in a very different way. We all know the litany: atmospheric carbon dioxide is at the highest level in the last hundred thousand years, the nitrogen cycle is unbalanced, and global warming is a reality demonstrated by scientific data and weather events all around our planet. Our planet is currently going through a great environmental degradation that is threatening not only biodiversity or our supply of clean water or human health but also the very existence of humankind and other living beings. As E.O. Wilson puts it, "We are inside a bottleneck

---

JOSÉ GALIZIA TUNDISI is an ecologist and president of the International Institute of Ecology in São Carlos, Brazil. He has served as a professor at the Federal University of São Carlos and the University of São Paulo and as president of the Brazilian National Research Council. His awards include the Gold Medal "Augusto Ruschi" Award from the Brazilian Academy of Sciences, the Moinho Santista Prize in Ecology, and the UN's Boutros Ghali Prize for Environment and Development. He is a member of the Brazilian and São Paulo Academies of Sciences.

of overpopulation and wasteful consumption."[1] In the simplest of terms, as we scientists have documented over and over again, this situation is dire and difficult.

Given this reality, this reality of which science has made us aware, how can we scientists, we gatherers and disseminators of knowledge, help to change this course of events? What is our role in remediation?

Certainly, striving to understand the world and then disseminating our findings to the public is more important than ever. In fact, this effort has taken on a tone of moral responsibility that is perhaps historically unprecedented. In our fulfillment of this moral responsibility, however, we scientists also need to think carefully about how to go from vision to action as fast as we can: in other words, we not only need to be very particular and selective about what the world most needs to know to avert at least the most egregious of today's environmental problems; we also need to suggest ways to move from that information to the implementation of policies aimed at averting those harms.

Science is, and has always been, a method for social transformation. Scientific activities and discovery can be an important part of change, from the local level to the international. Science can be an integral part of the noble work to improve the quality of life. Environmental science can have a great impact by helping to provide access to clean water, information about the cause of the loss of biodiversity, information about what might mitigate such loss, and so on.

I am a water scientist and have worked my entire life in South America on the limnology of lakes and reservoirs, water resources management, and the recovery of aquatic ecosystems. I have also worked professionally on water resources in forty countries. During that time I have witnessed the misery and poverty of the world's poorest areas and have noted a direct correlation between that misery and a lack of access to clean water. The effect on human health—the devastation wrought by waterborne diseases, high rates of infant mortality, and generally a poor quality of life—is always dramatic, often sickening, and probably preventable.

This reality prompted me to spend part of my energy as a scientist

---

1. E. O. Wilson, *The Future of Life* (New York: Knopf, 2002).

on other, more tangible efforts. First, colleagues and I started the Water School at the International Institute of Ecology in São Carlos. The Water School is a facility (a house, a van, a truck) where we strive to teach the general public, the schoolteachers, the children, anyone we can, everything we can about water: the water cycle, the reality of waterborne diseases, how to save water in our daily activities, how to clean water, how to get access to clean water. The Water School is a way to mobilize society. Our work began with small towns in Brazil (about five to ten thousand inhabitants) but is now moving to larger cities (sixty thousand inhabitants) as well as to the periurban areas of large cities and metropolitan areas. This is an independent, self-financed activity aimed at sustainable permanent action.

Second, I work on the IAP (Inter Academy Panel) water project. Given that poor water management is one of the bottlenecks of the water crisis, we need to improve the capacity of midlevel managers to do a better job of providing clean water, saving water, and educating the community. Sixty Academies of Science have enrolled in this project, and we have partnerships in six countries—Brazil, Poland, Jordan, Russia, China, and South Africa—where we conduct seminars, workshops, and field activities for hundreds of technicians and managers. This effort is creating a network of people engaged in improving the quality of water and life in their own regions.

Third, I regularly give presentations about water to the general public. I try to interact as much as I can with the audience in order to promote empathy, empowerment, and participation. I believe this public outreach has positive results.

Are these activities sufficient to solve the enormous water problems we face? Of course not. We need to do much more! And there are many other examples of what we can do to mobilize society: we can all give public seminars; we can all write booklets, maybe even children's books in collaboration with colleagues who specialize in children's literature; we can all organize unique and interdisciplinary seminars; we can all put together demonstrations and joint activities with political representatives and community organizers; we can all work to make sure that our institutions are strong, creative, and open enough to allow us to act as caring and informed members of society. Each one of us scientists can do a bit. And I would suggest that as knowledgeable citizens in a world in peril, each of us scientists has an obligation to do at least a bit, maybe more. Indifference from us can lead to disaster.

Wulf and Jones argue that the research community owes to the society a decisive engagement in the dissemination of knowledge *and* in the interaction with the administration in such a way that the best policy decisions can be made.[2] When considering what we know about the state of our world, when considering that society looks to us for leadership, when considering our moral responsibility as citizens of this planet, can we honestly say that we owe nothing to the effort to remedy our planet's problems?

E. O. Wilson points out that *Homo sapiens* has become a geophysical force on the planet. To change the tragic influence of that force, our moral commitment as scientists is essential. To publish papers and books is most certainly necessary for every professional scientist. But such effort is not sufficient to change our societies. We need to do more! We have to actively disseminate our knowledge to all the different components of society. We have to work to improve the environment at local and regional levels. We have to actively promote access to clean water; initiate restoration projects; engage in global conservation projects; educate the general public on environmental problems; and try to influence decision makers, administrations, and politicians providing advice for better policy decisions. We owe it to society today and to society in the future.

Twenty years ago, I wrote a paper titled "Ecology and Development: Perspectives for a Better Society."[3] In that paper I discussed the importance of using ecological knowledge at local and regional levels to improve society and promote a better quality of life for all. I still believe in this. I am convinced that our moral responsibility and our engagement with society can help to save our planet. I believe Mahatma Gandhi had it right when he declared, "We must become the change we want to see in the world." Environmental scientists should, right now in this place and time, decide to be active participants in the effort to heal the world—to be that change they want to see in the world.

2. A. W. Wulf and A. K. Jones, "A Call to Serve" (editorial), *Science* 322 (2008), 1025.
3. J. G. Tundisi, "Ecology and Development: Perspectives for a Better Society," *Physiological Ecology* 27, special number (1990), 93–130.

How to Be Hopeful

Barbara Kingsolver

The world shifts under our feet. The rules change. Not the Bill of Rights, or the rules of tenting, but the big unspoken truths of a generation. Exhaled by culture, taken in like oxygen, we hold these truths to be self-evident: You get what you pay for. Success is everything. Work is what you do for money, and that's what counts. How could it be otherwise? And the converse of that last rule, of course, is that if you're not paid to do a thing, it can't be important. If a child writes a poem and proudly reads it, adults may wink and ask, "Think there's a lot of money in that?" Being a good neighbor, raising children: the road to success is not paved with the likes of these. Some workplaces actually quantify your likelihood of being distracted by family or volunteerism. It's called your coefficient of drag. The ideal number is zero. This is the Rule of Perfect Efficiency.

Now, the rule of "success" has traditionally meant having boatloads of money. But we are not really supposed to put it in a boat. A house would be the customary thing. Ideally it should be large, with a lot of bathrooms and so forth, but no more than four people. If two friends come over during approved visiting hours, the two chil-

---

BARBARA KINGSOLVER is a novelist, essayist, and nonfiction writer whose books have earned major literary awards at home and abroad. In 2000 she received the National Humanities Medal, the highest honor for service through the arts in the United States. Her books include *Prodigal Summer* (2000), *Small Wonder* (2002), *Animal, Vegetable, Miracle* (2007), and *The Lacuna: A Novel* (2009). In 1997 she established the Bellwether Prize, which awards outstanding literary quality and a commitment to literature as a tool for social change.

dren have to leave. The bathroom-to-resident ratio should at all times remain greater than one. I'm not making this up, I'm just observing; it's more or less my profession. As Yogi Berra told us, you can observe a lot just by watching. I see our dream houses standing alone, the idealized life taking place in a kind of bubble. So you need another bubble, with rubber tires, to convey yourself to places you must visit, such as an office. If you're successful, it will be a large, empty-ish office you don't have to share. If you need anything, you can get it delivered. Play your cards right and you may never have to come face to face with another person. This is the Rule of Escalating Isolation.

And so we find ourselves in the chapter of history I would entitle "Isolation and Efficiency, and How They Came Around to Bite Us in the Backside." Because it's looking that way. We're a world at war, ravaged by disagreements, a bizarrely globalized people in which the extravagant excesses of one culture wash up as famine or flood on the shores of another. Even the architecture of our planet is collapsing under the weight of our efficient productivity. Our climate, our oceans, migratory paths, things we believed were independent of human affairs. Twenty years ago, climate scientists first told Congress that unlimited carbon emissions were building toward a disastrous instability. Congress said, we need to think about that. About ten years later, nations of the world wrote the Kyoto Protocol, a set of legally binding controls on our carbon emissions. The United States said, we still need to think about it. Now we can watch as glaciers disappear, the lights of biodiversity go out, the oceans reverse their ancient orders. A few degrees looked so small on the thermometer. We are so good at measuring things and declaring them under control. How could our weather turn murderous, pummel our coasts, and push new diseases like dengue fever onto our doorsteps? It's an emergency on a scale we've never known. We've responded by following the rules we know: Efficiency, Isolation. We can't slow down our productivity and consumption, that's unthinkable. Can't we just go home and put a really big lock on the door?

Not this time. Our paradigm has met its match. The world will save itself, don't get me wrong. The term "fossil fuels" is not a metaphor or a simile. In the geological sense, it's over. The internal combustion engine is so twentieth-century. Now we can either shift away from a carbon-based economy, or find another place to live. Imagine it: we raised you on a lie. Everything you plug in, turn on, or drive, the out-of-season foods you eat, the music in your ears. We gave you this world and promised you could keep it running on *a fossil substance*. Dinosaur slime, and it's running out. The geologists only disagree on how much is left, and the climate scientists are now saying they're sorry, but that's not even the point. We won't get time to use it all. To stabilize the floods and firestorms, we'll have to reduce our carbon emissions by 80 percent, within a decade.

Heaven help us get our minds around that. We're still stuck on a strategy of bait-and-switch: Okay, we'll keep the cars but run them on ethanol made from corn! But . . . we use petroleum to grow the corn. Even if you like the idea of robbing the food bank to fill up the tank, there is a math problem: it takes nearly a gallon of fossil fuel to render an equivalent gallon of corn gas. By some accounts, it takes more. Think of the Jules Verne novel in which the hero is racing around the world in eighty days, and finds himself stranded in the mid-Atlantic on a steamship that's run out of coal. It's day 79. So Phileas Fogg convinces the captain to pull up the decks and throw them into the boiler. "On the next day the masts, rafts, and spars were burned. The crew worked lustily, keeping up the fires. There was a perfect rage for demolition." The captain remarked, "Fogg, you've got something of the Yankee about you." Oh, novelists. They always manage to have the last word, even when they are dead.

How can we get from here to there without burning up our ship? That will be the central question of your adult life: to escape the wild rumpus of carbon-fuel dependency, in the nick of time. You'll make rules that were previously unthinkable, imposing limits on what we can use and possess. You will radically reconsider the power relationship between humans and our habitat. In the words of my esteemed colleague and friend Wendell Berry, the new Emancipation Proclamation will not be for a specific race or species, but for life itself. Imagine it. Nations have already joined together to rein in global consumption. Faith communities have found a new point of agreement with student activists, organizing around the conviction that caring for our planet is a moral obligation. Before the last UN

Climate Conference in Bali, thousands of U.S. citizens contacted the State Department to press for binding limits on carbon emissions. We're the 5 percent of humans who have made 50 percent of all the greenhouse gases up there. But our government is reluctant to address it, for one reason: it might hurt our economy.

For a lot of history, many nations said exactly the same thing about abolishing slavery. We can't grant humanity to all people, it would hurt our cotton plantations, our sugar crop, our balance of trade. Until the daughters and sons of a new wisdom declared: We don't care. You have to find another way. Enough of this shame.

Have we lost that kind of courage? Have we let economic growth become our undisputed master again? As we track the unfolding disruption of natural and global stabilities, you will be told to buy into business as usual: You need a job. Trade your future for an entry-level position. Do what we did—preserve a profitable climate for manufacture and consumption, at any cost. Even at the cost of the other climate, the one that was hospitable to life as we knew it. Is anyone thinking this through? In the awful moment when someone demands at gunpoint, "Your money or your life," that's not supposed to be a hard question.

A lot of people, in fact, are rethinking the money answer. Looking behind the cash price of everything to see what it cost us elsewhere: to mine and manufacture, to transport, to burn, to bury. What did it harm on its way here? Could I get it closer to home? Previous generations rarely asked about the hidden costs. We put them on layaway. You don't get to do that. The bill has come due. Some European countries already are calculating the "climate cost" on consumer goods and adding it to the price. The future is here. We're examining the moralities of possession, inventing renewable technologies, recovering sustainable food systems. We're even warming up to the idea that the wealthy nations will have to help the poorer ones, for the sake of a reconstructed world. We've done it before. That was the Marshall Plan. Generosity is not out of the question. It will grind some gears in the machine of efficiency. But we can retool.

We can reimagine an unconventionally communal sense of how life may be. This could be the key to a new order: you don't need so much stuff to fill your life, when you have people in it. You don't need jet fuel to get food from a farmer's market. You could invent a new kind of success that includes children's poetry, butterfly migrations, butterfly kisses, the Grand Canyon, eternity. If somebody says "Your money or your life," you could say "Life." And mean it. You'll see things collapse in your time, the big houses, the empires of glass. The new green things that sprout up through the wreck—those will be yours.

The arc of history is longer than human vision. It bends. We abolished slavery, we granted universal suffrage. We have done hard things before. And every time it took a terrible fight between people who could not imagine changing the rules, and those who said, "We already did. We have made the world new." The hardest part will be to convince yourself of the possibilities, and hang on. If you run out of hope at the end of the day, rise in the morning and put it on again with your shoes. Hope is the only reason you won't give in, burn what's left of the ship, and go down with it—the ship of your natural life and your children's only shot.

Hope: An Owner's Manual

Look, you might as well know, this thing
is going to take endless repair: rubber bands,
Krazy Glue, tapioca, the square of the hypotenuse.
Nineteenth-century novels. Heartstrings, sunrise:
all of these are useful. Also, feathers.

To keep it humming, sometimes you have to stand
on an incline, where everything looks possible;
on the line you drew yourself. Or in
the grocery line, making faces at a toddler
secretly, over his mother's shoulder.

You might have to pop the clutch and run
past all the evidence. Past everyone who is
laughing or praying for you. Definitely you don't
want to go directly to jail, but still, here you go,
passing time, passing strange. Don't pass this up.

In the worst of times, you will have to pass it off.
Park it and fly by the seat of your pants. With nothing
in the bank, you'll still want to take the express.
Tiptoe past the dogs of the apocalypse that are sleeping
in the shade of your future. Pay at the window.
Pass your hope like a bad check.

You might still have just enough time. To make a deposit.

To a Future Without Hope
Michael P. Nelson

I don't hope for anything. I don't fear anything. I'm free.
NIKOS KAZANTZAKIS

We are bombarded with a contradiction. Many of our environmental leaders go to great lengths, masterfully and persuasively, to recount our many and looming and depressing and overwhelming environmental problems. We will suffer the various effects of massive global climate change, hundreds or thousands of species will be wiped from the face of this good green Earth, and there will be tremendous and disproportionate human suffering along the path to this dimmer future. This is to say, our leaders convince us that the future is not only in jeopardy, but is essentially hopeless. These leaders then do something amazing, even audacious: they turn around and assert that there is hope, almost as a sort of un-reflected-upon reflex, an utterance of seemingly obligatory expectation. I realize this may sound terrible, but this gesture toward hope has begun to make me angry. I am told a story convincing me there is no hope, and then I am told to have hope. You are not fooling anyone: to quote Leonard

---

MICHAEL P. NELSON (coeditor) is the Ruth H. Spaniol Chair of Natural Resources and a professor of environmental ethics and philosophy at Oregon State University. He is the coeditor of *The Great New Wilderness Debate* (1998) and *The Wilderness Debate Rages On* (2008) and the coauthor of *American Indian Environmental Ethics: An Obijwa Case Study* (2003), all with J. Baird Callicott.

Cohen, "Everybody knows that the ship is sinking, everybody knows that the captain lied."

I worry that not only is hope a throwaway term—a vacuous sentiment we dole out at the end of a paragraph, or a book, or a film, or an interview—but also that it is dangerous and counterproductive. To hope, quite literally, is "to desire with expectation of obtainment" or "to expect with confidence." But the portrait our environmental leaders paint is one where there is no reasonable expectation of obtaining a desired end. Come on—it is either dishonest or lazy to tell me both that I cannot reasonably expect some future condition, and that I can confidently expect some future condition.

I am not saying that I dislike all uses or forms of hope. Sometimes I admit there seems to be so little at stake. Someone says or writes that we need to maintain hope in the face of great harms, great sorrows; the rest of us nod and resolve to do just that; and it is not clear that any of us—the speaker, the writer, the listener—really even knows what it is that we are hoping for, or what work hope is supposed to be doing. What really worries me—terrifies me, truth be told—is the use of hope as a motivator for healing our wounded and warped relationship with the natural world. I worry that hope will actually stifle, not aid, our resolve. I worry that hope can be, and often is, a distraction, an excuse for not getting on with the work at hand.

If you Google the words "philosophy" and "hope" together, one of the first hits is a cosmetics company, oddly called Philosophy, that peddles a skin moisturizer product even more oddly called "Hope in a Jar." Jarred hope is pricey—28 bucks for a single ounce. The company tagline reads, "Where there is hope there can be faith. Where there is faith miracles can occur." I have to fight my gag reflex here for the same reason that I fight it when I hear hope used as a motivator for action aimed at averting environmental harms.

There is another story of jarred hope: a story equally, though differently, sickening. In ancient Greek mythology, Pandora, sister-in-law of that fire thief Prometheus, was given a dowry and ordered by Zeus to keep it sealed. But curious Pandora (or her curious husband Epimetheus) could not resist. She opened the dowry and unleashed the scourges of humanity: greed, vanity, slander, envy, pining, and other diseases. The last scourge, the last evil let loose, was hope ("hope" in

Greek, Ελπις or *elpis,* is translated as "anticipation of misfortune"). Even as early as 700 BCE, Hesiod too expressed doubts about hope: "Hope is a bad companion for the man in need who sits in an idle place, when he has no sufficient livelihood."

I think I understand what is going on here. Our environmental leaders are looking to provide a motivation to act. And sometimes people act in desperation; sometimes they do amazing, creative, and wondrous things. But sometimes, many times, they do not. I see it in my students nearly every day—the desire to do good, and just, and beautiful things in the world so easily quashed by the realization that what they decide to do won't make a difference, or that the world is beyond repair. Some other student (or professor!) flippantly points out that their individual decisions and action do not make any real difference in the world, and the student is gutted, mouth agape, no response in sight. So they give up before they start. And I blame that on hope. As Francis Bacon once put it, "Hope is a good breakfast, but a bad supper." Hope is sugary cereal, quick yet vacuous energy for the masses—hope is not protein, meat that will suffice in a world gone awry.

How did we set this trap for ourselves? This, it seems to me, is the nub of the problem: hope results from a fixation on consequences—on judging right and wrong actions, picking professions, and even justifying or evaluating the value of a lived life on what those actions, those professions, and that life produce. We have built a society fixated on the future, perpetually risking all the attendant problems of justifying means by their ends, and forever flirting with endorsing the hedonistic instincts of the masses. We have therefore built a society that can be readily disempowered.

It is time for a new form of motivation. Turning our backs on hope might be the best thing we can do at this moment in time. To be motivated by hope is to be stripped naked, to be vulnerable, to be disempowered. To be motivated by a sense of obligation, a commitment to virtue, is to put on a Kevlar bodysuit. "Lack of power consists only in this," Baruch Spinoza points out, "that a man allows himself to be guided by things outside him, and to be determined by them." Psychopathic serial killers tell us that the way they control their victims is to give them little tastes, little slivers of hope—but when their victims lose hope, they can no longer be controlled.

The writer Derrick Jensen nails it when he proclaims that he does not "have much hope. But I think that's a good thing. Hope is what keeps us chained to the system, the conglomerate of people and ideas and ideals that is causing the destruction of the Earth. . . . A WONDERFUL THING happens when you give up on hope, which is that you realize you never needed it in the first place."[1] You are free to act rightly, because it is the right way to act and not because your action will move you or the world toward some future state.

So here's what I want, what I think we need so desperately. I want us to replace "I hope" with "I resolve to do the work" or "I will be this kind of person, I will live this kind of life" or any sort of utterance that focuses on virtue rather than on consequence. Any sort of commitment that is not subject to the fickle and fragile focus on the results of our actions and commitments. This, I think, is the new ethic in the face of a future without hope. This is the only moral anchor imaginable in the sea change rolling our way.

I am calling for us to satisfy our obligation to the future by suggesting that we ignore, or at least greatly downplay, the consequences of our actions. I am suggesting instead that our obligation to the future is most properly satisfied when we act rightly and virtuously, and when our motivation stands stubbornly apart from, not held hostage to, the consequences of our actions.

If we are really going to resolve to act on behalf of the future, we need a motivation as steadfast and tough as the one Wendell Berry's character Dorie Catlett shows toward her forever-drunken Uncle Peach. Referring to Dorie's moral resolve, Berry writes, "She had long ago given up hope for Uncle Peach. She cared for him without hope, because she had passed the place of turning back or looking back. Quietly, almost submissively, she propped herself against him, because in her fate and faith she was opposed to his ruin."

What we need more than anything, need to save us from our own destructive fixation on the future, is no less than an individual and

---

1. Derrick Jensen, "Beyond Hope," *Orion*, May/June 2006.

collective moral revolution: a revolution that includes abandoning hope, caring without hope, and a commitment where we quietly, almost submissively, prop ourselves against those forces in the world that are working to bring ruin. We often hear that people only change their ideas, and therefore their behavior, in the face of crisis. But we forget that a crisis can be a moral crisis as well, a sense of revulsion for a life that we are living, a commitment to live differently and to be a different kind of person. We need The Great "Yuck!" Yuck, what we are doing is repulsive. Yuck, this is not the way a responsible person lives. The Great "Yuck!" can be followed by The Great "No!" No, I will not live this way. No, I will not be this kind of a person, this kind of an agent in the world. Finally, The Great "No!" will give way to The Great "Yes!" Yes, I will live a life of respect, of humility, empathy, care, and attentiveness. Yes, I will choose to live with dignity and grace, no matter what. But none of this—the yucks or nos or yeses—is held hostage by the attainment of some future state. Each of us, right now, at this exact moment in time, has the power to choose to live the moral life, to live a life that is indeed worth living.

The Most Amazing Challenge

Paul Hawken

The evening before I gave the commencement speech, I threw the original away. That night, at a dinner given for those receiving honorary degrees, I got the distinct impression that some of the trustees and officials were not happy about my being the commencement speaker. I was introduced as Paul "Hawker," and someone read a desultory bio. I had a crisis of confidence and wondered if I should even be there. I decided that my mandate was to talk to the eight hundred young people who were graduating the next day, not to try to please the bishops and faculty and alumni. I discarded the speech I had brought and stayed up until midnight and drafted a new one. When I got up in the morning and read it, I didn't like it any better than the first one, so I tore it up and started over once again. I decided to get right to the point and say what needed to be said. Gone was the customary greeting to the president, faculty, distinguished alumni, and so on. I wanted to speak directly to the students from the first line. I was sitting at a desk in my shorts eating off a tray at 11:55 a.m. when I realized I had to stop. People were downstairs to pick me up. I copied what I written onto a thumb drive, put on my suit, printed the speech in the lobby, placed it in my suit pocket, and rushed to

---

PAUL HAWKEN is an author, activist, economist, and environmentalist. His coauthored book *Natural Capitalism: Creating the Next Industrial Revolution* (2008) has been called by former president Bill Clinton one of the five most important books in the world today. He is author and coauthor of dozens of other articles, op-eds, papers, and books, including *The Next Economy, The Ecology of Commerce* (1994), and *Blessed Unrest: How the Largest Movement in the World Came into Being and Why No One Saw It Coming* (2008).

the commencement. The first time I read it all the way through was when I was on the podium, as if someone else had written it and I was asked to be the reader.

When I was invited to give this speech, I was asked if I could give a simple short talk that was "direct, naked, taut, honest, passionate, lean, shivering, startling, and graceful." No pressure here.

Let's begin with the startling part. Class of 2009: you are going to have to figure out what it means to be a human being on Earth at a time when every living system is declining, and the rate of decline is accelerating. Kind of a mind-boggling situation . . . but not one peer-reviewed paper published in the last thirty years can refute that statement. Basically, civilization needs a new operating system, you are the programmers, and we need it within a few decades.

This planet came with a set of instructions, but we seem to have misplaced them. Important rules like "Don't poison the water, soil, or air," "Don't let the Earth get overcrowded," and "Don't touch the thermostat" have been broken. Buckminster Fuller said that Spaceship Earth was so ingeniously designed that no one has a clue that we are on one, flying through the universe at a million miles per hour, with no need for seatbelts, lots of room in coach, and really good food—but all that is changing.

There is invisible writing on the back of the diploma you will receive, and in case you didn't bring lemon juice to decode it, I can tell you what it says: You are brilliant, and the Earth is hiring. The Earth couldn't afford to send recruiters or limos to your school. It sent you rain, sunsets, ripe cherries, night-blooming jasmine, and that unbelievably cute person you are dating. Take the hint. And here's the deal: Forget that this task of planet-saving is not possible in the time required. Don't be put off by people who know what is not possible. Do what needs to be done, and check to see if it was impossible only after you are done.

When asked whether I am pessimistic or optimistic about the future, I always give the same answer: If you look at the science about what is happening on Earth and aren't pessimistic, you don't understand the data. But if you meet the people who are working to restore this Earth and the lives of the poor, and you aren't optimistic, you haven't got a pulse. What I see everywhere in the world are ordinary people willing to confront despair, power, and incalculable odds in order to restore some semblance of grace, justice, and beauty to this world. The poet Adrienne Rich wrote, "So much has been destroyed I have cast my lot with those who, age after age, perversely, with no extraordinary power, reconstitute the world." There could be no better description. Humanity is coalescing. It is reconstituting the world, and the action is taking place in schoolrooms, farms, jungles, villages, campuses, companies, refuge camps, deserts, fisheries, and slums.

You join a multitude of caring people. No one knows how many groups and organizations are working on the most salient issues of our day: climate change, poverty, deforestation, peace, water, hunger, conservation, human rights, and more. This is the largest movement the world has ever seen. Rather than control, it seeks connection. Rather than dominance, it strives to disperse concentrations of power. Like Mercy Corps, it works behind the scenes and gets the job done. Large as it is, no one knows the true size of this movement. It provides hope, support, and meaning to billions of people in the world. Its clout resides in idea, not in force. It is made up of teachers, children, peasants, businesspeople, rappers, organic farmers, nuns, artists, government workers, fisherfolk, engineers, students, incorrigible writers, weeping Muslims, concerned mothers, poets, doctors without borders, grieving Christians, street musicians, the president of the United States of America, and as the writer David James Duncan would say, the Creator, the One who loves us all in such a huge way.

There is a rabbinical teaching that says if the world is ending and the Messiah arrives, first plant a tree, and then see if the story is true. Inspiration is not garnered from the litanies of what may befall us; it resides in humanity's willingness to restore, redress, reform, rebuild, recover, reimagine, and reconsider. "One day you finally knew what you had to do, and began, though the voices around you kept shouting their bad advice" is Mary Oliver's description of moving away from the profane toward a deep sense of connectedness to the living world.

Millions of people are working on behalf of strangers, even if the evening news is usually about the death of strangers. This kindness of strangers has religious, even mythic origins, and very specific eighteenth-century roots. Abolitionists were the first people to create a national and global movement to defend the rights of those they did not know. Until that time, no group had filed a grievance except on behalf of itself. The founders of this movement were largely unknown—Granville Sharp, Thomas Clarkson, Josiah Wedgwood—and their goal was ridiculous on the face of it: at that time three out of four people in the world were enslaved. Enslaving each other was what human beings had done for ages. And the abolitionist movement was greeted with incredulity. Conservative spokesmen ridiculed the abolitionists as liberals, progressives, do-gooders, meddlers, and activists. They were told they would ruin the economy and drive England into poverty. But for the first time in history a group of people organized themselves to help people they would never know, from whom they would never receive direct or indirect benefit. And today tens of millions of people do this every day. It is called the world of nonprofits, civil society, schools, social entrepreneurship, nongovernmental organizations, and companies who place social and environmental justice at the top of their strategic goals. The scope and scale of this effort is unparalleled in history.

The living world is not "out there" somewhere, but in your heart. What do we know about life? In the words of biologist Janine Benyus, life creates the conditions that are conducive to life. I can think of no better motto for a future economy. We have tens of thousands of abandoned homes without people and tens of thousands of abandoned people without homes. We have failed bankers advising failed regulators on how to save failed assets. We are the only species on the planet without full employment. Brilliant. We have an economy that tells us that it is cheaper to destroy Earth in real time than to renew, restore, and sustain it. You can print money to bail out a bank, but you can't print life to bail out a planet. At present we are stealing the future, selling it in the present, and calling it "gross domestic product." We can just as easily have an economy that is based on healing the future instead of stealing it. We can either create assets for the future or take the assets of the future. One is called "restoration," and the other "exploitation." And whenever we exploit the Earth, we exploit people and cause untold suffering. Working for the Earth is not a way to get rich; it is a way to *be* rich.

The first living cell came into being nearly 40 million centuries ago, and its direct descendants are in all of our bloodstreams. Literally, you are breathing molecules this very second that were inhaled by Moses, Mother Teresa, and Bono. We are vastly interconnected. Our fates are inseparable. We are here because the dream of every cell is to become two cells. And dreams come true. In each of you are one quadrillion cells, 90 percent of which are not human cells. Your body is a community, and without those other microorganisms you would perish in hours. Each human cell has 400 billion molecules conducting millions of processes between trillions of atoms. The total cellular activity in one human body is staggering: 1 septillion actions at any one moment, a one with twenty-four zeros after it. In a millisecond, our body has undergone ten times more processes than there are stars in the universe, which is exactly what Charles Darwin foretold when he said science would discover that each living creature was a "little universe, formed of a host of self-propagating organisms, inconceivably minute and as numerous as the stars of heaven."

So I have two questions for you all: First, can you feel your body? Stop for a moment. Feel your body. One septillion activities going on simultaneously, and your body does this so well you are free to ignore it, and wonder instead when this speech will end. You can feel it. It is called life. This is who you are. Second question: Who is in charge of your body? Who is managing those molecules? Hopefully not a political party. Life is creating the conditions that are conducive to life inside you, just as in all of nature. Our innate nature is to create the conditions that are conducive to life. What I want you to imagine is that collectively humanity is evincing a deep innate wisdom in coming together to heal the wounds and insults of the past.

Ralph Waldo Emerson once asked what we would do if the stars only came out once every thousand years. No one would sleep that night, of course. The world would create new religions overnight. We would be ecstatic, delirious, made rapturous by the glory of God. Instead, the stars come out every night and we watch television.

This extraordinary time when we are globally aware of each other and the multiple dangers that threaten civilization has never happened, not in a thousand years, not in ten thousand years. Each of us is as complex and beautiful as all the stars in the universe. We have done great things, and we have gone way off course in terms of honoring creation. You are graduating to the most amazing, stupefying challenge ever bequeathed to any generation. The generations before you failed. They didn't stay up all night. They got distracted and lost sight of the fact that life is a miracle every moment of your existence. Nature beckons you to be on her side. You couldn't ask for a better boss. The most unrealistic person in the world is the cynic, not the dreamer. Hope makes sense only when it doesn't make sense to be hopeful. This is your century. Take it and run as if your life depends on it.

ETHICAL ACTION

What does personal integrity ask of us?

Polls tell us that a large percentage of people aren't alarmed by global warming or ecological collapse. Don't worry about them. Time will teach them. Focus on the people who do understand the enormity of the threat and the urgency of the emergency. If the actions of these people were actually consistent with their convictions, we could quickly resolve the climate crisis and redesign lifeways that support ongoing life. The problem is that there is no direct connection between knowing what is right and doing it. *Akrasia,* the ancient Greeks called it—weakness of the will. Nowadays we call it "hypocrisy."

"So convenient a thing it is to be a *reasonable creature,*" Benjamin Franklin wrote, "since it enables one to find or make a reason for everything one has a mind to do." And he feasted on the roasted fish he had vowed not to eat. And so we—good people all, conscientious, civic-minded, generous—find good reasons to continue to fly, to drive long distances, to consume beyond measure, to power up our electronics with coal-fired electricity, thinking we'll stop next year, or when our neighbors do, or when the good we do on the lecture circuit no longer outweighs the harm, or when the solar cells are installed on the roof, or when somebody invents an affordable electric car, even as we know better. Even as we are ashamed and frightened.

The times call for integrity, which is the consistency of belief and action. The times call for the courage to refute our own bad arguments and call ourselves on our own bad faith. We are called to live lives we believe in—even if a life of integrity is very different, let us suppose radically different, from how we live now. Knowledge imposes responsibility. Knowledge of a coming threat requires action to avert it. There is no way around it, if our lives are to be worthy of our view of ourselves as moral beings.

How to begin? Maybe with four lists. List 1: These are the things I value most in my life (good health for my children, the love and respect of my family, and so on). List 2: These are the things I do that are supportive of those values. List 3: These are the things I do that are destructive of those values. List 4: These are the things I am going to do differently. From now on. No matter what.

The odd fact is that those lifeways that are most destructive of the world often turn out to be exactly those lifeways that are most destructive of the spirit. You'll know you are achieving integrity when you are surprised by happiness and the unexpected joy of freedom.

Acknowledgments

We extend our deepest thanks:

To the team of wonderful people who helped put this book together—Carly Johnson Lettero, for diligent and well-informed research; Carol Mason, for formatting and assembling the book and helping in a thousand other ways; and Frank L. Moore, for managing the massive amount of information and communication the book entailed.

To the scholars and friends who helped us identify and contact writers for the book—H. Emerson Blake, James Blumenthal, Rick Brown, Alison Hawthorne Deming, Jane Elder, Peter Forbes, Eddin Khoo, Robin Kimmerer, Jane Lubchenco, Curt Meine, Christopher Merrill, Lani Roberts, Scott Slovic, Kim Stafford, Bron Taylor, Bryan Tilt, John Vucetich, Al-Yasha Williams. Thank you for your generous help with this project, and especially for your good work for the world.

To Barbara Ras at Trinity University Press, a high-powered and helpful editor and a valued friend. And to Laura Blake Peterson, of Curtis Brown Literary Agency, with thanks for believing so strongly in *Moral Ground*.

To Rachel Myron and Steve Lewis, and to Charles King and Davi Richards, who generously shared their Internet access in remote Alaska, enabling the book project to continue throughout the summer.

To Jonathan W. Moore, University of California at Santa Cruz, and Erin E. Moore, University of Oregon, for deeply insightful conversations about the direction of this book, and of the world. To Frank Moore, who gave up so much for *Moral Ground,* and gave so much to it.

To colleagues at Michigan State University who have been amazingly supportive, especially Kelly Millenbah, Elizabeth Simmons, Mike Jones, William Taylor, Mark Largent, Lissy Goralnik, and

John Waller. And to Oregon State University for writing support, and to OSU colleagues Charles Goodrich and Fred Swanson.

To the Kendeda Fund and the Germeshausen Foundation, for supporting the book's efforts to enliven and inform public discussion of our moral obligations to the future. To the Shotpouch Foundation; the Spring Creek Project for Ideas, Nature, and the Written Word at Oregon State University; and the Long-Term Ecological Reflections Project funded by the USDA Forest Service, whose collaborative conference "What Is the Work of a Philosopher in a Wounded World?" in part inspired this book.

A special thanks to all those who work for the sake of a future that is beautiful, flourishing, and fair, with gratitude and celebration.

Credits

Ecumenical Patriarch Bartholomew I, "To Commit a Crime Against the Natural World Is a Sin." From the address of His All-Holiness Ecumenical Patriarch Bartholomew at the Environmental Symposium, Saint Barbara Greek Orthodox Church, Santa Barbara, California, November 8, 1997. Reprinted by permission.

Thomas Berry, "The Great Work." Excerpts from "Religion in the Ecozoic Era" from *The Sacred Universe*, edited by Mary Evelyn Tucker. Copyright © 1993 by Thomas Berry. Copyright © 2009 by Columbia University Press. Reprinted with the permission of Columbia University Press. Excerpt from *The Great Work*. Copyright © 1999 by Thomas Berry. Used by permission of Bell Tower, a division of Random House, Inc.

Wendell Berry, "A Promise Made in Love, Awe, and Fear." Copyright © 2005 by Wendell Berry, from *Standing by Words: Essays*. Reprinted by permission of Counterpoint Press.

The Dalai Lama, "A Question of Our Own Survival." Excerpt from the Dalai Lama, "A Universal Task," *EPA Journal: A Magazine on National and Global Environmental Perspectives* 17, no. 4 (September/October 1991). Excerpt from *My Tibet*, by his Holiness the Fourteenth Dalai Lama of Tibet. Copyright © 1990 by the Regents of the University of California. Reprinted by permission of the University of California Press. Excerpt from "Universal Responsibility and the Good," in *Freedom in Exile: The Autobiography of the Dalai Lama*. Copyright © 1990 by Tenzin Gyatso, His Holiness, The Fourteenth Dalai Lama of Tibet. Reprinted by permission of HarperCollins Publishers.

Thomas L. Friedman, "Who We Really Are," adapted from *Hot, Flat, and Crowded: Why We Need a Green Revolution—and How It Can Renew America*. Copyright © 2008. Adapted by permission of Farrar, Straus & Giroux LLC.

CREDITS

Thich Nhat Hanh, "The Bells of Mindfulness," from *The World We Have*. Copyright © 2008. Reprinted with the permission of Parallax Press, Berkeley, California.

bell hooks, "Touching the Earth." Excerpt from *Belonging, a Culture of Place*. Copyright © 2009 by bell hooks. Reprinted with the permission of Routledge/Taylor & Francis Books, Inc.

John Paul II and the Ecumenical Patriarch Bartholomew I, "Steering the Earth Toward Our Children's Future." Copyright © Libreria Editrice Vaticana. Reprinted with permission.

Barbara Kingsolver, "How to Be Hopeful: Duke University Commencement Address," 11 May 2008. Reprinted by permission.

Chief Oren Lyons, "Keepers of Life." From his address to the delegates to the United Nations Organization, "The Year of the Indigenous Peoples," 10 December 1992. Reprinted by permission of the author.

Sallie McFague, "A Manifesto to North American Middle-Class Christians." From *Life Abundant: Rethinking Theology and Ecology for a Planet in Peril*, copyright © 2000 Fortress Press. Reproduced by special permission of Augsburg Fortress Publishers. All rights reserved.

Wangari Maathai, "We Are Called to Help the Earth to Heal." Copyright © 2010 by Wangari Maathai. Reprinted with permission. Excerpt from Nobel Lecture (Oslo, Norway, December 10, 2004). Copyright © 2004 by The Nobel Foundation. Reprinted by permission. "On Climate Change" and "Africa Must Be Heard on Climate Change." Both reprinted by permission of the author.

N. Scott Momaday, "An Ethic of the Earth." Excerpt from "An American Land Ethic" from *The Man Made of Words: Essays, Stories, Passages*. Copyright © 1997 by N. Scott Momaday. Reprinted with the permission of the author.

Ernest Partridge, "Moral Responsibility Is the Price of Progress," adapted from "Future Generations" from *A Companion to Environmental Philosophy*, edited by Dale Jamieson. Copyright © 2001. Adapted and reprinted with the permission of the author.

Daniel Quinn, "The Danger of Human Exceptionalism." Copyright © 2010 by Daniel Quinn.

CREDITS

Libby Roderick, "Winter Wheat" (Turtle Island Records, Anchorage, AK). Words and music by Libby Roderick, www.libbyroderick.com. Copyright © Libby Roderick Music. All rights reserved.

Carl Safina, "The Moral Climate," from *Orion* (September–October 2008). Copyright © 2008. Reprinted with the permission of the author.

Scott Russell Sanders, "We Bear You in Mind" (originally titled "For the Children"). Excerpt from *A Conservationist Manifesto* (Bloomington: Indiana University Press, 2009). Copyright © 2009. Reprinted by permission of the author.

Peter Singer, "A Fair Deal." Excerpts from Peter Singer, "A Fair Deal on Climate Change," copyright © 2007 Project Syndicate, www.project-syndicate.org/commentary/singer24; and "Will the Polluters Pay for Climate Change?" copyright © 2006 Project Syndicate, www.project-syndicate.org/commentary/singer14. Both reprinted with permission.

Gary Snyder, "For the Children," from *Turtle Island*. Copyright © 1974 by Gary Snyder. Reprinted by permission of New Directions Publishing Corp.

Quincy Troupe, "The Architecture of Language, Parts 9 and 10," from *The Architecture of Language*. Copyright © 2006 by Quincy Troupe. Reprinted with the permission of Coffee House Press, www.coffeehousepress.org.

Brian Turner, "Sky," in *Just This* (Wellington, NZ: Victoria University Press, 2010). Reprinted with permission.

Sheila Watt-Cloutier, "The Inuit Right to Culture Based on Ice and Snow," adapted from *Transcripts from Indigenous Peoples' Resistance to Economic Globalization: A Celebration of Victories, Rights and Cultures* (23 November 2006, New York, NY). Reprinted by permission of the author.

E. O. Wilson, "The Fate of Creation Is the Fate of Humanity." Excerpt from *The Creation: An Appeal to Save Life on Earth*. Copyright © 2006 by E. O. Wilson. Used by permission of W. W. Norton & Company, Inc.

List of Contributors

Allendorf, Fred W., 202
Bartholomew I, 51, 133
Bateson, Mary Catherine, 211
Berry, Thomas, 396
Berry, Wendell, 387
Borg, Marcus J., 250
Callicott, J. Baird, 356
Campbell, Courtney S., 146
Chapin, F. Stuart, III, 75
Collin, Robin Morris, 82
Crow, Michael M., 301
Dalai Lama, 15
Deming, Alison Hawthorne, 347
Doyle, Brian, 167
Duncan, David James, 434
Ebtekar, Massoumeh, 178
Fink, Jesse M., 230
Foreman, Dave, 100
Friedman, Thomas L., 189
Garvey, James, 279
Hanh, Thich Naht, 79
Hawken, Paul, 463
Heinrich, Bernd, 331
Hogan, Linda, 407
hooks, bell, 363
Jamieson, Dale, 183
Jensen, Derrick, 60
John Paul II, 51
Kaplan, Martin S., 263
Kawagley, Angayuqaq Oscar, 217
Kellert, Stephen R., 373
Kimmerer, Robin W., 141

Kingsolver, Barbara, 452
Krech, Shepard, III, 108
Le Guin, Ursula K., 158
Lentfer, Hank, 343
Lettero, Carly, 103
Lyons, Oren, 42
Maathai, Wangari, 271
McFague, Sallie, 243
McKibben, Bill, 174
McShane, Katie, 369
Meine, Curt, 403
Ming, Xu, 275
Momaday, N. Scott, 400
Moore, Kathleen Dean, xiii, 390
Murray-Philipson, Hylton, 55
Nabhan, Gary Paul, 115
Nasr, Seyyed Hossein, 254
Nelson, Michael P., xiii, 458
Obama, Barack, 30
Partridge, Ernest, 421
Perry, John, 169
Pister, Edwin P., 220
Pope, Carl, 294
Pyle, Robert Michael, 123
Quammen, David, 119
Quinn, Daniel, 9
Rawles, Kate, 88
Robinson, Tri, 260
Roderick, Libby, 269
Rolston, Holmes, III, 70
Rose, Deborah Bird, 154
Rose, Jonathan F. P., 207

LIST OF CONTRIBUTORS

Safina, Carl, 324
Sanders, Scott Russell, 45
Savoy, Lauret, 312
Selvamony, Nirmal, 137
Serageldin, Ismail, 317
Singer, Peter, 321
Sivaraksa, Sulak, 283
Snyder, Gary, 50
Speth, James Gustave, 3
Swimme, Brian, 410
Taylor, Bron, 379
Thompson, Paul B., 440
Tinker, George, 196

Tremmel, Joerg Chet, 444
Troupe, Quincy, 286
Tucker, Mary Evelyn, 410
Tundisi, José Galizia, 448
Turner, Brian, 69
Vanderheiden, Steve, 306
Vucetich, John A., 337
Wade-Benzoni, Kimberly A., 225
Watt-Cloutier, Sheila, 25
Weisman, Alan, 32
Williams, Terry Tempest, 429
Wilson, E. O., 21
Xin, Wei, 275

KATHLEEN DEAN MOORE is Distinguished Professor of Philosophy at Oregon State University, where she directs the Spring Creek Project for Ideas, Nature, and the Written Word. She is an environmental philosopher and essayist whose most recent books are *Wild Comfort: The Solace of Nature* and *The Pine Island Paradox: Making Connections in a Disconnected World*. She serves on the board of directors for the Orion Society and the Island Institute. She lives in Corvallis, Oregon.

MICHAEL P. NELSON is the Ruth H. Spaniol Chair of Natural Resources and a professor of environmental ethics and philosophy at Oregon State University, where he also serves as the lead principle investigator for the HJ Andrews Long-Term Ecological Research program in the Oregon Cascades. He is the coeditor, with J. Baird Callicott, of *The Great New Wilderness Debate* and *The Wilderness Debate Rages On*. He is also the cofounder and codirector of the Conservation Ethics Group, an award-winning environmental consultancy group fusing ethics with social and ecological science. He lives in Corvallis, Oregon.

ARCHBISHOP EMERITUS DESMOND TUTU has received the Nobel Peace Prize, the Gandhi Peace Prize, the Albert Schweitzer Prize for Humanitarianism, and the U.S. Presidential Medal of Freedom. In 1986 he was elected archbishop of Cape Town, the highest position in the Anglican Church in South Africa. In 1995, after the end of apartheid and the election of Nelson Mandela, Tutu was appointed chair of South Africa's Truth and Reconciliation Commission. He is currently chair of the Elders, an independent group of global leaders who offer their influence and experience to support peace building.